New Perspectives in Thermodynamics

Edited by J. Serrin

New Perspectives in Thermodynamics

Edited by J. Serrin

With 10 Figures

Springer-Verlag
Berlin Heidelberg New York Tokyo

James Serrin

Department of Mathematics, University of Minnesota,
Minneapolis, MN, USA

ISBN 3-540-15931-2 Springer-Verlag Berlin Heidelberg New York Tokyo
ISBN 0-387-15931-2 Springer-Verlag New York Heidelberg Berlin Tokyo

Library of Congress Cataloging in Publication Data. Main entry under title: New perspectives in thermodynamics. 1. Thermodynamics. I. Serrin, J. (James), 1926-. QC311.N53 1986 536'.7 85-22269

Printed in Germany

Typesetting: K + V Fotosatz, 6124 Beerfelden
Offset printing and bookbinding: Beltz Offsetdruck, 6944 Hemsbach
2153/3130-543210

Foreword

The material included in this book was first presented in a series of lectures delivered at the University of Minnesota in June 1983 in connection with the conference "Thermodynamics and Phase Transitions". This conference was one of the principal events in the first year of operation of the Institute for Mathematics and its Applications (IMA) at the University of Minnesota.

The Institute was founded under the auspices of the National Science Foundation of the United States and the University of Minnesota and is devoted to strengthening and fostering the relation of mathematics with its various applications to problems of the real world.

The present volume constitutes an important element in the continuing publication program of the Institute. Previous publications in this program have appeared as lecture notes in the well-known Springer series, and future ones will be part of a new series "IMA Volumes in Applied Mathematics".

Preface

Until recently it was believed that thermodynamics could be given a rigorous foundation only in certain restricted circumstances, particularly those involving reversible and quasi-static processes. More general situations, commonly arising in continuum theories, have therefore been treated on the assumption that internal energy, entropy and absolute temperature are a priori given quantities, or have been dealt with on a more or less ad hoc basis, with emphasis for example on various types of variational formulations and maximization rules.

At the same time, the last decade has seen a unity of method and approach in the foundations of thermodynamics and continuum mechanics, in which rigorous laws of thermodynamics have been combined with invariance notions of mechanics to produce new and deep understanding. Real progress has been made in finding a set of appropriate concepts for classical thermodynamics, by which energy conservation and the Clausius inequality can be given well-defined meanings for arbitrary processes and which allow an approach to the entropy concept which is free of traditional ambiguities. There has been, moreover, a careful scrutiny of long established but nevertheless not sharply defined concepts such as the Maxwell equal-area rule, the famous Gibbs phase rule, and the equivalence of work and heat.

The thirteen papers in this volume accordingly gather together for the first time the many ideas and concepts which have raised classical thermodynamics from a heuristic and intuitive science to the level of precision presently demanded of other branches of mathematical physics.

The basic notions of thermodynamics already appear in the work of the founders of the subject – Carnot, Clausius, Kelvin and Gibbs – though their ideas now appear with greater brilliance than ever as they are seen in a pure and abstract light. Perhaps more unexpected is the diminished role played by Carathéodory's principle of accessibility. Indeed in an important essay opening Part II of this volume Clifford Truesdell argues with great force that Carathéodory's contribution has turned the subject away from its mainstream, and that it is more appropriate to return to classic directions.

The first two papers in Part I, by Serrin and Šilhavý, carry out this approach. They are firmly based on the idea that *classical thermodynamics is precisely the study of heat, work and hotness:* in other words, *thermodynamics is concerned with the general structure of systems which exchange work and heat with their environment.* Though these papers differ in method and technique they both adopt the primitive concepts of heat and work, and develop mathematical structures in which the laws of thermodynamics can be stated with exactness and clarity. Equally, both papers clarify the role of entropy and energy and their relation with heat and work, leading to new insight into various formulations of entropy for irreversible thermodynamics and continuum mechanics.

In a third paper on the foundations of thermodynamics, Feinberg and Lavine present, in a tour de force of invention, a theory in which hotness and temperature are *defined* concepts, thus providing great generality to our views of the meaning of temperature as a measure of equilibrium.

The central position of cyclic processes in statements of the laws of thermodynamics is addressed in papers of Coleman and Owen, Truesdell, and Ricou. In recent years it has been realized with increasing force that there are numerous materials which have few, if any, cyclic processes available to them (hereditary systems, hardening cements, and so forth). For these systems the traditional laws provide little guidance. Coleman and Owen accordingly have introduced the notion of approximate cycles and have proposed a detailed topological structure for state spaces, so that for large classes of non-cyclic materials one can obtain a meaningful concept of entropy. Following a different line of argument, Truesdell in his paper in Part I takes thermodynamic efficiency as the guiding principle and obtains a number of estimates for this quantity in non-cyclic situations.

Manuel Ricou's paper considers the same problem from a new point of view. It is not possible to summarize his ideas here, but his short paper promises to fundamentally alter our conception of energy as a state function and is destined to become one of the classics of the subject. The astute reader will even find in this paper the beginnings of an axiomatic treatment of the third law of thermodynamics!

Part II of the volume deals with the thermodynamics of Gibbs and Carathéodory, and their legacy to later generations. The essay of Professor Truesdell which opens this part is concerned particularly with Gibbs's contribution to thermodynamics (as opposed to thermostatics). His remarks are especially important and appropriate today, not only because we now see thermodynamics and thermostatics as different subjects but also because students continue to this day to reread and rethink the ideas of Gibbs. It goes without saying that this essay confirms once again the lasting place which Gibbs holds in thermodynamic science.

It has repeatedly been emphasized, most recently in a paper of Martin Klein in "Springs of Scientific Creativity" (University of Minnesota Press, 1984), that Gibbs was a writer of notorious scientific difficulty in spite of the lucidity of his English style. The papers of Fosdick and Man in Part II are specifically concerned with the elucidation of difficult but key ideas introduced by Gibbs. In particular, Fosdick addresses the question of the structure and existence of equilibrium states in thermostatics, together with the dynamical significance of Gibbsian stability. His work provides important new insight into these elusive concepts, and should be widely read.

Man's contribution presents a clear and vigorous discussion of the empirical and philosophical meaning of the Gibbs phase rule. Although this rule has been criticized for its ambiguity, Man shows that Gibbs himself cannot be held responsible for the many inaccurate statements of the rule. Even more, he provides a context in which the phase rule becomes generically and rigorously valid.

The series of papers in Part III are concerned with special material systems. While quite different from each other, and from the earlier papers, they nevertheless possess an important underlying unity in that each uses ideas generated by modern thermodynamical research. The paper of Coleman, Fabrizio and Owen introduces and analyses a generalization of Cattaneo's theory of the propagation of heat. Their conclusions are of value not only in rationalizing the well-known paradox of the infinite speed of heat transmission (a consequence of the parabolic nature of the heat equation) but also in the theory of heat conduction in dielectric crystals at low temperatures.

The paper of J. E. Dunn studies a new type of work, important in the transfer of mechanical energy between a material system and its environment. This quantity, called *interstitial working*, does not appear in standard theories of energy conservation: on the other hand, absence of such a term places severe limitations on the types of materials which can be studied by the methods of continuum mechanics, at least when the second law of thermodynamics is taken into account. The new term accordingly can prove of particular value in the study of Korteweg fluids as well as chemical mixture theory.

The contribution of Richard James deals with phase transformations in certain types of crystals. Using arguments which are brilliant in their simplicity, he is able to enumerate all the possible transformations which can exist and thus to clarify a previously unknown situation. Underlying his approach is the purely thermodynamic concept of minimizing the free energy of the crystal, furnishing thereby a direct connection with the results of Fosdick.

Landau and Lifschitz have stated in their classic treatise on fluid mechanics that a shock wave can never compress a gas sufficiently to liquefy it. However recent experiments of Thompson and his coworkers have shown that, at least for certain gases, this belief is inaccurate. The final paper in the volume provides a theoretical model for these experiments, and a classification of possible gas-liquid shock waves (the actual computations are concerned with the special case of a van der Waals fluid, but the method is quite general). This work employs the concept of interstitial working, thus providing a link with Dunn's paper and a new application of the second law of thermodynamics.

This brief overview can only hint at the material covered, but it is hoped that readers will find much of lasting value among the ideas which are introduced here.

The preparation of the volume was greatly aided by the professional efforts of Kathleen Pericak-Spector, who edited many of the manuscripts, and by George Sell, whose helpful guidance even in difficult moments was always in evidence. Finally the entire work owes its existence to the good auspices of the Institute for Mathematics and its Applications at the University of Minnesota.

Minneapolis, January 1986 *James Serrin*

Contents

Part I Foundations of Thermodynamics

Contributors

Coleman, Bernard D.

Department of Mathematics, Carnegie-Mellon University, Pittsburgh, PA, USA

Dunn, J. E.

Sandia National Laboratories, Albuquerque, NM, USA

Fabrizio, Mauro

Istituto di Matematica, Universitá di Bologna, Bologna, Italy

Feinberg, Martin

Department of Chemical Engineering, University of Rochester, Rochester, NY, USA

Fosdick, Roger L.

Department of Aerospace Engineering and Mechanics, University of Minnesota, Minneapolis, MN, USA

Hagan, Robert

Lockheed Corporation, Palo Alto, CA, USA

James, Richard D.

Department of Aerospace Engineering and Mechanics, University of Minnesota, Minneapolis, MN, USA

Lavine, Richard

Department of Mathematics, University of Rochester, Rochester, NY, USA

Man, Chi-Sing

Department of Mathematics, University of Kentucky, Lexington, KY, USA

Owen, David R.

Department of Mathematics, Carnegie-Mellon University, Pittsburgh, PA, USA

Ricou, Manuel

Departamento de Matemática, Instituto Superior Técnico, Lisbon, Portugal

Serrin, James

Department of Mathematics, University of Minnesota, Minneapolis, MN, USA

Šilhavý, Miroslav

Mathematical Institute, Czechoslovak Academy of Sciences, Žitná ulice 25, Prague, Czechoslovakia

Truesdell, Clifford

The Johns Hopkins University, Baltimore, MD, USA

Part I

Foundations of Thermodynamics

Chapter 1
An Outline of Thermodynamical Structure

J. Serrin

1.1 Introduction

It is my purpose to formulate the laws of classical thermodynamics in a clear and precise way, sufficiently general to include not only traditional applications to reversible and quasi-static processes but also irreversible theories of continuum mechanics. I then go on to express the Clausius inequality in a precise form, independent of the structure of any particular system.

Finally, I shall give a general definition for the concepts of internal energy and entropy, and in particular elucidate the logical position of the Clausius-Duhem inequality in continuum mechanics.

Since the earliest days of thermodynamical science, it has always been recognized that the conclusions of the subject were to be obtained deductively from general laws. At the same time, in contrast with the case of other sciences, these laws have not been expressed in any standard or usual mathematical formalism, thus making the deductions appear different from those in other branches of physics. Indeed, *Buchdahl* expressed the situation well when he wrote "There is no doubt that part of the difficulty of the classical arguments lies in the subtlety with which mathematical notions and ostensibly physical notions are almost inextricably interwoven". Finally, since the early days of the subject there have been repeated calls for rigor in proofs. Thus, along with the discovery of an appropriate mathematical structure in which to carry out the deductions, modern research is also concerned with the struggle for precision. It is not that nothing correct has been available, but rather that rigor has been only occasional and confined to special circumstances. These facts themselves contributed to the general mysteries of the subject, since they caused further confusion between mathematical derivations and physical thinking.

The purpose of this chapter is to provide clarity and understanding for phenomenological thermodynamics, not to give "prescriptions" for calculating entropy or internal energy. Nevertheless, by introducing precise definitions one at least is presented with a definite problem to be solved rather than mystical statements about vague operational procedures. The precision which is obtained may eventually lead to a revision of present physical beliefs as to the primacy of energy and entropy, although for the moment such a view is speculative.

The basis of thermodynamics in fundamental restrictions about cyclic processes has also been reexamined in recent years, and alternative points of view are discussed in Sect. 1.6. Our presentation is partly motivated by this work, as well as by important new researches of Šilhavý. In particular, Šilhavý's central discovery that one can state the general First Law without the intervention of the concept of the mechanical equivalent of heat is crucial to the formulation of the First Law given here.

This latter direction of research has produced as well a generalized version of thermodynamics, which gives up the precise equivalence of work and heat but retains in all other respects the structure of the subject. Because this aspect of modern work is even more speculative, one cannot yet determine its future impact, but certainly it leads to interesting possibilities for explaining material behavior in non-classical irreversible systems.

Let me emphasize again that this work is not to be construed as providing answers to specific thermal problems. On the contrary, it concerns definite formulational questions which have their own implications and interrelations with the general subject.

I wish to thank Professor Kathleen Pericak-Spector for her great help in the preparation of this paper.

1.2 Classical Thermal Structure

> ● The subject of thermodynamics seems to present peculiar difficulties.
>
> *A. H. Wilson* [1.2]

Thermodynamics is a subject which has been studied for well over a century. For just as long a period there have been, as there still are, skeptics and critics. *Kelvin* [1.3] has written that "A mere quicksand has been given as a foundation for thermometry, by building from the beginning on an ideal substance called the perfect gas, with none of its properties realised rigorously by any real substance". *Cardwell* [1.4] states that "The student is usually introduced to the concepts of thermodynamics – the Carnot cycle, the principle of reversibility, the idea of entropy – in a way which does violence to credibility". There appears to be a special difficulty encountered with the second law. *Fong* [1.5] writes "The second law of thermodynamics is the most profound, yet the most elusive, fundamental principle in physics. ... As a college student, I was dissatisfied with many of the arguments involved in classical thermodynamics. It was a great relief to find that Max Born and P. W. Bridgman have expressed similar views".

It will be useful to begin by considering the traditional thermal structure of processes. Here the first law is frequently stated as follows:

I) For a cyclic process of an arbitrary thermal system, the net heat $\bar{Q}$ supplied to the system equals the net work $\bar{W}$ done by the system, that is

$$\bar{W} = \bar{Q} .$$

Correspondingly, the (traditional) second law can be stated analytically in three different ways:

II) For a cyclic process of a system the sum of the heats supplied divided by their absolute temperatures is non-positive:

$$\int \frac{dQ}{T} \leqslant 0 .$$

II_A) For any process of an isolated system the entropy S must either stay fixed or increase, that is

$$\Delta S \geqslant 0 .$$

II_B) The absolute temperature T is an integrating divisor for the heat.

All these statements appear unconnected; moreover, although all are analytical, the level of abstraction rises while the level of mathematical clarity falls. The Clausius inequality II and the entropy inequality II_A in fact are little more than heuristic descriptions, the formulae being mnemonic devices written in suggestive symbolism. Also, whether these are axioms or theorems depends on the writer, even assuming that the meaning is clarified. Fong has written further that "The feeling of uneasiness in thermodynamics is as old as thermodynamics. ... The real difficulty has always been in the basic concepts. The axiomatists made a very great effort to build elaborate and complicated structures. Unfortunately, equal care was not taken in selecting satisfactory construction materials, i.e., basic concepts".

In the following section we shall use the classical notions of *heat, work,* and *hotness* as primitive elements, and set up a thermodynamical structure in which the laws of thermodynamics can be expressed in a clear and concise way.

That heat is an appropriate and natural primitive for thermodynamics was already accepted by Carnot. Its continued validity as a primitive element of thermodynamical structure is due to the fact that it synthesizes an essential physical concept, as well as to its successful use in recent work to unify different constitutive theories. Similarly, by taking work as a second primitive not only do we follow the classical tradition but at the same time we avoid reference to specific mechanical theories which might otherwise reduce the generality of the treatment. Finally, "hotness" represents the abstract physical content of ordinary temperature readings, and as such again constitutes a reasonable and natural primitive element, intuitively understood from the historical beginnings of the subject and clearly expressed by a number of writers in the nineteenth century (see the quoted paragraph in [1.11], where *Mach* carefully notes that "temperature is ... nothing else than the characterization, the mark of a hotness level by a number").

1.3 The Formal Structure of Thermodynamics

- Science constantly seeks for separable entities which can either be perceived in the outside world, or, more often, have to be inferred speculatively in the outside world.

J. Bronowski [1.6]

It is convenient to begin our treatment by introducing the concept of hotness, which will here be considered a primitive notion within the theory. It is represented by a *thermal manifold* $\mathscr{H}$ consisting of the set of *hotness levels* L open to material systems.

At the simplest level, the manifold $\mathscr{H}$ should be a *totally ordered set*, with the order relation $\succ$ corresponding to increasing levels of hotness. In particular, if L_1 and L_2 are any two different hotness levels in $\mathscr{H}$, then either $L_1 \succ L_2$ or $L_2 \prec L_1$ (but not both). Moreover, if $L_1 \succ L_2$ and $L_2 \succ L_3$ then $L_1 \succ L_3$, that is, the order is transitive. The relation $L_1 \succ L_2$ will be read "L_1 is hotter than L_2" or alternatively "L_2 is colder than L_1"; we write also $L_1 \succcurlyeq L_2$ to indicate that either $L_1 \succ L_2$ or $L_1 = L_2$.

A *temperature scale* is a strictly increasing map from $\mathscr{H}$ into the reals $\mathbb{R}$. If ψ is a temperature scale then $\psi(L)$ is called the *temperature* of L in the scale ψ. At this stage of the theory there is no reason to prefer any one temperature scale over any other.

Fundamental to thermodynamical structure is the concept of a *thermodynamical system*, examples of which might be a body of gas or an elastic solid, to name two particularly simple cases. Every thermodynamical system $\mathscr{S}$ comes endowed with a set $\mathbb{P}(\mathscr{S})$ of *processes*, denoted by P, R, S, etc., which the system may undergo, together with a subset $\mathbb{P}_{\text{cyc}}(\mathscr{S})$ of *cyclic processes* of the system.[1] To every process $P \in \mathbb{P}(\mathscr{S})$ there correspond real numbers $\bar{W}(P)$ and $\bar{Q}(P)$, respectively the *total work* done by the process P and the *total heat* used by the process P (of course, any appropriate and agreed set of units may be used to "measure" heat and work, say calories and joules). Formally

$$\bar{W}: \mathbb{P}(\mathscr{S}) \to \mathbb{R}$$
$$\bar{Q}: \mathbb{P}(\mathscr{S}) \to \mathbb{R}.$$

We adopt the standard sign convention that $\bar{W}(P) > 0$ if work is done by the system on the exterior environment and $\bar{W}(P) < 0$ if the exterior environment does work on the system. Similarly $\bar{Q}(P) > 0$ if heat is supplied to the system, while $\bar{Q}(P) < 0$ means that the system has supplied heat to the environment. How the functions $\bar{W}$ and $\bar{Q}$ are to be computed for a given system is a *constitutive* matter

1 A scientist naturally takes it for granted that a thermodynamic system may move in various ways. The mathematician, wishing to quantify this idea, requires a listing (set) of the various things the system can do. Some systems may have relatively few available processes – for example, a gas *confined to a container with a movable piston* – but other systems – *the body of gas itself*, thought of as a separate system – may have many more possible motions open to it. In Sect. 1.8 we shall discuss several examples of thermodynamic systems in detail.

for the system, a matter which naturally must be discussed in detail when one turns to the specific study of particular systems. Suffice it to say for the moment that any material system $\mathscr{S}$ of physical interest certainly admits a definite value for both the total heat and the total work corresponding to each of its processes P.

This is not quite all. In 1875 *Gibbs* [1.7] had emphasized an important fact:

> "In thermodynamic problems, heat received at one temperature is by no means the equivalent of the same amount of heat received at another temperature. For example, a supply of a million calories at 150° is a very different thing from supply of a million calories at 50°. Hence, in thermodynamic problems, it is generally necessary to distinguish between the quantities of heat received or given out by the body at different temperatures, while as far as work is concerned, it is generally sufficient to ascertain the total amount performed."

In order to make the ideas expressed by Gibbs more concrete, we require a method for discriminating between the kinds of heat supplied to a system at different hotness levels. To do this, we make the basic observation (structural axiom) that to *every* process $P \in \mathbb{P}(\mathscr{S})$ and *every* hotness level $L \in \mathscr{H}$ there is associated a real number $Q(P, L)$ representing the total or net heat transferred to the system during the process *P at hotness levels L' not exceeding L*. Formally we have

$$Q: \mathbb{P}(\mathscr{S}) \times \mathscr{H} \rightarrow \mathbb{R} .$$

The function $Q(P, \cdot)$ is called the *accumulation function* of the process P. We emphasize that it is a mapping from the hotness manifold $\mathscr{H}$ into the reals $\mathbb{R}$.

The accumulation function expresses analytically the essential properties of the relation between heat and hotness for a given process P. For example, during a process P the total heat added *between* the hotness levels L_1 and L_2 (with $L_1 \prec L_2$ say) is given by $Q(P, L_2) - Q(P, L_1)$. It follows in particular that the accumulation function of an *isothermal process P*, operating at a single hotness level L_0, is constant except for a single jump at L_0, the jump being positive if $\bar{Q}(P) > 0$ and negative if $\bar{Q}(P) < 0$. Similarly if the system only absorbs heat during a process P – but never emits heat – then $Q(P, \cdot)$ is monotonically increasing. Finally, if P is *adiabatic* – that is, exchanges no heat whatsoever with its environment – then $Q(P, \cdot) \equiv 0$.

The reader should note that in actual physical processes it is possible for heat to be added at one time at a hotness level L_0, and then to be emitted later in the same amount and at the same hotness level. In such a case, the two amounts of heat cancel as far as the accumulation function is concerned, a property of this function which vitally contributes to its usefulness. Because of this kind of cancellation it is clear that one could have $Q(P, \cdot) \equiv 0$ without P being adiabatic, even though (as observed above) it is necessarily true that $Q(P, \cdot) \equiv 0$ when *P is* adiabatic.

The accumulation function is assumed to have the following structural property:

1) For every $P \in \mathbb{P}(\mathscr{S})$ there exists a *lower* hotness level, denoted by $L_l(P)$, or simply by L_l, such that

$$Q(P, L) = 0 \quad \text{when} \quad L \prec L_l$$

and an *upper* hotness level, denoted by L_u, such that

$$Q(P, L) = \bar{Q}(P) \quad \text{when} \quad L \succ L_u .$$

Condition (1) reflects the fact that, for any given process, heat is supplied to or emitted from the system only on some *bounded* range of hotnesses. In particular, if no heat is added at any hotness level below L_l then clearly $Q(P, L) \equiv 0$ for $L \prec L_l$, while similarly if no heat is added at hotness levels above L_u then obviously $Q(P, L) \equiv \bar{Q}(P)$ (the total heat) when $L \succ L_u$. The levels L_l and L_u of course may be different for different processes of the system. For a given process P with $Q(P, \cdot) \not\equiv 0$ it is convenient to specify L_l and L_u uniquely as the highest and lowest levels respectively with the given properties.

Two further structural assumptions are required, the first guaranteeing a minimal degree of regularity for the accumulation function $Q(P, \cdot)$, the second providing an axiomatization for the notion of cycle.[2] These assumptions can be stated as follows:

2) For every $P \in \mathbb{P}(\mathscr{S})$ the function $Q(P, \cdot)$ is bounded and right-continuous, and has at most a denumerable number of discontinuities.

3) To every process $P \in \mathbb{P}_{\text{cyc}}(\mathscr{S})$ and to every positive integer m there corresponds another process $P^{(m)} \in \mathbb{P}_{\text{cyc}}(\mathscr{S})$, called the *m-times repeated cycle of P*, with the properties

$$\bar{W}(P^{(m)}) = m\,\bar{W}(P)$$
$$\bar{Q}(P^{(m)}) = m\,\bar{Q}(P)$$
$$Q(P^{(m)}, \cdot) = m\,Q(P, \cdot) .$$

We define a *thermodynamical universe* to be a set $\mathscr{U}$ of thermodynamical systems $\mathscr{S}$.

In the sequel we shall need several further structural concepts. A particularly valuable idea is that of *products* of thermodynamical systems. The well-known heuristic arguments presented in standard treatments of thermodynamics to justify the classical efficiency theorem, arguments which ultimately go back to Carnot, involve comparing Carnot cycles for two different systems by forming a third (union) system for which the heat and work are found by adding the corresponding quantities for the original systems. In effect, the union idea involves taking the heat emitted by one body and transferring it by some unspecified mechanism, frequently involving a central heat reservoir, directly to a second body, with a corresponding reduction of the heat supplied to the second system from its other surroundings. These well-known but nevertheless somewhat vague ideas require a formal description in order to be useful.

---

2 The notion of cycle also can be replaced as primitive by the idea of "follower", see [1.8, 9].

Let $\mathscr{S}_1$ and $\mathscr{S}_2$ be a pair of physical systems. The *product system*, $\mathscr{S}_1 \oplus \mathscr{S}_2$, is characterized by its processes and their work and heat functions, which are required to satisfy the following conditions:

$$
\begin{array}{lll}
\text{(i)} & \mathbb{P}(\mathscr{S}_1 \oplus \mathscr{S}_2) = \mathbb{P}(\mathscr{S}_1) \times \mathbb{P}(\mathscr{S}_2) & \\
\text{(ii)} & \mathbb{P}_{\mathrm{cyc}}(\mathscr{S}_1 \oplus \mathscr{S}_2) = \mathbb{P}_{\mathrm{cyc}}(\mathscr{S}_1) \times \mathbb{P}_{\mathrm{cyc}}(\mathscr{S}_2) & \\
\text{(iii)} & \bar{W}(P_1 \oplus P_2) > 0 & \text{if} \quad \bar{W}(P_1) + \bar{W}(P_2) > 0 \\
\text{(iv)} & \bar{Q}(P_1 \oplus P_2) < 0 & \text{if} \quad \bar{Q}(P_1) + \bar{Q}(P_2) < 0 \\
\text{(v)} & Q(P_1 \oplus P_2, \cdot) \geqslant 0 & \text{if} \quad Q(P_1, \cdot) + Q(P_2, \cdot) \geqslant 0 .
\end{array}
$$
[3]

Here $P_1 \oplus P_2$ denotes the union process (in $\mathbb{P}(\mathscr{S}_1 \oplus \mathscr{S}_2)$) corresponding to the pair of processes $P_1 \in \mathbb{P}(\mathscr{S}_1)$, $P_2 \in \mathbb{P}(\mathscr{S}_2)$ and $\times$ denotes the Cartesian product.

It is open to question whether the concept of a product system should be meaningful for all conceivable pairs of thermodynamical systems. To avoid such metaphysical points, we shall henceforth restrict the formation of product systems only to special and distinguished pairs of system, which will be called *thermodynamically compatible systems* (or simply *compatible* systems). Thus if $\mathscr{S}_1$ and $\mathscr{S}_2$ are a pair of compatible systems, then the product system $\mathscr{S}_1 \oplus \mathscr{S}_2$ is itself assumed to be a meaningful thermodynamical system satisfying the laws of thermodynamics.

If $\mathscr{U}$ is a collection of thermodynamical systems, we shall say that $\mathscr{U}$ is compatible with a thermodynamical system $\mathscr{T}$ if and only if $\mathscr{T}$ is in $\mathscr{U}$ and each system $\mathscr{S}$ in $\mathscr{U}$ is compatible with $\mathscr{T}$.

Finally we shall say that a process P of a thermodynamic system $\mathscr{S}$ is *weakly reversible* if there exists at least one associated process P' of $\mathscr{S}$ such that

$$
\begin{array}{llll}
\text{(a)} & \bar{W}(P') & = -\bar{W}(P) & \\
\text{(b)} & \bar{Q}(P') & = -\bar{Q}(P) & \\
\text{(c)} & Q(P', \cdot) & = -Q(P, \cdot) & \\
\text{(d)} & P' \in \mathbb{P}_{\mathrm{cyc}}(\mathscr{S}) & \text{if} & P \in \mathbb{P}_{\mathrm{cyc}}(\mathscr{S}) .
\end{array}
$$

The process P' will be called a *weak reversal* of P. Note that there is no requirement that P' be unique for a given reversible process P or that it should follow some "path" reverse to the "path" of P; indeed in general this will not be the case.

Note. The formal structure presented here was first developed during the period 1977–1979 in papers of the author [1.10, 11, 12]. A similar structure was found slightly later but entirely independently by *Šilhavý* [1.13, 14, 15], Šilhavý's development requires considerably deeper topological and measure theoretic considerations, however, and accordingly we follow the treatment in [1.10, 11, 12]. Another concurrently developed and philosophically related approach to the foundations of thermodynamics is due to *Feinberg* and *Lavine*. In their treat-

[3] This is a weaker formulation of the union axiom than is usually stated. For the strong version of the axiom one requires that $\bar{W}(P_1 \oplus P_2) = \bar{W}(P_1) + \bar{W}(P_2)$, and $\bar{Q}(P_1 \oplus P_2) = \bar{Q}(P_1) + \bar{Q}(P_2)$.

ment, beginning in 1978 [1.16] and ultimately published as reference [1.17], it is not necessary to include the notion of hotness as an explicit primitive concept, though this gain is partially balanced by the need to apply fairly deep measure and function theoretic ideas.

Further bibliographical notes are contained in later sections. These notes are intended for accuracy of presentation, but are not necessary for understanding the text.

1.4 The First Law

> • One of the principal objects of theoretical research in any department of knowledge is to find the point of view from which the subject appears in its greatest simplicity.
>
> *J. W. Gibbs* [1.8]

Even since the work of James Joule in the mid-nineteenth century, a first principle in thermodynamic theory has been the basic interconvertibility of heat and work as forms of energy. If one is to state this principle without recourse to special assumptions regarding state spaces and internal energy it appears necessary to use, in one way or another, the general concept of cyclic processes. The principle of interconvertibility of heat and work then asserts that there exists a universal constant $\mathcal{J} > 0$ such that $\bar{W}(P) = \mathcal{J}\bar{Q}(P)$ for any cyclic process P of any physical system.

A particular feature of this formulation which may strike one as unusual is the appearance of the universal constant $\mathcal{J}$. In developing the theory of absolute temperature, for example, the existence of this canonical scale is not postulated, but rather is *derived* from more basic laws. It would therefore seem more appropriate to state the first law without reference to an absolute equivalent of work and heat, and to demonstrate within the theory that such an absolute equivalent must exist. Once one turns in this direction, however, a number of alternatives present themselves, and it is not immediately clear which of these should be taken as the fundamental expression of the relation between work and heat.

The first step in this direction was due to *Truesdell* [1.19], who set up such an axiomatic structure for *reversible systems* and who proved in this framework the existence of internal energy, entropy and an absolute equivalent of work and heat for certain two-variable systems. *Truesdell* and *Bharatha* [1.20] then derived stronger statements from somewhat weaker assumptions of the same kind (see also [1.21, 22]).

Though limited to reversible systems, Truesdell's work nevertheless strikingly shows that a fundamental axiom structure of thermodynamics need not involve a direct postulation that heat and work are interconvertible. The axioms used by Truesdell derive from Carnot's ideas; whether these can lead to a satisfactory theory of irreversible systems is uncertain.

In formulating here an appropriate and general set of laws for thermodynamics we shall seek statements which (following the above discussion) do not give *a priori* significance either to a mechanical equivalent of heat or to the existence of a canonical absolute temperature scale. Moreover, the statements should express only the most certain and secure of our intuitive beliefs about heat and work, should apply to general thermodynamical systems, and above all should yield a comprehensive and satisfying theory. In particular if we give some thought to the gist of the first law, namely that work can only be produced at the expense of heat energy, we are led to the following formulation, due to Šilhavý.

Weak First Law. *If* $\bar{W}(P) > 0$ *for a cyclic process P of a thermodynamical system* $\mathscr{S}$, *then also* $\bar{Q}(P) > 0$.

The weak first law formalizes the idea that positive work can be obtained from a cyclically operating process only when a positive total amount of heat is supplied to the system during the process. While representing a generally weaker requirement than the strict interconvertibility of heat and work, it nevertheless carries great conviction and provides all the normal conclusions drawn from the stronger statement. Of course the weak first law, as stated above, is logically consistent with the strict interconvertibility of work and heat in the sense that if the latter is asserted to hold, then the weak first law is an obvious consequence.

To obtain strict interconvertibility, certainly a desideratum, we shall also consider a stronger version of the first law, again due to Šilhavý.

Strong First Law. *For a cyclic process P of a thermodynamical system* $\mathscr{S}$, *the conditions* $\bar{W}(P) > 0$ *and* $\bar{Q}(P) > 0$ *are equivalent.*

The principal goal of elementary thermodynamics is to provide analytic tools for studying thermal systems. We shall say that a thermodynamic system $\mathscr{R}$ is a *reversible heat engine* if there exists a weakly reversible cyclic process R of $\mathscr{R}$ such that $\bar{W}(R) \neq 0$.[4] The following analytic result then follows as a demonstrated consequence of the Weak First Law.

The Energy Inequality. *Let* $\mathscr{U}$ *be a thermodynamical universe which is compatible with a reversible heat engine. Then there exists a unique universal constant* $\mathscr{J} > 0$ *such that for every cyclic process P of every system* $\mathscr{S}$ *in* $\mathscr{U}$ *we have* $\bar{W}(P) \leqslant \mathscr{J}\bar{Q}(P)$.

Proof of the Energy Inequality. Let $\mathscr{R}$ be a reversible heat engine which is compatible with $\mathscr{U}$, and let R be a reversible cyclic process of $\mathscr{R}$ with $\bar{W}(R) \neq 0$. We may assume that $\bar{W}(R) > 0$, if necessary by interchanging the roles of R and its associated reversal process R'. By the Weak First Law, it follows that $\bar{Q}(R) > 0$. We may thus define the positive quantity

$$\mathscr{J} = \frac{\bar{W}(R)}{\bar{Q}(R)} .$$

Let P be a cyclic process of an arbitrary system $\mathscr{S}$ in $\mathscr{U}$, and suppose for contradiction that

4 An example of such a system is a perfect gas (see footnote 8).

$$\bar{W}(P) > \mathscr{J}\bar{Q}(P) .$$

Let $P^{(m)}$ denote the cycle P repeated m times (see Sect. 1.3) and $R^{(n)}$ the cycle R repeated n times, and consider the joint process $P^{(m)} \oplus R^{(m)}$ of the system $\mathscr{S} \oplus \mathscr{R}$. By (iii) in Sect. 1.3

$$\bar{W}(P^{(m)} \oplus R^{(n)}) > 0 \quad \text{if} \quad \bar{W}(P^{(m)}) + \bar{W}(R^{(n)}) = m\bar{W}(P) + n\bar{W}(R) > 0$$

and similarly by (iv)

$$\bar{Q}(P^{(m)} \oplus R^{(n)}) < 0 \quad \text{if} \quad \bar{Q}(P^{(m)}) + \bar{Q}(R^{(n)}) = m\bar{Q}(P) + n\bar{Q}(R) < 0 .$$

Now define $R^{(-n)}$ when $n > 0$ to be the cycle R' repeated n times, that is $R^{(-n)} = R'^{(n)}$. Then

$$\bar{W}(R^{(-n)}) = \bar{W}(R'^{(n)}) = n\bar{W}(R') = -n\bar{W}(R)$$

and similarly $\bar{Q}(R^{(-n)}) = -n\bar{Q}(R)$. It follows that in the above inequalities n can be allowed to take negative as well as positive integer values.

Consider the vectors A, B in $\mathbb{R}^2$ given by

$$A = \bar{W}(P)i + \bar{W}(R)j , \qquad B = \mathscr{J}\bar{Q}(P)i + \mathscr{J}\bar{Q}(R)j .$$

Their second components are equal and positive, and their first components satisfy $\bar{W}(P) > \mathscr{J}\bar{Q}(P)$. Hence they bear the geometric relationship shown Fig. 1.1.

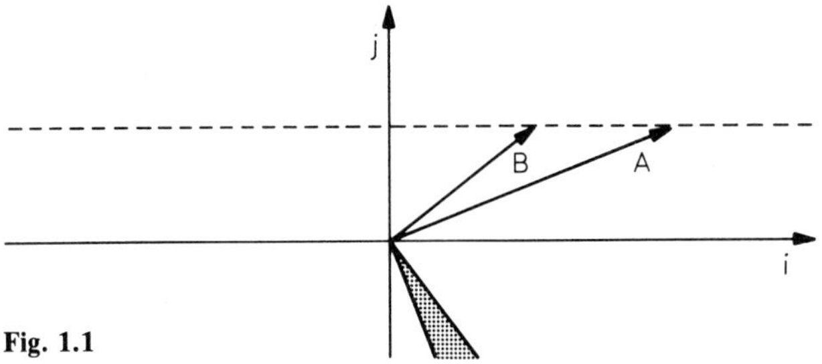

Fig. 1.1

Consequently there exists a vector $C = mi + nj$, with m and n integers, $m > 0$, $n \neq 0$, such that

$$0 < A \cdot C = m\bar{W}(P) + n\bar{W}(R) ,$$
$$0 > B \cdot C = m\mathscr{J}\bar{Q}(P) + n\mathscr{J}\bar{Q}(R) .^{5}$$

Thus $\bar{W}(P^{(m)} \oplus R^{(n)}) > 0$ and yet $\bar{Q}(P^{(m)} \oplus R^{(n)}) < 0$. This contradicts the First Law since $P^{(m)} \oplus R^{(n)}$ is a cyclic process of the thermodynamic system $\mathscr{S} \oplus \mathscr{R}$ by (ii). This proves the condition $\bar{W}(P) \leqslant \mathscr{J}\bar{Q}(P)$.

---

[5] Indeed any lattice point in the shaded area defined by vectors respectively orthogonal to A and B suffices for the endpoint of C. This elegant geometric construction of the pair (m, n) is due to *Robert Hummel*.

To prove that $\mathcal{J}$ is unique, suppose that $\tilde{\mathcal{J}}$ is a second constant with the property that

$$\bar{W}(P) \leqslant \tilde{\mathcal{J}} \bar{Q}(P)$$

for all cyclic processes P of systems $\mathcal{S}$ in $\mathcal{U}$. We choose for P the special cyclic processes R, R' of $\mathcal{R}$. Then, in particular,

$$\bar{W}(R) \leqslant \tilde{\mathcal{J}} \bar{Q}(R) \,, \qquad \bar{W}(R') \leqslant \tilde{\mathcal{J}} \bar{Q}(R') \,.$$

On the other hand, by the properties of R and R' we have

$$\frac{\bar{W}(R)}{\bar{Q}(R)} = \frac{\bar{W}(R')}{\bar{Q}(R')} = \mathcal{J} \,,$$

where $\bar{Q}(R) > 0$, $\bar{Q}(R') < 0$. Thus the preceding two inequalities imply both $\mathcal{J} \leqslant \tilde{\mathcal{J}}$ and $\tilde{\mathcal{J}} \geqslant \mathcal{J}$. Hence $\mathcal{J} = \tilde{\mathcal{J}}$ as required.

The energy inequality of course applies only to thermodynamic systems $\mathcal{S}$ in some universe $\mathcal{U}$ which is compatible with $\mathcal{R}$. Since we may assume, realistically, that any system $\mathcal{S}$ of interest belongs to such a universe it follows that the relation $\bar{W}(P) \leqslant \mathcal{J} \bar{Q}(P)$ can be presumed to hold for cyclic processes of arbitrary thermodynamic systems. The constant $\mathcal{J}$ is called the *mechanical equivalent of heat.*

When the *Strong First Law* is posited instead of the *Weak Law*, and the strong union axiom is assumed (see footnote 3), a similar proof yields the conclusion

$$\bar{W}(P) = \mathcal{J} \bar{Q}(P)$$

for all cyclic processes P of systems $\mathcal{S}$ in the universe $\mathcal{U}$. That is, the Strong First Law is equivalent to the interconvertibility of heat and work for arbitrary cyclic processes.

In what follows, if a conclusion is valid whenever *either* the Weak or Strong First Law is assumed to hold, we shall simply say that it follows from the "First Law". If the Strong First Law is *necessary* to obtain the conclusion we shall remark this separately.

Note. Both the weak and strong versions of the First Law were given by *Šilhavý* in his fundamental paper [1.13], though in his context some additional topological considerations appear which are extraneous to our purposes. Šilhavý moreover does not emphasize the independent importance of these two versions, as we do here: see [1.13], Part II, Theorem 2.2.1; [1.14], Part 1, Sects. 4.6, 4.7; and [1.14], Part II, Section 4.12.

The energy inequality, as a possible axiomatic expression of the first law, was first noted by *Fosdick* and the author [1.23]. The inequality, as a theorem based on the Weak First Law, is due to *Šilhavý* ([1.13], Part II, Sect. 2.1). The present proof of the energy inequality is simpler than that given in [1.13], because of the different underlying thermodynamic structure used here.

1.5 The Second Law

- The scientist must order; science is made out of facts as a house out of stones, but an accumulation of facts is no more a science than a heap of stones is a house.

H. Poincaré

The second law of thermodynamics involves more subtle ideas than the first since it deals with the quality of heat at different hotness levels. Moreover, the physical notions which originally motivated the various nineteenth century statements of the second law are fairly obscure, requiring some effort to phrase clearly.

The first essentially correct formulation of the second law is due to Rudolf Clausius. Here is his original formulation of 1850, as paraphrased by *Kelvin* [1.24]:

It is impossible for a self-acting machine, unaided by any external agency, to convey heat from one body to another at a higher temperature.

Later *Clausius* [1.25] restated this in the form:

A passage of heat from a colder to a hotter body cannot take place without compensation.

While this is not at all precisely stated, we may consider it to mean that if a cyclic process absorbs heat at a hotness level L_0 and emits heat at a hotness level $L_1 \succ L_0$ then necessarily $\bar{W}(P) < 0$. In terms of the accumulation function $Q(P, \cdot)$ this takes the form

Second Law (Clausius). *The accumulation function of a cyclic process P cannot have a single positive jump followed by a single negative jump, unless* $\bar{W}(P) < 0$.

That is, the accumulation function of a cyclic process with $\bar{W}(P) \geqslant 0$ cannot have any of the forms shown:

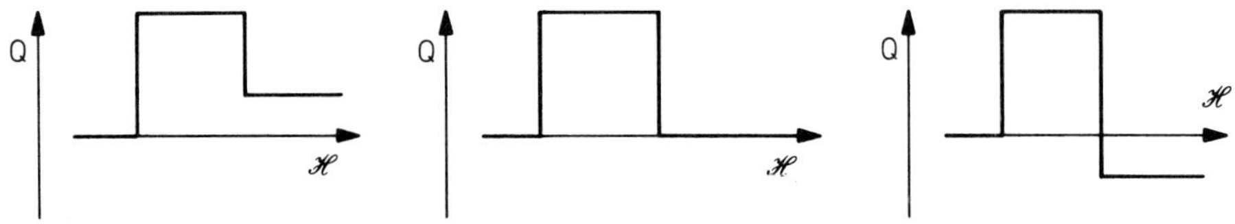

This version of the second law is not easily applicable to the case of general thermal processes without the further intervention of sophisticated topological notions. On the other hand, a slightly stronger formulation of essentially the same idea can be given which avoids this difficulty. In particular we observe that when $\bar{W}(P) \geqslant 0$ the third picture above cannot apply because of the First Law. Now, if graphs of the first two types are disallowed by Clausius's version of the second law, it seems equally the case that no *linear combination* of such graphs could occur as the accumulation function of a cyclic process with $\bar{W}(P) \geqslant 0$. Indeed, such an accumulation function would represent a process which raises various low temperature heat supplies to various higher temperatures, without the need of doing work on the system. In the same way, any *closure* of such linear combinations would also appear impossible for cyclic processes (by continuity considerations) at least if $\bar{W}(P) > 0$. But the set of such closures coincides with the set of *non-negative* accumulation functions. We are thus led to the following general version of the second law, proposed first by Serrin.

Second Law. *The condition $Q(P, \cdot) \geqslant 0$ can occur for a cyclic process P of a thermodynamical system $\mathscr{S}$ only in the exceptional case when $Q(P, \cdot) \equiv 0$.*

The second law as stated obviously implies the Clausius Law. Its plausibility further derives from the physical idea that should heat be *so strongly* added to a system that the accumulation function is non-negative at every hotness level, and positive at least at some hotness levels, then the system must necessarily move away from its initial condition.

The reader may observe an interesting duality between the First and Second Laws, namely that when $\bar{W}(P) > 0$ for a cyclic process P the former requires a *positive* value for $\bar{Q}(P)$ while the latter implies a *negative* value for $Q(P, \cdot)$ at some hotness level. This result can be stated formally as follows.

Combined Laws. *For any cyclic process P with $\bar{W}(P) > 0$ there holds*

$$\begin{aligned} Q(P,L) &> 0 \quad \textit{for some} \quad L \succ L_u \\ Q(P,L) &< 0 \quad \textit{for some} \quad L \prec L_u . \end{aligned}$$

The Second Law is an *intrinsic* statement about the relation between heat and hotness in cyclic processes. In parallel with the discussion of the First Law in the previous section, the Second Law also has an equivalent *analytical* formulation of great usefulness. We state this as follows.

The Accumulation Theorem. *Let $\mathscr{U}$ be a thermodynamical universe which is thermodynamically compatible with a perfect gas $\mathscr{G}$. Then there exists an (absolute) temperature scale $\hat{T}$ on the hotness manifold $\mathscr{H}$, with $\hat{T}(\mathscr{H}) \equiv \mathbb{R}^+$, such that for every cyclic process P of every thermodynamic system $\mathscr{S}$ in $\mathscr{U}$ we have*

$$\int_0^\infty \frac{Q(P,L)}{T^2}\, dT \leqslant 0 ,$$

where $L = \hat{L}(T)$ is the hotness level associated with the temperature T in the scale $\hat{T}$. Any temperature scale $\hat{T}$ with the above property either agrees with the perfect gas scale of $\mathscr{G}$ or is a positive constant multiple of this scale.

The accumulation theorem immediately accomplishes two purposes: *It establishes the concept of absolute temperature without ambiguity, and it characterizes once and for all the allowable behavior of the accumulation function of any cyclic process.*

Indeed the accumulation theorem implies the Second Law, for if $Q(P, \cdot) \geqslant 0$ in a cyclic process P then the accumulation integral above will of necessity be positive unless $Q(P, \cdot) \equiv 0$.

The reader should also observe that the accumulation integral is well-defined and finite, as follows easily from properties (1) and (2) of the accumulation function given in Sect. 1.3.

The accumulation inequality is a generalization of the Clausius inequality noted in Sect. 1.2. In particular, should the function $Q(P, \cdot)$ be continuously differentable with respect to T, then an integration by parts shows that

$$\int_0^\infty \frac{Q(P,L)}{T^2} dT = \int_0^\infty \frac{dQ(P,L)}{dT} \frac{dT}{T} ,$$

the latter integral representing the "sum" of the heats added divided by their absolute temperatures.[6] The advantage of the accumulation integral compared to the Clausius integral is that it can be expressed analytically in terms of clearly formulated primitive concepts, and at the same time is applicable to a broader class of processes since its existence relies only on the structural properties (1) and (2) of the accumulation function.

Proof of the Accumulation Theorem. We first define the required temperature scale $\hat{T}$. By hypothesis the universe $\mathscr{U}$ is compatible with a perfect gas $\mathscr{G}$. Let $\theta: \mathscr{H} \to \mathbb{R}^+$ be the gas scale defined by $\mathscr{G}$, by Charles' Law an invertible mapping of $\mathscr{H}$ onto $\mathbb{R}^+$. We shall prove below that θ is a *temperature* scale, that is, that θ is strictly monotone increasing. Assuming this for the moment, we name this scale $\hat{T}$ and shall show that it suffices for the validity of the accumulation inequality.

Thus suppose for contradiction that in the scale $\hat{T}$ there is some cyclic process C of some system $\mathscr{S}$ in $\mathscr{U}$ such that

$$\int_0^\infty \frac{Q(C,L)}{T^2} dT > 0 .$$

Our purpose will be to construct an auxiliary cyclic process of $\mathscr{G}$ which, *when considered together with the cyclic process C,* provides a violation of the second law. To do this we state the following

Principal Lemma. *There is a cyclic process G of $\mathscr{G}$ with the properties*

$$\bar{W}(G) + \bar{W}(C) > 0 , \qquad Q(G, \cdot) + Q(C, \cdot) \geqslant 0 .^7$$

---

[6] More generally, for the analytically minded reader, should $Q(P, \cdot)$ be of bounded variation then we can use Stieltjes integration to write

$$\int_0^\infty \frac{Q(P,L)}{T^2} dT = \int_0^\infty \frac{dQ(P,L)}{T} .$$

[7] For a proof of this lemma see *Serrin* [1.12] or *Coleman, Owen* and *Serrin* [1.32]. In these papers it is also shown how to replace the perfect gas $\mathscr{G}$ in the proof by less special model materials – essentially those with a suitably rich supply of Carnot cycles – thus avoiding Kelvin's criticism (at the expense, however, of a longer proof).

In standard thermodynamic arguments results having essentially the same physical content as the Principal Lemma are usually taken for granted.

Using this result, consider the union process $C \oplus G$ of the thermodynamic system $\mathscr{S} \oplus \mathscr{G}$. Conditions (iii) and (v) in Sect. 1.3 imply that $\bar{W}(C \oplus G) > 0$ and $Q(C \oplus G, \cdot) \geqslant 0$ respectively. By (ii) the process $C \oplus G$ is cyclic. Hence from the First Law we have $\bar{Q}(C \oplus G) > 0$, while by the Second Law $Q(C \oplus G, \cdot) \equiv 0$. This is impossible, however, in view of the second part of property (1) of the accumulation function. Consequently we have established that the temperature scale $\hat{T} = \theta$ serves as an *absolute* temperature in which the accumulation inequality is valid.

Now assume that $\tilde{T}$ is a another scale such that

$$\int_0^\infty \frac{Q(P,L)}{T^2}\, dT \leqslant 0, \qquad L = \tilde{L}(T),$$

for cyclic processes. We must show that $\tilde{T}$ is a constant multiple of the gas scale θ. To this end let us choose for P in the preceding inequality a Carnot cycle R of the perfect gas $\mathscr{G}$. Since R is reversible it follows easily that

$$\int_0^\infty \frac{Q(R,L)}{T^2}\, dT = 0, \qquad L = \tilde{L}(T).$$

(cf. also Sect. 1.7). Now by well known properties of Carnot cycles for perfect gases it is clear that there is such a cycle R operating between any two hotness levels L_1 and L_2 ($L_1 \prec L_2$) and that for such a cycle[8]

$$\frac{Q_2}{Q_1} = \frac{\theta_2}{\theta_1},$$

where Q_2 is the heat absorbed at L_2, Q_1 is the heat emitted at L_1, and θ_1, θ_2 are the gas temperatures at L_1 and L_2. The accumulation function for R then has the form

$$Q(R,L) = \begin{cases} 0 & \text{if} \quad L \prec L_1 \\ -Q_1 & \text{if} \quad L_1 \preccurlyeq L \prec L_2 \\ Q_2 - Q_1 & \text{if} \quad L_2 \preccurlyeq L\,. \end{cases}$$

Consequently the integral relation above (in the scale $\tilde{T}$) yields the equality

$$-Q_1\left(\frac{1}{T_1} - \frac{1}{T_2}\right) + (Q_2 - Q_1)\frac{1}{T_2} = 0$$

where T_1 and T_2 are the absolute temperature of L_1 and L_2 *in the scale* $\tilde{T}$. Rearranging this result gives

---

[8] The argument for this is briefly as follows. The heat form (see Sect. 1.8) of a perfect gas $\mathscr{G}$ is given by

$$q = du + p\,dv$$

where $u = u(\theta)$ and $p = R\theta/v$. Hence

$$q/\theta = d(R \log v + \textstyle\int du/\theta)\,,$$

an exact differential, from which the structure of the Carnot cycles of $\mathscr{G}$ is easily deduced.

$$\frac{Q_2}{Q_1} = \frac{T_2}{T_1},$$

whence

$$\frac{T_2}{T_1} = \frac{\theta_2}{\theta_1}.$$

By fixing L_1 and varying L_2 it is now clear that $\tilde{T} = \text{const}\ \theta$.

We observe that if $L_2 \succ L_1$ as above, then necessarily $Q_2 > Q_1$ – for otherwise the Second Law would be violated by the reverse of the Carnot cycle in question. Hence in turn $\theta_2 > \theta_1$, showing that θ is strictly increasing and thus completing the proof of the accumulation theorem.

From the proof it follows that we can physically recognize the direction of increasing hotness by means of a perfect gas thermometer.

Another method to determine this direction involves the use of general Carnot cycles; if a Carnot cycle operates between hotness levels L_1 and L_2, absorbing heat at L_2 and emitting heat at L_1, and does positive work, then necessarily $L_1 \prec L_2$. Indeed, if this were not so then (since $\bar{Q}(P) > 0$ by the First Law) the accumulation function $Q(P, \cdot)$ for the process would have the graph shown:

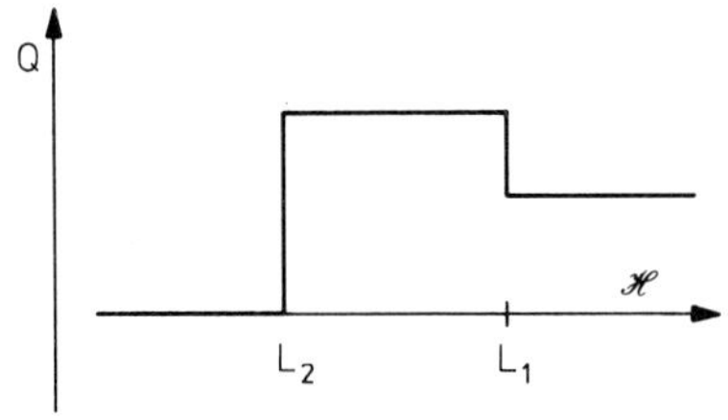

But this is impossible according to the Second Law, and so $L_1 \prec L_2$.

The analytical results which we have obtained so far provide a rigorous setting for phenomenological thermodynamics. In particular, the Weak First Law takes the analytical form

$$\bar{W}(P) \leqslant \mathscr{J}\bar{Q}(P) \quad \text{for all} \quad P \in \mathbb{P}_{\text{cyc}}(\mathscr{S}),$$

the Strong First Law the form

$$\bar{W}(P) = \mathscr{J}\bar{Q}(P) \quad \text{for all} \quad P \in \mathbb{P}_{\text{cyc}}(\mathscr{S}),$$

and the Second Law the form

$$\int_0^\infty \frac{Q(P, L)}{T^2}\, dT \leqslant 0 \quad \text{for all} \quad P \in \mathbb{P}_{\text{cyc}}(\mathscr{S}).$$

Here $\mathscr{S}$ of course refers to an arbitrary thermodynamic system.

For convenience in the sequel we shall suppose that the units used to measure heat are normalized so that the constant $\mathscr{J}$ has the value 1.

Appendix to Sect. 1.5: The Direction of Increasing Hotness. From the point of view of foundational axiomatics one might wish to weaken the hypotheses defining the hotness manifold, giving up the condition that there is a distinguished direction of increasing hotness and retaining only the one dimensional topological structure. If this is done, however, several unavoidable consequences ensue, each having definite costs. To begin with, one must give up the possibility of any second law reflecting Clausius' idea that cyclic processes cannot raise hotness levels without the expenditure of work for if there is no a priori direction of increasing hotness such a law would lose its very meaning. But if one cannot use a Clausius-type second law then reliance must be placed on a second law of Kelvin's form. As experience has shown, however, this in turn forces the introduction of fairly deep topological considerations into the fundamental structure in order to obtain a useful theory.

In any case, the physically relevant issue in a discussion of increasing hotness levels is not *whether* such a direction exists. Indeed we are convinced by experience that there is such a direction, and we would be shocked by a theory which neither supposed nor proved this fact. From a physical point of view, rather, what *is* important is whether a theory can provide some experimental or observational method for determining this direction. That is, even though we believe (by sensation) that there is a direction of increasing hotness, we require an objective method for recognizing it and, even more, a method for distinguishing which of two hotness levels is the hotter one. But all modern theories do provide such a method; the theory in this paper, for example, shows that a perfect gas thermometer does the job, or alternately, at a deeper level, the use of Carnot cycles as discussed above.

The principal difference at this level between a theory which assumes an a priori direction and one which does not, is that in the former type a model material (such as a perfect gas) can be proved to be an empirical thermometer, while in the latter some model material must be endowed with a directional character – after which a direction of increasing hotness can be proved to exist on the basis of the fundamental structure.

Note. The Second Law and the Accumulation Theorem in the form stated here appeared first in references [1.10, 1.11]. The proof of the Accumulation Theorem was originally given in [1.12]. For other related points of view, see [1.8, 9, 13–17].

1.6 State Structure and Potentials

In order to provide a concrete framework for the notions of internal energy and entropy it is necessary to introduce the idea of a state space, and an associated state structure. At the simplest (and most general) level this may be defined as follows.

A *state structure* for a system $\mathscr{S}$ consists of a set Σ, whose elements are called *states* of the system, and a corresponding family of processes $\mathbb{P}_\Sigma(\mathscr{S}) \subset \mathbb{P}(\mathscr{S})$, with each process $P \in \mathbb{P}_\Sigma(\mathscr{S})$ having a well-defined initial state $P_i \in \Sigma$ and final state $P_f \in \Sigma$.[9] Moreover if P is a cyclic process in $\mathbb{P}_\Sigma(\mathscr{S})$ then $P_i = P_f$. [In practice, a state structure should also be compatible with the notion of a process P *following* another process P', and should include in this case the axiom $P_i = P'_f$. Similarly, the structure may include (at least in some weak sense) the intuitive notion that if two processes P, P' satisfy $P_f = P'_i$ then P' can follow P.]

We shall say that a system $\mathscr{S}$ has an *internal energy* corresponding to the state structure $(\Sigma, \mathbb{P}_\Sigma)$ if there exists a function

$$U\colon \Sigma \to \mathbb{R}$$

such that

$$\Delta U \leqslant \bar{Q}(P) - \bar{W}(P)$$

for each $P \in \mathbb{P}_\Sigma(\mathscr{S})$. Here ΔU denotes the difference between U evaluated at the final state and the initial state of P, that is

$$\Delta U \equiv U(P_f) - U(P_i)\ .$$

Roughly speaking, then, a function U is an internal energy for a system if it is a lower potential for the difference $\bar{Q}(P) - \bar{W}(P)$.

If $P \in \mathbb{P}_{\mathrm{cyc}}(\mathscr{S}) \cap \mathbb{P}_\Sigma(\mathscr{S})$ then necessarily $P_i = P_f$ and in turn $\Delta U = 0$. Consequently we recover the energy inequality

$$\bar{W}(P) \leqslant \bar{Q}(P)$$

from the above formula, this in fact being the motivation for the definition of internal energy.

In parallel with the Strong First Law, we may also introduce the idea of a strong internal energy, in which the inequality $\Delta U \leqslant \bar{Q}(P) - \bar{W}(P)$ is replaced by the stronger requirement

$$\Delta U = \bar{Q}(P) - \bar{W}(P)\ .$$

Turning to the Second Law it is natural to proceed in a similar way, but now based on the accumulation inequality. For convenience in formulation, we introduce the abbreviation $\bar{A}(P)$ for the integral appearing in the accumulation theorem; thus

$$\bar{A}(P) \equiv \int_0^\infty \frac{Q(P, L)}{T^2}\, dT\ ;$$

naturally, once one has a definite absolute temperature scale in hand one can define $\bar{A}(P)$ whether or not the process P is cyclic. This being understood, we shall say that a system $\mathscr{S}$ has an *entropy* corresponding to the state structure $(\Sigma, \mathbb{P}_\Sigma)$ if there exists a function

$$S\colon \Sigma \to \mathbb{R}$$

such that

$$\Delta S \geqslant \bar{A}(P)$$

[9] Formally, the assignment of initial and final states can be considered as a pair of mappings

$$i\colon \mathbb{P}_\Sigma(\mathscr{S}) \to \Sigma\ , \quad i(P) = P_i$$
$$f\colon \mathbb{P}_\Sigma(\mathscr{S}) \to \Sigma\ , \quad f(P) = P_f\ .$$

for each $P \in \mathbb{P}_\Sigma(\mathscr{S})$. To express this in succinct form, one can say that *entropy is an upper potential for the accumulation integral* $\bar{A}(P)$.

If $P \in \mathbb{P}_{cyc}(\mathscr{S}) \cap \mathbb{P}_\Sigma(\mathscr{S})$ then of course $\Delta S = 0$ so that we recover the cyclic condition

$$\bar{A}(P) \leqslant 0$$

stated in the accumulation theorem. Another case of interest is that of an *adiabatic process*, for which $Q(P, \cdot) \equiv 0$. In this situation one has $\bar{A}(P) = 0$, whence in turn the entropy hypothesis yields

$$\Delta S \geqslant 0 ,$$

the condition $\mathrm{II_A}$ noted at the beginning of the paper. Other direct consequences of the formulas $\Delta U \leqslant \bar{Q} - \bar{W}$, $\Delta S \geqslant \bar{A}$, including the classical efficiency theorem of Kelvin, are noted in my appended remarks to Professor Truesdell's paper following in this volume.

A system $\mathscr{S}$ is said to satisfy the *energy-entropy hypothesis* for a state structure if there exists an internal energy function and an entropy function corresponding to this structure.

It is one of the principal conclusions of elementary thermodynamics that simple reversible systems necessarily possess both an entropy and an internal energy. As we shall see in Sect. 1.8, this result is an easy, almost immediate consequence of the general structure exhibited above.[10]

On the other hand, reversible systems by no means exhaust the range of thermodynamic interest: it is not self-evident that a *given* system, with its given preassigned structure, will possess either an entropy or an internal energy. Thus the construction of an appropriate state structure becomes a matter of paramount importance. The state space Σ must be large enough to accommodate a non-trivial set of processes $\mathbb{P}_\Sigma$, but at the same time not so large as to be uninteresting or carry superfluous information. Different process classes $\mathbb{P}_\Sigma$ may be of importance for different purposes, and accordingly a given system might have more than one useful state structure. Finally a state structure can serve not only as a vehicle for the concept of internal energy and entropy, but also as a way of formulating the constitutive structure of the system itself, that is, the dependence of working and heating in the system upon changes of state. This idea, of funda-

[10] Historically, the existence of internal energy for reversible systems is due to Clausius for the special case of a perfect gas and to Kelvin for two-variable reversible materials (see [1.34], p. 227). Kelvin's argument applies unchanged to n-variable reversible systems, and is essentially the one given below in Sect. 1.8. The first derivation of existence of entropy for reversible systems is due to Rankine (see [1.34], pp. 215, 333), though his derivation was neither particularly clear nor applicable beyond two dimensional state spaces. A rigorous derivation (of the two dimensional result) was indicated by Reech a little later, and carried to its conclusion by *Truesdell* and *Bharatha* [1.20].

A number of different methods have since been devised for extending the Rankine-Reech theorem to n-variable state spaces, the most recent being due to *Pitteri* [1.29] on the basis of the axiomatic structure introduced in [1.27]. The simple and direct method here adopted in Sect. 1.8 appeared first in [1.11].

A thorough discussion and evaluation of the various proofs now available, their interrelation and their advantages and disadvantages, would be an important contribution to the subject though no light task.

mental importance to the discussion of actual systems, can be given a strict formal development as we shall see in Sect. 1.8.

For any particular state structure $(\Sigma, \mathbb{P}_\Sigma)$ the *existence* of internal energy and entropy is a desideratum. In addition, it is equally important to know whether they are *unique*, or under what circumstances one can expect this to be the case. A partial answer to the latter question is contained in the theorem at the end of Sect. 1.7.

All these issues are closely related to the First and Second Laws, as is of course apparent. Not so apparent is the specific role played by cycles. If the class $\mathbb{P}_\Sigma(\mathscr{S})$ contains a *rich* supply of cycles, then the cyclic laws alone may suffice to settle the question of existence and uniqueness. But in other cases, as emphasized in the important work of *Day* [1.26] and *Coleman* and *Owen* [1.27 – 31], proving the existence of entropy can require more subtle versions of the Second Law in which statements are made about non-cyclic as well as cyclic processes. Further work in this direction, due to *Coleman, Owen,* and *Serrin* [1.32], *Owen* [1.33], and *Šilhavý*, is discussed elsewhere in this volume.

Finally, in his dissertion *Ricou* [1.8] has generalized these ideas to obtain the remarkable result that if a state structure is deterministic for a given system (that is, if the condition $P_i' = P_f$ implies that the process P' can follow P) then the system must have an internal energy and an entropy.

1.7 Reversible Processes

Recall that a process P of a thermodynamic system $\mathscr{S}$ is weakly reversible if there exists an associated process P' of $\mathscr{S}$ such that

$$\begin{aligned} &\text{(a)} \quad \bar{W}(P') &&= -\bar{W}(P) \\ &\text{(b)} \quad \bar{Q}(P') &&= -\bar{Q}(P) \\ &\text{(c)} \quad Q(P',\cdot) &&= -Q(P,\cdot) \\ &\text{(d)} \quad P' \in \mathbb{P}_{\text{cyc}}(\mathscr{S}) &&\quad \text{if} \quad P \in \mathbb{P}_{\text{cyc}}(\mathscr{S})\,. \end{aligned}$$

Suppose in addition that $P, P' \in \mathbb{P}_\Sigma(\mathscr{S})$ for some state structure $(\Sigma, \mathbb{P}_\Sigma)$ and that

$$\text{(e)} \quad P_i' = P_f\,, \qquad P_f' = P_i\,.$$

Then the process is called *strongly reversible* with respect to the state structure.

Now consider the energy inequality for a weakly reversible cyclic process. We have first of all the inequality $\bar{W}(P) \leqslant \bar{Q}(P)$. From (a) and (b) and the energy inequality for the reversal process P' it follows that $\bar{W}(P) \geqslant \bar{Q}(P)$. Hence, *for any cyclic weakly reversible process P of a thermodynamical system there holds*

$$\bar{W}(P) = \bar{Q}(P)\,,$$

That is, for such processes, heat and work are strictly interconvertible – even without the intervention of the Strong First Law!

Next suppose that P is strongly reversible for a state structure $(\Sigma, \mathbb{P}_\Sigma)$ in which $\mathscr{S}$ has an internal energy function U. Then, proceeding exactly as before, we obtain the conclusion

$$\Delta U = \bar{Q}(P) - \bar{W}(P) ,$$

one of the standard results of classical thermodynamics, again proved without recourse to the Strong First Law.

We turn next to the accumulation theorem. If P is a weakly reversible cyclic process then

$$\int_0^\infty \frac{Q(P,L)}{T^2} dT \leqslant 0 \quad \text{and so} \quad \int_0^\infty \frac{Q(P',L)}{T^2} dT \geqslant 0 .$$

But the second integral must also be non-positive by the accumulation theorem. Therefore, *for any cyclic weakly reversible process P of any thermodynamic system $\mathscr{S}$ we have*

$$\bar{A}(P) = \int_0^\infty \frac{Q(P,L)}{T^2} dT = 0 .$$

Similarly if the process P is strongly reversible for a state structure in which $\mathscr{S}$ has an entropy, we obtain the conclusion

$$\Delta S = \bar{A}(P) .$$

The above results immediately imply the following interesting

Theorem. *Let $\mathscr{S}$ be a thermodynamic system satisfying the energy-entropy hypothesis for a state structure $(\Sigma, \mathbb{P}_\Sigma)$. Suppose that there exists a "base" state $\sigma_0 \in \Sigma$ such that to each $\sigma \in \Sigma$ there corresponds at least one strongly reversible process P with initial state σ_0 and final state σ. Then both the internal energy and the entropy are unique up to arbitrary constants (namely the values $U(\sigma_0)$ and $S(\sigma_0)$).*

This result supplies the underlying logic for textbook statements that entropy can be calculated by means of reversible processes joining states in question. (Unfortunately, the textbook derivations are not formulated with any degree of explicitness, which can mislead readers who fail to notice the qualifications that are an integral part of the theorem.)

Partial converses of this result, and extensions to quasi-reversible processes, have been given by *Coleman* and *Owen* [1.27], *Feinberg* and *Lavine* [1.17], and by *Ricou* and *Serrin*.

1.8 Special Systems

> • If we review the classical treatments, we see that they did not distinguish between the general laws of thermodynamics and the constitutive equations defining particular thermodynamic bodies.
>
> *C. Truesdell* [1.35]

It is of considerable interest to apply the preceding ideas to several concrete systems in order to see the advantages of a rigorous approach to thermodynamic structure. We shall consider particularly the classical reversible systems of standard thermodynamics, and the structure of classical single constituent continuum mechanics. As has always been apparent in the thermodynamic literature, a clear understanding of these cases, especially the first, is crucial to the development of the subject. Indeed, in light of these examples it becomes possible to give a strict definition of the notion of constitutive structure in thermodynamics. This we do at the end of the section.

As has already been emphasized, our main purpose is to describe a general thermodynamical structure, rather than to apply this to particular systems. Accordingly it almost goes without saying that the following discussions will be brief and given without accompanying motivations, and that they will leave more specialized considerations aside.

1.8.1 Reversible Systems

A classical reversible system $\mathscr{S}$ is, in the first instance, endowed with a finite dimensional state space Σ (a connected open subset of $\mathbb{R}^k$) with the property that each process $P \in \mathbb{P}(\mathscr{S})$ has a unique corresponding *path* in Σ,

$$\Gamma\colon I \to \Sigma\,,$$

where $I = [a, b]$ is a closed time interval and the function $t \to \Gamma(t)$ is piecewise smooth.[11] Naturally the path of any cyclic process is assumed to be closed, that is $\Gamma(a) = \Gamma(b)$.

It is assumed that there are two differential forms w and q (with continuous coefficients) defined on Σ, with the property that

$$\bar{W}(P) = \int_\Gamma w\,, \qquad \bar{Q}(P) = \int_\Gamma q\,,$$

Γ being the path associated with P.

The accumulation function $Q(P, \cdot)$ is determined through the assignment of a hotness level $\mathscr{L}(\sigma)$ to each state $\sigma \in \Sigma$, that is

$$\mathscr{L}\colon \Sigma \to \mathscr{H}\,.$$

Assuming that $\mathscr{L}$ is continuous, we then put

[11] We do not study quasi-static processes, though in many ways their theory is parallel or even identical.

$$Q(P,L)=\int_{\Gamma(L)} q$$

where $\Gamma(L)$ denotes that "part" of the path Γ where the associated hotness level $\mathscr{L}$ does not exceed L (i.e. that part of the path where $\mathscr{L}\circ\Gamma \preccurlyeq L$).

To complete the formulation of a reversible system we assume that to each path Γ there corresponds at least one process P with the path Γ, and in particular that this process is cyclic if Γ is closed. Note, from a mathematical point of view, that the association from processes to paths is *onto* but not necessarily *one-to-one.*

The system can be given a "natural" state structure, with $\mathbb{P}_\Sigma(\mathscr{S})\equiv\mathbb{P}(\mathscr{S})$, by defining the initial and final states of any process in the obvious way,

$$P_i=\Gamma(a)\ ,\qquad P_f=\Gamma(b)\ .$$

This being the case, it is easy to see that *every process* $P\in\mathbb{P}(\mathscr{S})$ *is strongly reversible.*[12] It follows moreover from the results of the preceding section that, provided an entropy and an internal energy exist for $\mathscr{S}$, they are unique up to additive constants and satisfy the conditions

$$\Delta U=\bar{Q}(P)-\bar{W}(P)\ ,\qquad \Delta S=\bar{A}(P)\ .$$

Of course this is not the whole story: we must still deal with the question of *existence* of U and S as well as their calculation in terms of the assigned structure of the system, namely the differential forms q and w and the hotness function $\mathscr{L}$.

The case of internal energy is fairly simple. By the First Law and reversibility there holds, for all cyclic processes P,

$$\bar{Q}(P)=\bar{W}(P)$$

(recall $\mathscr{J}=1$). Choosing an arbitrary closed piecewise smooth path Γ, and letting P be some associated cyclic process, it follows from the formulas for $\bar{Q}$ and $\bar{W}$ that

$$\int_\Gamma q=\int_\Gamma w\ .$$

The differential form $q-w$ is thus exact, and accordingly there exists a continuously differentable function U such that

$$dU=q-w\ .$$

In turn, for any process P of $\mathscr{S}$ we have

$$\bar{Q}(P)-\bar{W}(P)=\int_\Gamma q-\int_\Gamma w=\int_\Gamma dU=\Delta U\ ,$$

proving that U is an internal energy for $\mathscr{S}$.

---

12 It is for this reason that *systems* of the kind under discussion are called reversible. The reader should bear in mind that while every process of a reversible system is strongly reversible, nevertheless not every reversible process belongs to a reversible system.

The procedure to establish existence of entropy is more involved, but uses similar ideas (see [1.11]). We observe that, for any process P of the system $\mathscr{S}$,

$$\bar{A}(P) = \int_0^\infty \frac{1}{T^2} \left\{ \int_{\Gamma(L)} q \right\} dT .$$

To this iterated integration we apply Fubini's theorem, noting that the inner integral can be expressed as a time integration over the set of values $t \in I$ where $(\hat{T} \circ \mathscr{L} \circ \Gamma)(t) \leqslant T$. Thus we find

$$A(P) = \int_\Gamma \frac{q}{T} ,$$

where $T = \hat{T} \circ \mathscr{L}$ is the absolute temperature associated with a state point σ. Since $\bar{A}(P) = 0$ for any cyclic process, by virtue of the Second Law and reversibility (note the closely parallel treatment of the First Law), it follows as before that the differential form q/T is exact. Hence there exists an entropy function $S: \Sigma \to \mathbb{R}$ such that

$$dS = \frac{q}{T} .$$

This result is the precise formulation of condition $\mathrm{II_B}$ given at the beginning of the paper (Sect. 1.2).

From the above discussion, it is clear that $\mathrm{II_B}$ is not a general principle of thermodynamics but rather a special relation holding for a restricted class of systems, a fact which cannot be too strongly emphasized. To be fair however, one should add that $\mathrm{II_B}$ (namely, the fact that the absolute temperature is an integrating divisor for the heat form of a reversible system) *does express fully the content of the second law for reversible systems.* To see this, consider a cyclic process P of a reversible system. Assuming that q/T is exact we see that $\int_\Gamma q/T = 0$ and in turn $\bar{A}(P) = 0$. Thus if $Q(P, \cdot) \geqslant 0$ then necessarily $Q(P, \cdot) \equiv 0$, as required.

In order to calculate the internal energy and entropy of a reversible system in terms of its characteristic differential forms q and w one can use the conditions

$$dU = q - w , \qquad T\,dS = q$$

derived above. Moreover, as demonstrated in every text, these conditions yield those numerous interrelations between q, w and T which are the lifeblood of the classical subject.

1.8.2 Continuous Media

In continuum theory the basic assumption is that hotness is a well-defined field concept. Let the motion and the hotness of a body (or system) then be defined by

$$\begin{aligned} x &= \chi(X, t) \colon \mathscr{B} \times I \to \mathbb{R}^3 \\ L &= \lambda(X, t) \colon \mathscr{B} \times I \to \mathscr{H} , \end{aligned} \qquad (*)$$

where for each $t \in I$ the mapping $x = \chi(X, t)$ is one-to-one and smooth from the reference configuration $\mathscr{B}$ into $\mathbb{R}^3$. The reference configuration is itself a closed smoothly bounded subset of $\mathbb{R}^3$.

The configuration comes equipped with a reference density $\varrho_0\colon \mathscr{B} \to \mathbb{R}^+$, whence the mass density at any time t can be defined by

$$\varrho(X, t) = \frac{\varrho_0(X)}{\det F(X, t)},$$

where

$$F(X, t) = \operatorname{grad} \chi = \left(\frac{\partial \chi}{\partial X}\right)$$

is the Jacobian of χ; here it is natural to require that $\det F > 0$.

The velocity vector $v = \partial\chi/\partial t$ is a mapping from $\mathscr{B} \times I$ into $\mathbb{R}^3$. Define $\mathscr{S}(t) = \chi(\mathscr{B}, t)$ to be the point set in $\mathbb{R}^3$ occupied by the body at time t and let $\partial\mathscr{S}(t)$ denote its boundary. If f is the body force vector per unit mass, τ the surface force vector, r the body rate of heat supply per unit mass, h the surface rate of heat supply, and

$$\mathscr{S}_L(t) = \{x \in \mathscr{S}(t) \,|\, \lambda(x, t) \preccurlyeq L\}, \qquad \partial_L \mathscr{S}(t) = \{x \in \partial\mathscr{S}(t) \,|\, \lambda(x, t) \preccurlyeq L\},$$

then (see [1.10, 15]) we have the following expressions for the work, heat and accumulation functions of the process P given by the motion (*):

$$\bar{W}(P) = -\int_I \left\{ \int_{\mathscr{S}(t)} \varrho f \cdot v \, dx + \int_{\partial\mathscr{S}(t)} \tau \cdot v \, ds \right\} dt$$

$$\bar{Q}(P) = \int_I \left\{ \int_{\mathscr{S}(t)} \varrho r \, dx + \int_{\partial\mathscr{S}(t)} h \, ds \right\} dt \tag{**}$$

$$Q(P, L) = \int_I \left\{ \int_{\mathscr{S}_L(t)} \varrho r \, dx + \int_{\partial_L \mathscr{S}(t)} h \, ds \right\} dt.$$

Here it is of course assumed that the integrands are bounded and continuous as functions of x and t. An easy calculation using Fubini's theorem now shows that additionally

$$\bar{A}(P) = \int_I \left\{ \int_{\mathscr{S}(t)} \frac{\varrho r}{T} dx + \int_{\partial\mathscr{S}(t)} \frac{h}{T} ds \right\} dt$$

where $T = \hat{T} \circ \lambda\colon \mathscr{B} \times I \to \mathbb{R}^+$ is the absolute temperature associated with the motion.

The energy-entropy hypothesis requires somewhat more care in a dynamical setting because of the presence of kinetic energy. While this quantity could be treated formally as part of the internal energy, it is in fact more convenient to distinguish between these types of energy and to write the energy-entropy hypothesis in the form

$$\Delta(U + K) \leqslant \bar{Q}(P) - \bar{W}(P)$$
$$\Delta S \geqslant \bar{A}(P),$$

where K denotes the kinetic energy, given by the relation

$$K(t) = \int_{\mathscr{S}(t)} \frac{1}{2} \varrho v^2 dx \, .$$

When the Strong First Law is assumed, one naturally should replace the first inequality above by equality, e.g.

$$\Delta(U+K) = \bar{Q}(P) - \bar{W}(P) \, .$$

Finally, it should be noted that no particular state structure is placed in evidence, since we wish to maintain a maximum generality.

The classical energy balance equation and the Clausius-Duhem inequality are closely related to the above considerations. In particular, let us assume further that a "local" state can be associated with each material particle in the body, during the motion, and that an internal energy density ε and entropy density η exist, functionally dependent on the local state. The total internal energy and entropy at any instant t are then given by

$$U(t) = \int_{\mathscr{S}(t)} \varrho \varepsilon \, dx \, , \qquad S(t) = \int_{\mathscr{S}(t)} \varrho \eta \, dx \, .$$

The preceding inequalities now take the form

$$[U(b)+K(b)] - [U(a)+K(a)] \leqslant \bar{Q}(P) - \bar{W}(P)$$
$$S(b) - S(a) \geqslant \bar{A}(P) \, .$$

In general neither of the functions $U(t)$ or $S(t)$ can be presumed continuously differentiable. If this assumption is made, however, then assuming that the process P induces subprocesses corresponding to arbitrary subintervals of I, it follows immediately that

$$\frac{d}{dt} \int_{\mathscr{S}(t)} \varrho \left(\varepsilon + \frac{1}{2} v^2 \right) dx \leqslant \int_{\mathscr{S}(t)} (\varrho f \cdot v + r) \, dx + \int_{\partial \mathscr{S}(t)} (\tau \cdot v + h) \, ds$$
$$\frac{d}{dt} \int_{\mathscr{S}(t)} \varrho \eta \, dx \geqslant \int_{\mathscr{S}(t)} \frac{\varrho r}{T} dx + \int_{\partial \mathscr{S}(t)} \frac{h}{T} ds \, .$$

If the Strong First Law is assumed, the first inequality should be changed to equality.

These relations are to be interpreted in the sense that they must hold identically for every process (*) available to the body. The first of these relations, typically in its equality form, is the equation of energy balance, the second is the Clausius-Duhem inequality (see [1.35]). This derivation, based directly on the energy-entropy hypothesis (and certain differentiability assumptions), provides the underlying motivation for their validity.[13]

---

[13] If the body force f and the heat supply r are eliminated from the energy inequality and the Clausius-Duhem inequality (to eliminate f requires also Cauchy's equations of motion), one

In conclusion, I would like to signal a possibly useful generalization of the formula for work in (**). As discussed in the paper of Dunn in this volume, a body whose work $\bar{W}(P)$ is given by (**) may be unable to tolerate stress responses of certain useful types. For this reason, it is valuable to include a generalized surface working i (independent of velocity) in (**). Thus one can propose the formula

$$\bar{W}(P) = \int_T \left\{ \int_{\mathscr{S}(t)} \varrho f \cdot v \, dx + \int_{\partial\mathscr{S}(t)} (\tau \cdot v + i) \, ds \right\} dt$$

as being more appropriate to the case of a general continuous body. If internal spins are present, as in micropolar media or liquid crystals, one must naturally add further terms to account for their effects; considerations of this type are naturally far beyond the intentions of this opening essay.

1.8.3 Constitutive Structure[14]

A set of processes $\mathbb{C}(\mathscr{S}) \subset \mathbb{P}(\mathscr{S})$ of a system is said to possess a *constitutive structure subordinate to the state space* Σ if to each process $P \in \mathbb{C}(\mathscr{S})$ there corresponds a *path*

$$\Gamma: I \to \Sigma\,, \qquad I = [a, b] \subset \mathbb{R}$$

with the following properties:

(i) If two processes in $\mathbb{C}(\mathscr{S})$ have the same path then they have the same total work $\bar{W}$, total heat $\bar{Q}$, and accumulation function Q.

(ii) If $P \in \mathbb{C}(\mathscr{S})$ is cyclic then the corresponding path Γ is closed, that is $\Gamma(a) = \Gamma(b)$.

(iii) Let $P \in \mathbb{C}(\mathscr{S})$ with corresponding path Γ. Let $\tau \in (a, b)$ and put $I_\tau = [a, \tau]$, $I'_\tau = [\tau, b]$. Then there exist processes P_τ, P'_τ in $\mathbb{C}(\mathscr{S})$, with the respective paths Γ restricted to I_τ and Γ restricted to I'_τ, such that

$$\begin{aligned} \bar{W}(P) &= \bar{W}(P_\tau) + \bar{W}(P'_\tau) \\ \bar{Q}(P) &= \bar{Q}(P_\tau) + \bar{Q}(P'_\tau) \\ Q(P, \cdot) &= Q(P_\tau, \cdot) + Q(P'_\tau, \cdot)\,. \end{aligned}$$

Conditions (i – iii) are obviously satisfied in the case of reversible systems, (i) because the line integrals defining $\bar{W}$, $\bar{Q}$, and Q depend only on the path Γ (the differential forms q and w are fixed for the system), and (iii) because integration is additive. For a reversible system the state space is finite dimensional and simple in its structure. For continuum mechanics the state space is infinite dimensional,

---

obtains the so-called reduced Clausius-Duhem inequality, a key relation in the study of material behavior. It is interesting to note that this result, like so many others, relies only on the weak rather than the strong form of the First Law. Another aspect of this situation occurs in Gibbsian theory, where the various minimization and stability theorems are unchanged whether or not the Strong First Law is invoked.

[14] The presentation here is crucially dependent on the important ideas of *Coleman* and *Owen* [1.27].

generally a *field* of local states. For materials with memory the states are particularly sophisticated, involving the past history of the particles themselves.

If $\mathbb{C}(\mathscr{S})$ has a constitutive structure subordinate to Σ, then the pair $(\Sigma, \mathbb{C})$ can be given a "natural" state structure by putting, for any $P \in \mathbb{C}(\mathscr{S})$,

$$P_i = \Gamma(a) , \quad P_f = \Gamma(b) .$$

In turn, within this state structure, the problem of entropy and internal energy can be formulated as in the previous sections.

We remark that $\mathbb{C}(\mathscr{S})$ can easily be a proper subset of $\mathbb{P}(\mathscr{S})$, or even of $\mathbb{P}_\Sigma(\mathscr{S})$ – the set of processes with well defined initial and final states in Σ. In this way, in fact, the constitutive structure of the subset $\mathbb{C}(\mathscr{S})$ can influence the behavior of processes going from one equilibrium state in Σ to another, even though the processes themselves need not be describable by means of paths in the state space.

1.9 Concluding Remarks

- One may feel a security in the conclusions of a [postulational] analysis which is impossible in a less formal structure.

P. W. Bridgman [1.36]

Reviewing the structure of phenomenological thermodynamics, one finds a beautiful edifice of immense aesthetic and physical appeal. First and foremost, thermodynamics is the science of *heat*, and the relations existing on the one hand between heat and work and on the other between heat and hotness; see Table 1.1. Two intrinsic principles govern these relations – the First and Second Laws of Thermodynamics. Each states a reasonable, even if somewhat pessimistic, conviction about the physical world. There are, next, analytical formulations of

Table 1.1

General thermodynamic structure (Sects. 1.3 – 7)		
Primitive concepts	Heat and Work	Heat and Hotness
Intrinsic law	First Law	Second Law
Analytical formulation	Energy Inequality	Accumulation Inequality
Scale	Mechanical Equivalent of Heat	Absolute Temperature
Potential	Internal Energy	Entropy
Reversible systems (Sect. 1.8)		
Constitutive structure	Heat form q, Work form w	Heat form q, Temperature T
Associated potential	$dU = q - w$	$dS = q/T$

Note. When the Strong First Law is assumed, the listing "Energy Inequality" should be replaced by "Energy Balance". All other entries are the same whether or not the Strong First Law is used.

these laws – first, the energy inequality (or, if the Strong First Law is used, the interconvertibility of heat and work for cycles), and second, the accumulation inequality. These analytical formulations pave the way to all direct applications of the theory. They rely in turn on two derived *scale* concepts – the Joule mechanical equivalent of heat and the Kelvin absolute temperature scale, each among the greatest conceptions of Nineteenth Century physics. Finally, there are two fundamental potentials, or more accurately semi-potentials – the internal energy and the entropy – which extend the direct cyclic principles to much broader classes of thermal processes. On this structure hangs all thermal science, from the elementary theory of reversible systems, to Gibbs' magnificent conception of thermal and chemical equilibrium, to sophisticated theories of material dynamics.

The most far-reaching implication of the structure, however, is the fact that it is *not* limited to equilibrium. In fact, a dogmatism which would lay this restriction on thermodynamics would in turn invalidate a massive sector of thermal physics, including heat transfer theory, compressible fluid mechanics, shock wave theory, and combustion theory. Conversely, allowing thermodynamics a natural scope beyond equilibrium yields a powerful and far-reaching theory with which to attack those *dynamical* problems where hotness and heat play a crucial role.

References

1.1 H. A. Buchdahl: *The Concepts of Classical Thermodynamics* (Cambridge University Press, Cambridge 1966)

1.2 A. H. Wilson: *Thermodynamics and Statistical Mechanics* (Cambridge University Press, Cambridge 1957)

1.3 J. W. Thompson (Lord Kelvin): Encyclopedia Brittanica, 9th edition (1878)

1.4 D. S. L. Cardwell: *From Watt to Clausius* (Cornell University Press, Ithaca 1971)

1.5 P. Fong: "Second Thoughts on the Second Law of Thermodynamics, Its Past, Present and Future", in *A Critical Review of Thermodynamics* (Mono Book Company, 1970) p. 407

1.6 J. Bronowski: *The Common Sense of Science* (London 1951)

1.7 J. W. Gibbs: *Graphical Methods in the Thermodynamics of Fluids.* Collected Works, Vol. 1, p. 10

1.8 M. Ricou: Energy, entropy, and the laws of thermodynamics. Ph. D. Thesis, University of Minnesota (1983); see also the final paper in Part I of this volume

1.9 M. Ricou, J. Serrin: A general second law of thermodynamics (to appear)

1.10 J. Serrin: *The Concepts of Thermodynamics. Contemporary Developments in Continuum Mechanics and Partial Differential Equations* (North Holland, Amsterdam 1978) pp. 411 – 451

1.11 J. Serrin: Conceptual analysis of the classical second laws of thermodynamics. Arch. Rational Mech. Anal. **70**, 355 – 371 (1979)

1.12 J. Serrin: Lectures on thermodynamics. University of Naples (1979)

1.13 M. Šilhavý: On measures, convex cones, and foundations of thermodynamics. I. Systems with vector-valued actions; II. Thermodynamic systems. Czech. J. Phys. **B30**, 841 – 861, 961 – 991 (1980)

1.14 M. Šilhavý: On the second law of thermodynamics for cyclic processes. I. General framework; II. Inequalities for cyclic processes. Czech. J. Phys. **B32**, 987 – 1010, 1073 – 1099 (1982)

1.15 M. Šilhavý: On the Clausius inequality. Arch. Rational Mech. Anal. **81**, 221 – 243 (1983) (Work originally presented at Euromech 111 Symposium, September 1978)

1.16 M. Feinberg: Private communication (1978)
1.17 M. Feinberg, R. Levine: Thermodynamics based on the Hahn-Banach theorem: the Clausius inequality. Arch. Rational Mech. Anal. **82**, 203 – 293 (1983)
1.18 J. W. Gibbs: *Letter of Acceptance for the Rumford Medal* (1881)
1.19 C. Truesdell: "How to Understand and Teach the Logical Structure and the History of Classical Thermodynamics", in *Proceedings of the International Congress of Mathematicians* (Vancouver, 1974) pp. 577 – 586
1.20 C. Truesdell, S. Bharatha: *The Concepts and Logic of Classical Thermodynamics* (Springer, Berlin, Heidelberg, New York 1977)
1.21 C. Truesdell: Absolute temperatures as a consequence of Carnot's general axiom. Arch. Hist. Exact Sci. **20**, 357 – 380 (1980)
1.22 M. Pitteri: Classical thermodynamics of homogeneous systems based upon Carnot's general axiom. Arch. Rational Mech. Anal. **80**, 333 – 385 (1982)
1.23 R. L. Fosdick, J. Serrin: Global properties of continuum thermodynamic processes. Arch. Rational Mech. Anal. **59**, 108 – 109 (1975)
1.24 J. W. Thompson (Lord Kelvin): *Mathematical and Physical Papers* (Cambridge University Press, Cambridge 1882); see especially pages 178 – 181
1.25 R. Clausius: *Abhandlungen über die mechanische Wärmetheorie* (Braunschweig 1864, 1867) [Translated by W. R. Browne as *The Mechanical Theory of Heat* (London, 1879); see especially pages 76 – 79]
1.26 W. A. Day: *The Thermodynamics of Simple Materials with Fading Memory* (Springer, Berlin, Heidelberg, New York 1972)
1.27 B. Coleman, D. Owen: A mathematical foundation for thermodynamics. Arch. Rational Mech. Anal. **54**, 1 – 104 (1974)
1.28 B. Coleman, D. Owen: On thermodynamics and elastic-plastic materials. Arch. Rational Mech. Anal. **59**, 25 – 51 (1975)
1.29 B. Coleman, D. Owen: On thermodynamics and intrinsically equilibrated materials. Ann. Mat. Pura Appl. (IV) **108**, 189 – 199 (1976)
1.30 B. Coleman, D. Owen: On the thermodynamics of semi-systems with restrictions on the accessibility of states. Arch. Rational Mech. Anal. **66**, 173 – 181 (1977)
1.31 B. Coleman, D. Owen: On the thermodynamics of elastic-plastic materials with temperature-dependent moduli and yield stresses. Arch. Rational Mech. Anal. **70**, 339 – 354 (1979)
1.32 B. Coleman, D. Owen, J. Serrin: The second law of thermodynamics for systems with approximate cycles. Arch. Rational Mech. Anal. **77**, 103 – 142 (1981)
1.33 D. Owen: The second law of thermodynamics for semi-systems with few approximate cycles. Arch. Rational Mech. Anal. **80**, 39 – 55 (1982)
1.34 C. Truesdell: *The Tragicomical History of Thermodynamics, 1822 – 1854* (Springer, Berlin, Heidelberg, New York 1980)
1.35 C. Truesdell: *Rational Thermodynamics* (McGraw-Hill, New York 1969)
1.36 P. W. Bridgman: *The Nature of Thermodynamics* (Harvard University Press, London 1941)

Chapter 2
Foundations of Continuum Thermodynamics

M. Šilhavý

2.1 Introduction

There have been many attempts to derive the existence of energy, entropy, and absolute temperature from the statements of the first and second laws of thermodynamics which do not presuppose the existence of these quantities. Until very recently, the rigorous part of this research has been based essentially upon the use of analytic and differential geometric methods in finite dimensional space [1].

As I have explained elsewhere [2.4], this approach generally fails to provide a basis for the existing structure of continuum thermodynamics as presented, e.g., by *Truesdell* [2.5].

In the past, the attempts to make the foundations of thermodynamics of irreversible inhomogeneous processes precise were blocked by the lack of an appropriate tool to describe the exchange of heat between a body and its environment in terms suitable for the statement of the second law of thermodynamics. In [2.4, 6, 7] I introduced Borel measures on the set of empirical temperatures to describe the information about this exchange of heat. Explicitly, with each process there is associated a Borel measure called the heat measure; the value of this measure on a given set of empirical temperatures is the gain of heat of the system at empirical temperatures from that set. [2]

An important ingredient in the approach developed in [2.6] and [2.4] is the concept of a thermodynamic system. Each system has states and can undergo processes, i.e., time-evolutions of states. [3] With each process there is associated a number, the work done by the system in the process, and a Borel measure on

---

1 See, e.g., *Carathéodory* [2.1], *Arens* [2.2] and *Boyling* [2.3].

2 A similar idea was suggested independently by *Serrin* [2.8, footnote p. 427]. In fact it was only after the essential parts of Professor Serrin's and my researches were completed that we realized close similarities of our results, not published at that time. Early in 1979 I sent a preliminary but complete version of [2.7] to Professor *D. R. Owen*. In response he sent me a preprint of *Serrin* [2.8] of which I was unaware. A final version of [2.7] was sent to Professor *J. Serrin* in the summer of 1979 (a printed version of [2.7] with updated references, but without any change of results, has appeared recently as [2.9]. *Serrin's* ideas were meanwhile made more explicit in [2.10], and a complete structure of thermodynamics, similar to that of *Šilhavý* [2.4], appears in *Coleman, Owen* and *Serrin* [2.11]. The proofs in *Šilhavý* [2.4] are based on the extensive use of the separation theorems in topological a linear space of measures, and a recent paper by *Feinberg* and *Lavine* [2.12] employs essentially the same mathematical tools.

3 Description of dynamical systems within the state space formalism similar to that employed in [2.6] and [2.4] was developed within different contexts by *Noll* [2.13], *Coleman* and *Owen* [2.14] and *Gurtin* [2.15].

the set of empirical temperatures, the heat measure of the process. The class of thermodynamic systems admits a natural operation of a composition of two thermodynamic systems to produce a new thermodynamic system. The theory deals with thermodynamic universes which are collections of thermodynamic systems closed under the compositions of thermodynamic systems, and which contains sufficiently many ideal reversible Carnot processes. Within this framework, the basic program of foundations of thermodynamics is carried out:

The First Law. Various versions of the first law are given in [2.4] which imply the universal proportionality of work and heat in cyclic processes and the existence of energy satisfying the equation of balance of energy. Alternatively, a weaker postulate is employed in [2.4] which implies the energy inequality rather than the equation of balance of energy.

The Second Law. The verbal statements of the second law are converted into mathematically meaningful postulates (see *Šilhavý* [2.4, 6, 7, 16, 17]). Relations among these statements are established; the existence of the absolute temperature is shown to be a consequence of various forms of the second law under appropriate additional assumptions; the absolute temperature satisfies the Clausius inequality for cyclic processes. Moreover, using an earlier idea of *Day* [2.18], the Clausius inequality is shown to imply the existence of non-equilibrium entropy satisfying the abstract version of the Clausius-Duhem inequality[4].

It is also shown in *Šilhavý* [2.4] that the first and the second laws can be stated in a formally similar way and that the above described consequences of these laws can be deduced from a general theory developed in the first part of *Šilhavý* [2.4]. The universal proportionality of work and heat in cyclic processes and the existence of the absolute temperature are established there by an essentially geometrical method which sheds additional light on the nature of these concepts. A less technical description of this geometrical approach is given by *Kratochvil* and *Šilhavý* [2.20]. Another survey of the results of *Šilhavý* [2.4] is contained in an appendix to the forthcoming second edition of *Truesdell's* "Rational Thermodynamics" [2.17].

The approach to the foundation of thermodynamics described above is extended to more general systems (which do not necessarily satisfy the accessibility axiom of [2.4]) in *Šilhavý* [2.21]. For more detailed information about this approach see Sect. 2.2.2.

The abstract theory [2.4] and its more general version [2.21] apply to continuous bodies in inhomogeneous irreversible processes. In [2.22] I develop an approach to the foundations of thermodynamics of continuous bodies. The basic definition of a continuous body does not presuppose any equation or inequality such as the balance equations and the Clausius-Duhem inequality. The statement of the first law of thermodynamics then implies the existence of energy satisfying the equations of balance of energy; moreover, the equation of balance of energy

4 A slightly different condition implying the existence of entropy and absolute temperature for genuinely non-equilibrium material is given in *Šilhavý* [2.19].

together with the principle of material frame indifference then yield (cf. *Green* and *Rivlin* [2.29]) the equations of balance of linear and angular momentum. Taking the concept of empirical temperature for granted, one can also define the heat measure corresponding to a process in terms of the rest of the structure of continuous bodies and the second law employed in the general theory then leads to the existence of absolute temperature and entropy satisfying the Clausius-Duhem inequality.

In addition to this, the rich structure of continuous bodies, the explicit account of heat through the heat flux and the body heating, enables one to remove the empirical temperature from the list of primitive quantities. Roughly speaking it turns out that the empirical temperature is determined by the second law to within a universal change of scale, provided the universe of bodies has an additional structure. In particular, one does not need the so called zeroth law to introduce and determine the empirical temperature[5].

2.2 Work, Heat, and Empirical Temperature

The purpose of this section is to develop a conceptual structure appropriate for the discussion of the basic laws of thermodynamics and their basic consequences. Thus, after a preliminary discussion of the concepts of work and heat measure in Sect. 2.2.1, the concept of a thermodynamic system is defined in Sect. 2.2.2. Section 2.2.3 then deals with universes of thermodynamic systems, some basic postulates expressing the presence of sufficiently many ideal reversible Carnot processes to be used in the sequel are formulated here.

2.2.1 Work and Heat Measures

Basically, thermodynamics is a science of work, heat, and temperature. Within the abstract theory presented here, work is a primitive undefined concept, i.e., work is a number associated with each process.

Also the concepts of heat and empirical temperature are taken for granted here. They are combined, however, in a single primitive concept, the concept of a heat measure (*Šilhavý* [2.4, 6, 7]). Throughout, $I \subset R$ denotes a fixed open interval which is interpreted as the set of empirical temperatures. Assuming that the empirical temperature is meaningful in "non-equilibrium situations", it is always possible to determine, for a given process of a system, and a given set of empirical temperatures $A \subset I$, the net gain, $Q(A)$, of heat of the system at empirical temperatures from the set A. Thus, each process determines a set function $A \to Q(A)$, defined on certain subsets of empirical temperatures. It is clear from the interpretation of this function that if $A_1, A_2 \subset I$ are two disjoint sets of empirical temperatures, $A_1 \cap A_2 = \phi$, then

$$Q(A_1 \cup A_2) = Q(A_1) + Q(A_2) \; ,$$

[5] See *Šilhavý* [2.22]; a particular case of this result can be found in *Šilhavý* [2.24].

i.e., $Q(\cdot)$ is an additive set function. For continuous bodies, the function Q is defined explicitly in terms of the fields of empirical temperature, the body heat supply, and the heat flux vector. *Šilhavý* [2.7] shows that there is essentially no loss of generality in assuming that Q is a finite real-valued signed Borel measure with compact support contained in I, i.e., it turns out that $Q(\cdot)$ is additive also on infinite sequences of mutually disjoint Borel sets.

Henceforth we shall assume that the set function $Q(\cdot)$ associated with the process is a finite real valued signed Borel measure with compact support contained in I; Q will be referred to as the heat measure of the process.

We denote by $\boldsymbol{M}(I)$ the set of all finite real valued signed Borel measures with compact support contained in I. The elements Q of $\boldsymbol{M}(I)$ will be referred to as measures; $\boldsymbol{M}(I)$ is an infinite dimensional real linear space.

Let $w \in R$ be the work done by the system in the process, and $Q \in \boldsymbol{M}(I)$ be the heat measure corresponding to the process. For a given process of a thermodynamic system, the pair $(w, Q) \in R \times \boldsymbol{M}(I)$, contains all the information about the work and exchange of heat in the process necessary for the statement of the laws of thermodynamics and their consequences.

Furthermore, a topology on $R \times \boldsymbol{M}(I)$ is used to determine when two pairs (w_1, Q_1) and (w_2, Q_2) are close to each other. Namely, $\boldsymbol{M}(I)$ is endowed (*Šilhavý* [2.4, 16]) with the weakest topology which renders all the functionals

$$Q \rightarrow \int f \, dQ$$

continuous, where f is a fixed continuous function on I. $R \times \boldsymbol{M}(I)$ is endowed with the product topology. This particular choice of the topology can be justified by a set of simple and physically motivated postulates (*Šilhavý* [2.4, 16, 21]).

2.2.2 Thermodynamic Systems

Basic to the approach described in *Šilhavý* [2.4, 6] is the concept of a thermodynamic system. At this level of generality a thermodynamic system is specified by giving its set of states, the class of processes, and two mappings which associate with each process the work done by the system in the process and the heat measure corresponding to the process. Processes are viewed as time-evolutions of states. We will now make the discussion precise.

Let $S = (\Sigma, \Pi, \bar{w}, \bar{Q})$ be a four-tuple of objects, where Σ is a set; Π is a set of paths in Σ, i.e., a set of mappings π of the type

$$\pi\colon [0, d_\pi] \rightarrow \Sigma\,, \tag{2.1}$$

where $d_\pi > 0$ depends on π, $\bar{w}$ is a mapping

$$\bar{w}\colon \Pi \rightarrow R\,;$$

and Q is a mapping

$$\bar{Q}\colon \Pi \rightarrow \boldsymbol{M}(I)\,.$$

The four-tuple $S = (\Sigma, \Pi, \bar{w}, \bar{Q})$ is called a *thermodynamic system* if it satisfies the conditions (i – iii) formulated below. The set Σ is then called the *state space*, its elements $\sigma \in \Sigma$ *states*; Π is called the *class of processes* of the system, the elements $\pi \in \Pi$ are *processes*. For each process $\pi \in \Pi$ the number d_π occurring in (2.1) is called the *duration* of the process; the values $\pi^{\mathrm{i}} = \pi(0)$ and $\pi^{\mathrm{f}} = \pi(d_\pi)$ are then referred to as the *initial* and the *final states* of π, respectively. For a given process $\pi \in \Pi$, the number $\bar{w}(\pi)$ is the *work done by the system in the process*, and the measure $\bar{Q}(\pi) \in \boldsymbol{M}(I)$ is the *heat measure corresponding to the process* π.

i) If $\pi_1, \pi_2 \in \Pi$ and $\pi_1^{\mathrm{f}} = \pi_2^{\mathrm{i}}$, then the path $\pi_1 * \pi_2$, of duration $d_{\pi_1} + d_{\pi_2}$, given by

$$\pi_1 * \pi_2(t) = \begin{cases} \pi_1(t)\,, & t \in [0, d_{\pi_1}] \\ \pi_2(t - d_{\pi_1})\,, & t \in [d_{\pi_1}, d_{\pi_1} + d_{\pi_2}]\,, \end{cases}$$

belongs to Π, i.e.,

$$\pi_1 * \pi_2 \in \Pi\,;$$

moreover

$$\bar{w}(\pi_1 * \pi_2) = \bar{w}(\pi_1) + \bar{w}(\pi_2)\,,$$

$$\bar{Q}(\pi_1 * \pi_2) = \bar{Q}(\pi_1) + \bar{Q}(\pi_2)\,.$$

This assumption says that if the system undergoes two processes π_1 and π_2 such that the final state of the first process coincides with the initial state of the second process, then these two processes give rise to a new process $\pi_1 * \pi_2$ in which the system undergoes the process π_1 and then π_2 continuously. The process $\pi_1 * \pi_2$ is called the continuation of π_1 with π_2. Moreover, the work and the heat measure corresponding to the continuation $\pi_1 * \pi_2$ are the sum of the quantities corresponding to the original processes π_1 and π_2.

ii) There exists at least one state $\sigma \in \Sigma$ with the following properties: for each $d > 0$ the path $\sigma^{[d]}$, of duration d, given by

$$\sigma^{[d]}(t) = \sigma\,, \quad 0 \leqslant t \leqslant d$$

is a process, i.e.,

$$\sigma^{[d]} \in \Pi\,, \tag{2.2}$$

which satisfies

$$\bar{w}(\sigma^{[d]}) = 0 \tag{2.3}$$

and

$$\bar{Q}(\sigma^{[d]}) = 0\,. \tag{2.4}$$

Any state $\sigma \in \Sigma$ satisfying (2.2), (2.3), and (2.4) will be referred to as an equilibrium state of the system; thus, an equilibrium state of a system is a state in which the system can stay an arbitrarily long time without doing work or exchanging heat with its environment. Condition (ii) asserts the existence of at least one equilibrium state.

iii) If $\sigma_1, \sigma_2 \in \Sigma$, then there exists a $\pi \in \Pi$ with $\pi^{\mathrm{i}} = \sigma_1$ and $\pi^{\mathrm{f}} = \sigma_2$.

In other words, any two states σ_1, σ_2 can be connected by at least one process of the system. Elastic, viscous, and certain elastic-plastic bodies satisfy this condition, but certain bodies, such as the bodies made of material with fading memory, satisfy this condition only approximately (cf. *Coleman* and *Owen* [2.14]). I assume condition (iii) here to make the discussion simple; in *Šilhavý* [2.21] I extend my approach of [2.4] to systems satisfying (iii) only approximately. In a stimulating paper, *Coleman, Owen* and *Serrin* [2.11] also deal with the second law of thermodynamics for systems satisfying (iii) only approximately. However, the version of the Clausius inequality derived in [2.11] from a basic statement of the second law is not strong enough to ensure the existence of entropy for general systems. The existence of entropy is proved in [2.11] only for thermally bounded systems, which, roughly speaking, are the systems which exchange heat below a given a priori bound of empirical temperatures. No such restriction occurs in *Šilhavý* [2.4] and thus the generality gained in [2.11] in one respect is accompanied by a loss of generality in another respect. More importantly, the thermal boundedness restriction occurring in [2.11] is not inherent to the second law for systems which do not satisfy (iii) strictly, as I show in [2.21]. In that paper I also show that the entropy can be constructed under slightly more general circumstances than in the paper by *Coleman* and *Owen* [2.14]. Namely, it turns out that it suffices to assume the existence of small perturbations of processes; in these perturbations the final state of the process, the value of work, and of the Clausius integral are assumed to be close to their respective quantities corresponding to the original process. This allows me to relax some specific assumptions made in [2.14], and the resulting theory is more suitable for the discussion of the qualitative properties of global continuous bodies such as the uniqueness of the solutions of the corresponding equations, continuous dependence on initial data, and stability (*Šilhavý* [2.22])[6].

2.2.3 Universes of Thermodynamic Systems

The present theory deals with classes of thermodynamic systems which are closed under the operation of the sum of two thermodynamic systems. Roughly speaking, a sum of two thermodynamic systems is a new thermodynamic system which corresponds to viewing these two systems as a single system. This should not be confused with the operation of putting two systems into thermal contact. It was Carnot who recognized the significance of considering two thermodynamic systems as a single thermodynamic system. Explicit mathematical definitions of the sum of two systems occur frequently in the literature (cf. *Arens* [2.2]) within the framework of reversible homogeneous systems. We now proceed to a formal definition of the sum of two thermodynamic systems within the present framework (*Šilhavý* [2.4]).

---

[6] The generalizations described above were announced at this Workshop. Full details were presented by myself in a lecture at the Institute for Mathematics and its Applications, Minneapolis, MN, USA, on June 22, 1983. The author wishes to thank Professors Coleman, Owen and Serrin for their valuable comments during this lecture.

Let $S_\alpha = (\Sigma_\alpha, \Pi_\alpha, \bar{w}_\alpha, \bar{Q}_\alpha)$, $\alpha = 1, 2$ be two thermodynamic systems; their sum, $S_1 + S_2$, is the thermodynamic system $S = (\Sigma, \Pi, \bar{w}, \bar{Q})$, where the state space Σ is given by

$$\Sigma = \Sigma_1 \times \Sigma_2 ;$$

the class Π of processes consists of those paths $\pi: [0, d_\pi] \to \Sigma$, for which there exist processes $\pi_1 \in \Pi_1$, $\pi_2 \in \Pi_2$ with

$$d_{\pi_1} = d_{\pi_2} = d_\pi$$

such that

$$\pi(t) = (\pi_1(t), \pi_2(t)) , \tag{2.5}$$

for all $t \in [0, d_\pi]$; the values of $\bar{w}$ and $\bar{Q}$ on the process π are then given by

$$\bar{w}(\pi) = \bar{w}_1(\pi_1) + \bar{w}_2(\pi_2)$$

and

$$\bar{Q}(\pi) = \bar{Q}_1(\pi_1) + \bar{Q}_2(\pi_2) ,$$

where $\pi_1 \in \Pi_1$, and $\pi_2 \in \Pi_2$ are as in (2.5).

It turns out that (*Šilhavý* [2.4]) the four-tuple $S = (\Sigma, \Pi, \bar{w}, \bar{Q})$ really is a thermodynamic system, i.e., satisfies conditions (i – iii) of the preceding section.

A collection U of thermodynamic systems is called a universe of thermodynamic systems if it is closed under the operation of addition of thermodynamic systems, i.e., if $S_1, S_2 \in U$ imply $S_1 + S_2 \in U$.

Having now defined the concept of a universe of thermodynamic systems, we proceed to introduce various sets of pairs, $(w, Q) \in R \times M(I)$, associated with cyclic processes starting from equilibrium of systems within a given universe. Let $S = (\Sigma, \Pi, \bar{w}, \bar{Q})$ be a thermodynamic system. A process $\pi \in \Pi$ is said to be a cyclic process starting from equilibrium if

$$\pi^{\mathrm{i}} = \pi^{\mathrm{f}}$$

and if π^{i} is an equilibrium state in the sense of assumption (ii) of Sect. 2.2.2.

If U is a universe of thermodynamic systems, we denote by $A_0(U)$ the set of all pairs $(w, Q) \in R \times M(I)$ corresponding to cyclic processes starting from equilibrium of systems from U, i.e.,

$$\begin{aligned} A_0(U) = \{(w, Q) \in R \times M(I)\colon & \text{ there exists a system } S = (\Sigma, \Pi, \bar{w}, \bar{Q}) \in U \text{ and} \\ & \text{a cyclic process starting from equilibrium } \pi \in \Pi \text{ such that} \\ & (w, Q) = (\bar{w}(\pi), \bar{Q}(\pi))\} . \end{aligned}$$

It is also convenient to consider the closure, $A(U)$, of the set $A_0(U)$ in $R \times M(I)$ being endowed with the topology described in Sect. 2.2.1. The operation of the closure adds to $A_0(U)$ certain pairs (w, Q) which need not correspond to any real cyclic process starting from equilibrium of any system from U; these pairs are interpreted as pairs corresponding to the ideal cyclic processes starting from equilibrium.

Note that for reversible systems, such as the elastic systems in homogeneous situations, the pair $(\hat{w}, \hat{Q})$ corresponding to the time reversal of a process is the opposite of the pair (w, Q) corresponding to the original process. We interpret the elements $(w, Q) \in A(U)$ which satisfy

$$-(w, Q) \in A(U)$$

as pairs corresponding to ideal reversible processes starting from equilibrium. The set of all such elements (w, Q) is denoted by $L(U)$, i.e.,

$$L(U) = A(U) \cap (-A(U)) ,$$

where

$$-A(U) = \{(w, Q) \in R \times M(I): -(w, Q) \in A(U)\} .$$

The following proposition is crucial to the proofs of the proportionality of work and heat in cyclic processes and for the proof of the existence of an absolute temperature scale.

Proposition[7]. *If U is a universe of thermodynamic systems then the sets $A_0(U)$, $A(U)$, and $L(U)$ are additive.*
(A subset S of a vector space V is said to be additive if $x_1, x_2 \in S$ implies $x_1 + x_2 \in S$.)

We shall also need an assumption expressing that the universe U has sufficiently many ideal reversible homogeneous Carnot processes. There are many ways to formulate this assumption mathematically. We here use, for simplicity, a rather strong version of this assumption, noting, however, that many of the results to be given below hold under less restrictive assumptions (cf. *Šilhavý* [2.4, 16, 21]).

Axiom. *There exists a subset L_0 of $L(U)$ with the following three properties*

i) *If $(w, Q) \in L_0$ and $\lambda \in [0, 1]$, then*

$$\lambda(w, Q) \in L_0 .$$

ii) *There exists $(w, Q) \in L_0$ with*

$$w > 0, \qquad Q(I) > 0 .$$

iii) *If $\theta^+, \theta^- \in I$ and $\theta^+ > \theta^-$ then there exists a $(w, Q) \in L_0$ with*

$$w > 0$$

and with Q of the form

$$Q = c^+ \delta_{\theta^+} - c^- \delta_{\theta^-}$$

where $c^+ > 0$, $c^- > 0$ and δ_θ denotes the Dirac measure concentrated at $\theta \in I$.

Roughly speaking, L_0 can be interpreted as the set of all homogeneous ideal reversible cyclic processes starting from equilibrium associated with the universe U; condition (i) then expresses a homogeneity property which can be related to

---

[7] See *Šilhavý* [2.4].

the homogeneity of the underlying processes; (ii) simply asserts the existence of at least one element in L_0 for which the corresponding work is positive and the net gain of heat $Q(I)$ is positive; condition (iii) then says that L_0 contains pairs (w, Q) corresponding to Carnot processes at any pair of operating temperatures θ^+, θ^-.

Henceforth I will deal with a universe of thermodynamic systems which satisfies the above axiom.

2.3 The First Law of Thermodynamics

Two different basic statements of the first law are given in Sect. 2.3.1. Section 2.3.2 contains a discussion of the relations among the basic statements and their consequences. It is shown that the basic statements of the first law imply Joule's relation and the existence of the energy function satisfying the equation of balance of energy. In Sect. 2.3.3 a postulate weaker than the first law is exhibited; it is shown that it implies an energy inequality rather than the equation of balance of energy. Some further aspects of the first law in relation to the second law are discussed in Sect. 2.4.2.

2.3.1 Basic Statements of the First Law

Recall that it is assumed that a universe U of thermodynamic systems satisfying the Axiom of Sect. 2.2.3 is given. For such a universe, $A(U)$ denotes the set of all pairs, (w, Q), associated with real or ideal cyclic processes starting from equilibrium. To avoid repeated statements, we denote, for a generic pair $(w, Q) \in A(U)$, by q the net gain of heat in a process characterized by (w, Q); i.e.,

$$q = Q(I) .$$

(The net gain of heat for the set of all possible empirical temperatures I.)

The following two versions of the first law are given in *Šilhavý* [2.4].

The First Law (I). *Let* $(w, Q) \in A(U)$. *Then*

$$q = 0 \quad \textit{implies} \quad w = 0 .$$

Roughly speaking, this postulate says that it is impossible to produce non-zero work in a cyclic process starting from equilibrium in which the overall net gain of heat vanishes.

The First Law (II). *Let* $(w, Q) \in A(U)$. *Then*

$$w > 0 \quad \textit{if and only if} \quad q > 0 .$$

The "if" part of the implication in the above equivalence says that it is impossible to produce, in a cyclic process starting from equilibrium, a positive work unless the net gain of heat is positive. The converse implication assures us that w is strictly positive if q is.

2.3.2 Relations Among the Statements; Consequences: Equation of Balance of Energy

Theorem 1[8]. *The versions* (I) *and* (II) *of the first law given in the preceding section are equivalent.*

From now on we assume that the two equivalent versions of the first law are satisfied, and discuss their consequences.

Theorem 2[8]. *There exists a unique positive constant J such that*

$$w = Jq \tag{2.6}$$

for each $(w, Q) \in A(U)$.

Theorem 2 asserts a universal proportionality of work and net gain of heat in cyclic processes starting from equilibrium. The relation (2.6) will be called Joule's relation; the constant J is the mechanical equivalent of a unit of heat. With a suitable choice of units one can achieve that $J = 1$. It turns out that Joule's relation holds for all cyclic processes, not only the cyclic processes starting from equilibrium. This is also a consequence of Theorem 3 stated below. In that theorem the consequences of the first law of thermodynamics are extended also to general, not necessarily cyclic processes. We employ the following notation. If $S = (\Sigma, \Pi, \bar{w}, \bar{Q}) \in U$ is a thermodynamic system and $\pi \in \Pi$ its process, then $\bar{q}(\pi)$ denotes the net gain of heat in that process, i.e.,

$$\bar{q}(\pi) = \bar{Q}(\pi)(I) \ .$$

Theorem 3[8]. *For each system* $S = (\Sigma, \Pi, \bar{w}, \bar{Q}) \in U$ *there exists a function* $E\colon \Sigma \to R$ *such that*

$$E(\pi^{\mathrm{f}}) - E(\pi^{\mathrm{i}}) = J\bar{q}(\pi) - \bar{w}(\pi) \tag{2.7}$$

for each process $\pi \in \Pi$. *The function E is unique to within an additive constant.*

The function E is the energy function of the thermodynamic system. Equation (2.7) is the equation of balance of energy. To summarize, the basic statements of the first law lead to Joule's relation and to the equation of balance of energy. Conversely, these consequences guarantee the validity of the basic statements of the first law. Hence (2.7) is completely equivalent to the basic statements (I and II); equation (2.7), however, is more convenient in applications.

2.3.3 A Weaker Postulate

To clarify in more detail the logical relationship between the basic statements of the first law and its consequences, this section is devoted to establishing the consequences of a postulate which is weaker than the first law (II) namely, we consider the following

---

[8] See *Šilhavý* [2.4].

Postulate. *Let* $(w, Q) \in A(U)$. *Then*

$$w > 0 \quad \textit{implies} \quad q > 0 \, .$$

In other words, it is impossible to produce, in a cyclic process starting from equilibrium, a positive work unless a positive net amount of heat is gained. The postulate, however, does not exclude the possibility that $w \leqslant 0$ even if $q > 0$. The first consequence is established in

Theorem 4[9]. *There exists a unique positive constant J such that*

$$w \leqslant Jq \tag{2.8}$$

for each $(w, Q) \in A(U)$.

Inequality (2.8) establishes an upper bound for the work done in a cyclic process starting from equilibrium in terms of the net gain of heat. The constant $J > 0$ is the maximum work which can be extracted from a unit of heat. The next theorem extends the consequences of the Postulate to general non-cyclic processes. It turns out that it is possible to define a state funtion E whose change in a process is dominated by the difference $Jq - w$.

Theorem 5[9]. *For each system* $S = (\Sigma, \Pi, \bar{w}, \bar{Q}) \in U$ *there exists a function* $E\colon \Sigma \to R$ *such that*

$$E(\pi^{\mathrm{f}}) - E(\pi^{\mathrm{i}}) \leqslant J\bar{q}(\pi) - \bar{w}(\pi) \tag{2.9}$$

for each $\pi \in \Pi$.

In contrast to the energy occurring in equation (2.7), there may be systems for which there exist two functions E_1, E_2 both satisfying inequality (2.9), but whose difference is non-constant.

2.4 The Second Law of Thermodynamics

This section is devoted to the second law of thermodynamics. In Sect. 2.4.1 the classical verbal statements due to Carnot, Clausius, Kelvin, and Planck are converted into meaningful postulates. In Sect. 2.4.2 the relationship among the various versions of the second law are established and, more importantly, it is shown that they lead to the existence of the absolute temperature scale and entropy.

2.4.1 Basic Statements of the Second Law

While the first law of thermodynamics establishes the equality (in suitable units) of the work done by the system and the net gain of heat of the system in a cyclic process, the second law deals with a finer question that not all of the heat

---

9 See *Šilhavý* [2.4].

supplied to the system can be converted into useful work. To make this rough idea precise, one has to look more closely at the exchange of heat between the system and its environment. The heat measure provides sufficiently detailed information for this purpose.

Rather than using the heat measure directly, it is convenient to introduce certain quantities derived from the heat measure (see *Šilhavý* [2.4, 6, 7]). Accordingly, let $Q \in \boldsymbol{M}(I)$ be a heat measure of a process of a thermodynamic system. Generally, there may be sets of empirical temperatures on which the measure is positive as well as sets of empirical temperatures where the measure is negative. It is a standard result of measure theory (cf. *Rudin* [2.25]) that each measure $Q \in \boldsymbol{M}(I)$ can be written uniquely as a difference

$$Q = Q^+ - Q^-$$

of two non-negative measures $Q^+, Q^- \in \boldsymbol{M}(I)$ which satisfy certain minimality conditions. Such a decomposition of Q into Q^+ and Q^- is called the Jordan decomposition. If $A \subset I$ is a Borel set of empirical temperatures, then the value $Q^+(A) \geqslant 0$ is interpreted as the heat absorbed by the system at empirical temperatures from the set A, and, similarly, $Q^-(A) \geqslant 0$ is interpreted as the heat emitted at empirical temperatures from the set A. In particular the value of the measure Q^+ on the set of all empirical temperature, i.e., the number

$$q^+ = Q^+(I)$$

is called the heat absorbed by the system in the process, while the number

$$q^- = Q^-(I)$$

is called the heat emitted by the system in the process. The net gain q of heat of the system in the process, i.e., the number

$$q = Q(I) ,$$

then satisfies

$$q = q^+ - q^- .$$

The support of the measure Q^+, denoted by $\operatorname{supp} Q^+$, (the smallest closed subset of I on which the measure is non-zero) is interpreted as the set of all empirical temperatures on which the system absorbs heat. If $\operatorname{supp} Q^+ \neq \phi$, i.e., if the system absorbs a non-zero amount of heat, then the number

$$\theta^+ = \max \operatorname{supp} Q^+$$

is referred to as the maximum empirical temperature at which heat is absorbed. Similarly, the support $\operatorname{supp} Q^-$ of Q^- is interpreted as the set of empirical temperatures on which the system emits heat, if $\operatorname{supp} Q^- \neq \phi$, then the number

$$\theta^- = \min \operatorname{supp} Q^-$$

is referred to as the minimum empirical temperature at which heat is emitted by the system in the process.

The basic statements of the second law of thermodynamics are formulated below as assertions about the properties of real and ideal cyclic processes starting from equilibrium of a universe U of thermodynamic systems. To avoid repeated hypotheses, we use, for a generic pair $(w, Q) \in A(U)$, the notation $q^+, q^-, \theta^+, \theta^-$ introduced above.

The Second Law (I). *Let* $(w, Q) \in A(U)$. *Then*

$$w > 0 \quad \textit{implies} \quad q^- > 0 .$$

Roughly speaking, it is impossible for a thermodynamic system to produce a positive work in a cyclic process starting from equilibrium unless a positive amount of heat is emitted. Statements of this type can be attributed, loosely, to Carnot, Kelvin, and Planck.

The Second Law (II). *If* $(w, Q) \in A(U)$ *and* Q *is of the form*

$$Q = c\delta_\theta$$

for some $c \in R$ *and* $\theta \in I$, *then*

$$w \leqslant 0 .$$

That is, it is impossible for a thermodynamic system to produce a positive work in a cyclic process starting from equilibrium by extracting heat at just one value of empirical temperature (Kelvin).

The Second Law (III). *Let* $(w, Q) \in A(U)$. *Then*

$$w > 0 \quad \textit{implies} \quad q^+ > 0 , \quad q^- > 0 , \quad \text{and} \quad \theta^+ > \theta^- .$$

It is impossible for a system to undergo a cyclic process starting from equilibrium in wich positive work is done unless heat is absorbed (this follows also from the first law), heat is emitted, and the maximum empirical temperature at which heat is absorbed exceeds the minimum empirical temperature at which heat is emitted (Clausius).

The Second Law (IV). *Let* $(w, Q) \in A(U)$. *Then*

$$w > 0 , \quad q^+ > 0 , \quad \textit{and} \quad q^- > 0 \quad \textit{imply} \quad \theta^+ > \theta^- .$$

Note that this version of the second law differs from the version (III) by shifting the implication from one place to another place. In words, (IV) says that it is impossible for a thermodynamic system to undergo a cyclic process starting from equilibrium in which positive work is done, and heat is transferred from cooler to hotter temperatures (Clausius).

All the above statements are contained in *Šilhavý* [2.4], (I and II) occur also in [2.6, 7]. For a more complete list of the statements of the laws of thermodynamics see *Šilhavý* [2.16, Sect. 4], where also the ambiguities arising in the process of translating the verbal statements into precise mathematical language are analysed.

2.4.2 Relations Among the Statements, Consequences: Entropy Inequality

We now deal with a universe U satisfying the Axiom of Sect. 2.2.3 and the first law of thermodynamics in any of the equivalent versions given in Sects. 2.3.1, 2.

Theorem 6. *The versions* (I – IV) *of the second law of thermodynamics given in the preceding section are equivalent.*

The equivalence of (I – IV) holds only provided the first law holds and provided an assumption embodied in item (iii) of the Axiom in Sect. 2.2.3 holds. A discussion of the relationships among the statements for systems satisfying the first law under less restrictive assumptions is given in *Šilhavý* [2.4]. A discussion of the relationships among the statements for general systems not necessarily satisfying the first law is given in *Šilhavý* [2.16]; it turns out that the relations in this general case become rather complicated; the statements are far from equivalent in that case.

In the remainder of this section we assume that the four equivalent versions (I – IV) of the second law are satisfied.

Theorem 7[10]. *There exists a positive, increasing, and continuous function* $T: I \to R^{++}$ *such that*

$$\int \frac{dQ}{T} \leqslant 0 \tag{2.10}$$

for all $(w, Q) \in A(U)$. *The function* T *is unique to within a positive multiplicative constant.*

The function T is, of course, the absolute temperature scale, and (2.10) is just the Clausius inequality for cyclic processes.

Thus, Theorem 7 shows that any of the versions (I – IV) of the second law leads to the existence of the absolute temperature scale obeying the Clausius inequality. The existence of this scale establishes the consequences of the second law of cyclic processes (starting from equilibrium). For general processes, the Clausius inequality leads to the existence of a state function, called entropy, whose change in the process dominated the expression $\int (dQ)/T$ corresponding to that process.

Theorem 8[11]. *For each system* $S = (\Sigma, \Pi, \bar{w}, \bar{Q}) \in U$ *there exists a state function* $H: \Sigma \to R$ *such that*

$$H(\pi^{\mathrm{f}}) - H(\pi^{\mathrm{i}}) \geqslant \int \frac{dQ(\pi)}{T} \tag{2.11}$$

for each process $\pi \in \Pi$.

---

[10] See *Šilhavý* [2.4, 6, 7].

[11] *Šilhavý* [2.4], the proof of the theorem given in the cited work is based on an earlier idea by *Day* [2.18].

The inequality (2.11) is an abstract entropy inequality which in the case of a continuous body takes the form of the Clausius-Duhem inequality. In contrast to the energy function (see Theorem 3), the entropy function need not be unique to within an additive constant.

To summarize, Theorems 7 and 8 show that any of the basic statements of the second law leads to the existence of the absolute temperature and entropy obeying an abstract entropy inequality. This holds only when the universe satisfies the first law and the assumption embodied in item (iii) of the Axiom. In *Šilhavý* [2.16] I present a complete discussion of the consequences of the statements of various forms of the second law for general systems not necessarily obeying the first law. The consequences then become very complicated and differ from statement to statement. Note, however, [2.16], that if a universe satisfying the Axiom of Sect. 2.2.3 contains an ideal gas with constant specific heats, then statements (III and IV) are equivalent and imply the Clausius inequality and the energy inequality (2.8) for cyclic processes. For general processes then (2.9) and (2.11) hold. *Thus, for universes containing an ideal gas with constant specific heats, the second laws* (III and IV) *contain a considerable amount of the first law.*

Acknowledgement. The author wishes to thank Dr. Kathy Pericak-Spector for her valuable comments on a previous version of this paper and for her help with the English language of it.

References

2.1 C. Carathéodory: Untersuchungen über die Grundlagen der Thermodynamik. Math. Ann. **67**, 355 (1909)
2.2 R. Arens: An axiomatic basis for classical thermodynamics. J. Math. Anal. Appl. **6**, 207 (1963)
2.3 J. B. Boyling: An axiomatic approach to classical thermodynamics. Proc. Roy. Soc. London **A329**, 35 (1972)
2.4 M. Šilhavý: On measures, convex cones, and foundations of thermodynamics. I. Systems with vector-valued actions. II. Thermodynamic systems. Czech. J. Phys. **B30**, 841, 961 (1980)
2.5 C. Truesdell: *Rational Thermodynamics* (McGraw-Hill, New York 1969)
2.6 M. Šilhavý: "On the Clausius Inequality", in *Abstracts of Euromech 111 Symposium*, September 26 – 28 (1978) p. 68
2.7 M. Šilhavý: On the Clausius inequality (manuscript); published recently as Ref. 2.9
2.8 J. Serrin: "The Concepts of Thermodynamics", in *Contemporary Developments in Continuum Mechanics and Partial Differential Equation*, ed. by G. de la Penha, L. A. Medeiros (North-Holland, Amsterdam 1978) p. 411 – 451; lecture held in Rio de Janeiro (1977)
2.9 M. Šilhavý: On the Clausius inequality. Arch. Ration. Mech. Anal. **81**, 221 (1983)
2.10 J. Serrin: Conceptual analysis of the classical second laws of thermodynamics. Arch. Ration. Mech. Anal. **70**, 355 (1979)
2.11 B. D. Coleman, D. R. Owen, J. Serrin: The second law of thermodynamics for systems with approximate cycles. Arch. Ration. Mech. Anal. **77**, 103 (1981)
2.12 M. Feinberg, R. Lavine: Thermodynamics based on the Hahn-Banach theorem: The Clausius inequality. Arch. Ration. Mech. Anal. **81**, 221 (1983)
2.13 W. Noll: A new mathematical theory of simple materials. Arch. Ration. Mech. Anal. **48**, 1 (1972)

2.14 B. D. Coleman, D. R. Owen: A mathematical foundation for thermodynamics. Arch. Ration. Mech. Anal. **54**, 1 (1974)
2.15 M. E. Gurtin: Thermodynamics and stability. Arch. Ration. Mech. Anal. **59**, 63 (1975)
2.16 M. Šilhavý: On the second law of thermodynamics. I. General framework. II. Inequalities for cyclic processes. Czech. J. Phys. **B32**, 987, 1073 (1982)
2.17 M. Šilhavý: The thermodynamics of cyclic processes. Appendix G in C. Truesdell: *Rational Thermodynamics,* 2nd ed. (Springer, Berlin, Heidelberg, New York, Tokyo 1984)
2.18 W. A. Day: A theory for thermodynamics of materials with memory. Arch. Ration. Mech. Anal. **34**, 85 (1969)
2.19 M. Šilhavý: A condition equivalent to the existence of nonequilibrium entropy and absolute temperature for materials with internal variables. Arch. Ration. Mech. Anal. **68**, 299 (1978)
2.20 J. Kratochvíl, M. Šilhavý: On thermodynamics of non-equilibrium processes. J. Non-Equilibrium Thermodyn. **7**, 339 (1982)
2.21 M. Šilhavý: Foundations of thermodynamics for systems with appropriate accessibility (to appear)
2.22 M. Šilhavý: Introduction to continuum thermodynamics (unpublished manuscript of a book)
2.23 A. E. Green, R. S. Rivlin: On Cauchy's equations of motion. Z. Angew. Math. Phys. **15**, 290 (1964)
2.24 M. Šilhavý: How many constitutive functions are necessary to determine a thermoeleastic material? Czech. J. Phys. **B29**, 981 (1979)
2.25 W. Rudin: *Real and Complex Analysis,* 2nd ed. (McGraw-Hill, New York 1974)

Chapter 3
Foundations of the Clausius-Duhem Inequality*

M. Feinberg and R. Lavine

3.1 Introduction

To some extent modern continuum thermodynamics amounts to a collection of "thermodynamical theories" sharing common premises and common methodology. There are theories of elastic materials, of viscous materials, of materials with memory, of mixtures, and so on. It is generally the case that, in the context of each theory, one considers all processes (compatible with classical conservation laws) that bodies composed of the prescribed material might admit, and, moreover, one *supposes* that there exists for the theory a "Clausius-Duhem inequality". In very rough terms this amounts to an assertion of the following kind: *For any process suffered by any body composed of the material under study*

$$\begin{bmatrix}\text{The total entropy}\\ \text{of the body at the}\\ \text{end of the process}\end{bmatrix} - \begin{bmatrix}\text{The total entropy of the}\\ \text{body at the beginning}\\ \text{of the process}\end{bmatrix} \geqq \int \frac{dq}{\theta}\Bigg|_{\text{process}} \qquad (3.1)$$

"*dq denoting the element of heat received from external sources and θ the temperature of the part of the system receiving it.*" (This interpretation of the integral on the right side of (3.1) is taken from the opening page of Gibbs's "On the Equilibrium of Heterogeneous Substances").

In fact, one generally supposes something more: that in calculating the various quantities appearing in (3.1) one has available "functions of state" that give instantaneous values of the specific entropy (entropy/mass) and the temperature at a material point within a body once the instantaneous "state" of the material point is specified.[1] More precisely, in a theory of a prescribed material it is usually taken for granted that "states" of material points can be identified with elements of a Hausdorff space Σ (which depends upon the manner in which the notion of "state" is rendered concrete within the theory[2]) and that there exist

---

* With very minor differences this article has also appeared as an appendix in the second edition of C. Truesdell's *Rational Thermodynamics* (Springer, Berlin, Heidelberg, New York, Tokyo 1984).

1 A "state" will always be regarded here as an attribute of a material point within a body, never as an attribute of the body as a whole. We say that a material point is in a certain "state" while the body containing it is in a certain "condition".

2 The notion of "state" will vary from one theory to another. In a theory of an elastic material the state of a material point might, for example, be identified with a pair (ε, F), ε and F denoting instantaneous values of the specific internal energy (energy/mass) and the deformation gradient at that material point. In a theory of a particular gas the state of a material point might be identified with a pair (p, v), p and v denoting instantaneous values at that material point of the pressure and specific volume.

two functions – a *specific entropy function* η: $\Sigma \to \mathbb{R}$ (where $\mathbb{R}$ denotes the real numbers) and a *thermodynamic temperature scale* θ: $\Sigma \to \mathbb{P}$ (where $\mathbb{P}$ denotes the positive real numbers) – such that the specific entropy and temperature of a material point in state $\sigma \in \Sigma$ are given by $\eta(\sigma)$ and $\theta(\sigma)$. If, for a body suffering a particular process, one has a specification of the manner in which states of the various material points evolve in time, then the functions $\eta(\cdot)$ and $\theta(\cdot)$ permit local and instantaneous calculation of the temperature and specific entropy. In turn these calculations, supplemented with suitable information about the process, permit evaluation of the quantities appearing in (3.1), each by means of an appropriate integration procedure. The crucial supposition here is not only that the functions $\eta(\cdot)$ and $\theta(\cdot)$ exist within the context of a particular theory, *but also that they are compatible with* (3.1) *for all processes that bodies considered within the theory might admit.* Indeed, for thermodynamical theories of the kind we have in mind, *this supposition plays the role of the Second Law.*

The Second Law, however, is traditionally invoked in far more basic terms, usually as a simple prohibition against certain kinds of heat receipt by bodies suffering cyclic processes. No initial appeal is made to notions of "entropy" or "thermodynamic temperature", much less to "functions of state" that give instantaneous values of the thermodynamic temperature and specific entropy at a material point. To the extent that these functions exist within the context of a particular theory, their existence is generally expected to be a *consequence* of the Second Law rather than a precursor of it.

Classical arguments for the existence of specific entropy functions and thermodynamic temperature scales are – at least in spirit – constructive ones, based entirely on consideration of *reversible* (usually homogeneous) processes. These are generally depicted as hypothetical processes, suitably well-approximated by actual processes, which operate so slowly that a body suffering such a process might be regarded to be in an equilibrated condition at every instant. Based on one or another of the primitive statements of the Second Law (and the tacit presumption of a rich supply of reversible processes), these arguments deliver *both* the *existence* and the *uniqueness* (up to inconsequential scale changes) of "functions of state" which give the specific entropy and thermodynamic temperature. However, the very nature of *these* arguments suggests that the functions so constructed should contain in their domain only those states that might be exhibited within a body during the course of a reversible process.

On the other hand, the methodology of modern continuum thermodynamics requires that the Clausius-Duhem inequality apply to *all* processes that bodies of a particular material under study might suffer, even those processes which involve rapid deformation and heating. Accordingly, states within the domains of the *pre-supposed* functions $\eta(\cdot)$ and $\theta(\cdot)$ are generally *not* restricted to those that might be manifested during reversible processes. In effect, results of the classical arguments – in particular the existence of $\eta(\cdot)$ and $\theta(\cdot)$ – are lifted from and applied beyond the restricted setting within which they were derived. In discussing modern use of the Clausius-Duhem inequality, *Truesdell*[3] [3.1] has

[3] That physical theory often proceeds in the way Truesdell suggests has been argued elsewhere by *Feynman* [3.2].

written that

> "... we are following here the common, tried path of theorists: We have observed a property that summarizes a number of known and understood facts, and we ask if it will serve by itself, stripped of the restricting assumptions by which we were led to it, as an axiom on the basis of which further facts may be explained and further effects predicted."

Nevertheless, it is not unreasonable that more orthodox thermodynamicists should call into question the very premises upon which modern use of the Clausius-Duhem inequality depends. It is our purpose here to address issues of this kind.

In the sense suggested at the outset, we shall continue to view thermodynamics as a collection of thermodynamical theories, each appropriate to a particular material. The notion of "thermodynamical theory" will eventually be made precise. It suffices here to say that we shall regard such a theory to be characterized, first, by specification of the "state space" Σ – that is, of the manner in which the notion of "state of a material point" is rendered concrete within the theory – and, second, by a suitable description of those processes bodies composed of the prescribed material are deemed to suffer. Once the notion of "thermodynamical theory" is made mathematical we shall want to know

i) for *precisely* which theories there *exist* two "functions of state" – a *thermodynamic temperature scale* $\theta: \Sigma \to \mathbb{P}$ and a *specific entropy function* $\eta: \Sigma \to \mathbb{R}$ – such that inequality (3.1), made suitably precise, is satisfied for every process the theory contains; and

ii) for *precisely* which of those theories that admit such functions is it the case that this pair of functions is *unique* (up to inconsequential changes of scale).

Although we wish to address these questions in general terms, we shall circumscribe our considerations in one respect, primilarily so that subsequent discussion might have a substantially less technical character. For all theories considered hereafter we shall presume that Σ is a *compact* Hausdorff space. *In effect, for a particular material we are restricting attention to processes wherein no material point experiences a state outside a fixed compact set, perhaps very large.*[4] This restriction will be understood throughout.

In this context we show that, among all thermodynamical theories, *existence* of functions $\eta: \Sigma \to \mathbb{R}$ and $\theta: \Sigma \to \mathbb{P}$ (compatible with the Clausius-Duhem inequality) is a property of *precisely* those we call Kelvin-Planck theories – roughly, of those thermodynamical theories for which the processes comply with a version of the Kelvin-Planck Second Law. Our proof, which we sketch very briefly, is based on the Hahn-Banach theorem (as are proofs of virtually all other

[4] When Σ is merely required to be *locally* compact, one can obtain results similar but not identical to those obtained here. It should be mentioned, however, that when Σ is not even locally compact our work becomes more problematical. For materials with memory (in which states are identified with "histories" residing in an infinite-dimensional Banach space) the assumption that Σ be compact or even locally compact may be too restrictive to permit consideration of a suitably rich supply of processes.

theorems stated here) and relies not at all on notions of reversibility, slow processes, Carnot cycles, differential forms, equilibrium states or any other apparatus normally built into classical *existence* arguments. Thus, for theories that admit functions $\eta(\cdot)$ and $\theta(\cdot)$, elements of the domain Σ need *not* be restricted to those states that might be exhibited in a body during the course of a reversible process. On the other hand, further theorems indicate that, for theories in which these functions not only exist but are also essentially *unique*, it *must* be the case that *each* state in Σ is "exhibited" in some process – if not an actual process then one well approximated by actual processes – which is *reversible* (in a certain weak sense). Taken together, our theorems suggest that orthodox criticism of premises upon which modern thermodynamical methodology is based should perhaps focus less upon questions concerning *existence* of crucial state functions and more upon questions concerning the extent to which *uniqueness* of these functions plays a role in applications.

More extensive discussion of the work sketched here as well as additional results can be found in two articles by us[5] [3.3]. These articles had their origin in unpublished notes we wrote for James Serrin in 1978 to show how some of his early ideas, when used in conjunction with the Hahn-Banach theorem, could deliver results similar to his Accumulation Theorem in a rather different way. Our concerns then were much more limited than they are here. We concentrated only on how the Hahn-Banach theorem provides information about the existence and uniqueness of thermodyamic temperature scales suited for the *Clausius* inequality (i.e. (3.1) restricted to *cyclic* processes), and then only when the existence of an empirical temperature scale (or hotness manifold) is taken for granted. That the existence of such thermodynamic temperature scales emerges from the Hahn-Banach theorem (in approximately the same limited setting) was apparently observed by Šilhavý at about the same time. Readers are concouraged to consult the contributions to this volume by Serrin and Šilhavý, and also those by Coleman and Owen, for discussion of their research on foundations of the Clausius-Duhem inequality.

3.2 Thermodynamical Theories

In this section we make precise what we mean by a *thermodynamical theory* and, in particular, what we mean by a *Kelvin-Planck theory.* At the outset our discussion will be informal: words like "theory" and "process" will be used in an intuitive way only to provide motivation for definitions we shall eventually record.

Although thermodynamical theories of different materials may vary greatly in detail, answer to questions posed earlier depend less upon the fine structure of a particular theory than they depend upon certain of its coarse-grained features. For our purposes, then, we regard a thermodynamical theory to be characterized by just those features that bear upon issues connected with existence and properties of specific entropy functions and thermodynamic temperature scales com-

[5] The second article, "Thermodynamics based on the Hahn-Banach Theorem: the Clausius-Duhem inequality" is being prepared for the same journal.

patible with the Clausius-Duhem inequality. In fact, we regard a theory to be specified by two sets, Σ and $\mathscr{P}$, that carry precisely the information we require. The first of these is called the set of *states* for the theory, and the second is called the set of (basic) *processes*.

The set Σ serves to specify for a particular theory the manner in which the notion of "state of a material point" is made concrete. In effect, specification of Σ amounts to a description of the domain of states that material points might conceivably experience during processes admitted for consideration in the theory. As indicated earlier, we shall always presume that Σ is endowed with a compact Hausdorff topology. For example if, in a theory of a particular gas, a state is regarded to be a pair (p, v), with p the pressure and v the specific volume at a material point, then Σ would be identified with a closed and bounded region of $\mathbb{R}^2$, perhaps very large. In any case, it should be kept in mind that a "state" is an attribute of a material point within a body, not an attribute of the body as a whole.

To explain the nature of the set $\mathscr{P}$ and the sense in which it characterizes the processes a theory contains, we shall require a somewhat more extensive discussion. For this purpose we shall consider a theory concerned with all bodies composed of a particular material, and we shall suppose that states of material points are identified with elements of a compact Hausdorff space Σ.

We begin by making precise the idea that a body (as distinct from a material point within a body) reveals itself in a certain instantaneous "condition". Consider a body composed of the prescribed material. At some fixed instant each material point within the body manifests itself in some state contained in Σ. (We do not presume that all material points are in the same state, although it may in fact be the case that the body is homogeneous throughout.) By the (instantaneous) *condition* of the body we mean a positive Borel measure on Σ which we denote here by m and which we interpret in the following way: *For each Borel set $\Lambda \subset \Sigma$, $m(\Lambda)$ is the mass of the part of the body consisting of material points in states contained within Λ.* In rough terms we can imagine $m(\Lambda)$ to be determined by removing from the body just those material points in states contained within Λ and weighing the material so removed. Thus, the measure m describes the instantaneous distribution of material among the various states. Note that $m(\Sigma)$ is the mass of the entire body. Note also that if a body of total mass M is homogeneous throughout with all material points in state $\sigma \in \Sigma$, then the condition of the body is given by $M\delta_\sigma$, where δ_σ is the Dirac measure[6] concentrated at σ.

Now we consider a process suffered by some body composed of the prescribed material. During the course of the process the body may suffer deformation, and heat exchange may occur between the body and its exterior. Moreover, each material point within the body may experience a variety of states as the process advances. As a result, the *initial condition* m_i of the body (at the beginning of the process) may be very different from its *final condition*, m_f (at the end of the process). In fact, with the process we may associate the *change of condition*

$$\Delta m := m_\mathrm{f} - m_\mathrm{i} \tag{3.2}$$

---

[6] The Dirac measure δ_σ is defined as follows: For each Borel set $\Lambda \subset \Sigma$, $\delta_\sigma(\Lambda) = 1$ [$=0$] if σ is [is not] a member of Λ.

suffered by the body between the beginning of the process and its end. Here Δm is a (signed) Borel measure on Σ; that is, Δm may take negative values on some Borel sets. Note, however, that

$$\Delta m(\Sigma) = m_{\mathrm{f}}(\Sigma) - m_{\mathrm{i}}(\Sigma) = 0 \tag{3.3}$$

since both $m_{\mathrm{f}}(\Sigma)$ and $m_{\mathrm{i}}(\Sigma)$ are equal to the total mass of the body suffering the process.

In addition to the change of condition we shall associate with the process under consideration another (signed) Borel measure on Σ called the *heating measure* for the process. We denote the heating measure by q and give it the following interpretation: *for each Borel set $\Lambda \subset \Sigma$, $q(\Lambda)$ is the net amount of heat received (from the exterior of the body suffering the process) during the entire process by material points in states contained within Λ.* In rough terms, $q(\Lambda)$ is the net amount of heat one would observe being received by the body from its exterior if, at each instant during the process, one ignored heat receipt by parts of the body in states outside Λ. Note that q may be positive (corresponding to net heat absorption) on some Borel sets and negative (corresponding to net heat emission) on others.

Because the measures Δm and q for a process will play significant roles in what follows it is perhaps instructive to consider how these measures derive from a somewhat more traditional (and more detailed) process description:

Example 1. For the material under consideration we consider a particular process. With the process we associate

i) a body B that suffers the process. We regard B to be a set (of material points) taken together with a σ-algebra of subsets of B called the *parts* of B. Moreover, we presume that B comes equipped with a positive measure μ defined on its parts: for each part $P \subset B$, $\mu(P)$ is the *mass* of part P;
ii) a closed interval $I(=[t_{\mathrm{i}}, t_{\mathrm{f}}])$ to be interpreted as the time interval during which the process takes place;
iii) a measurable function $\hat{\sigma}: B \times I \to \Sigma$, where $\hat{\sigma}(X, t)$ is to be interpreted as the state of material point X at time t;
iv) a real-valued (signed) measure h on $B \times I$ with the following interpretation: for each part $P \subset B$ and each Lebesgue measurable set $J \subset I$, $h(P \times J)$ is the net amount of heat received (from the exterior of B) by part P during instants contained in J.

For the process so described the *heating measure* q is constructed as follows: for each Borel set $\Lambda \subset \Sigma$,

$$q(\Lambda) = h(\hat{\sigma}^{-1}(\Lambda)) . \tag{3.4}$$

To construct the change of condition for the process we proceed in the following way: Let the functions $\hat{\sigma}_{\mathrm{i}}: B \to \Sigma$ and $\hat{\sigma}_{\mathrm{f}}: B \to \Sigma$ be defined by

$$\hat{\sigma}_{\mathrm{i}}(\cdot) = \hat{\sigma}(\cdot, t_{\mathrm{i}}) \quad \text{and} \quad \hat{\sigma}_{\mathrm{f}}(\cdot) = \hat{\sigma}(\cdot, t_{\mathrm{f}}) . \tag{3.5}$$

Then the *initial condition* m_i of body B and its *final condition* m_f are defined as follows: for each Borel set $\Lambda \subset \Sigma$,

$$m_i(\Lambda) = \mu(\hat{\sigma}_i^{-1}(\Lambda)) \quad \text{and} \quad m_f(\Lambda) = \mu(\hat{\sigma}_f^{-1}(\Lambda)) \,. \tag{3.6}$$

The *change of condition* Δm for the process is then given by $m_f - m_i$.

We are now in a position to describe the set $\mathscr{P}$ for the theory under consideration. With a process admitted by a body composed of the prescribed material we can associate the pair $(\Delta m, q)$, where Δm is the change of condition and q is the heating measure for the process. *By $\mathscr{P}$ we shall mean the set of all $(\Delta m, q)$ pairs that derive from processes admitted by bodies composed of the prescribed material.*[7]

We let $\mathscr{M}(\Sigma)$ denote the vector space of finite signed regular Borel measures on Σ, and we let $\mathscr{M}^\circ(\Sigma)$ denote the linear subspace of $\mathscr{M}(\Sigma)$ defined by

$$\mathscr{M}^\circ(\Sigma) := \{v \in \mathscr{M}(\Sigma) \mid v(\Sigma) = 0\} \,. \tag{3.7}$$

It follows from (3.3) that the change of condition for any process is an element of $\mathscr{M}^\circ(\Sigma)$. Thus we can view $\mathscr{P}$ as a subset of $\mathscr{M}^\circ(\Sigma) \oplus \mathscr{M}(\Sigma)$. Hereafter, for a theory with state space Σ we give $\mathscr{M}(\Sigma)$ the weak-star topology[8], $\mathscr{M}^\circ(\Sigma)$ the topology it inherits as a subspace of $\mathscr{M}(\Sigma)$, and $\mathscr{M}^\circ(\Sigma) \oplus \mathscr{M}(\Sigma)$ the resulting product topology.

There is a certain amount of structure one might expect the set $\mathscr{P}$ to possess. For example, if $(\Delta m, q) \in \mathscr{P}$ derives from a process suffered by a body composed of the material under study and if $(\Delta m, q)' \in \mathscr{P}$ derives from a process suffered by another such body (perhaps a copy of the first), then one might expect $(\Delta m, q) + (\Delta m, q)'$ to be an element of $\mathscr{P}$ on the grounds that the two processes executed in separate locations constitute a third process suffered by the "union" of the two separate bodies. (In effect, we are invoking what Serrin has called the "union axiom".) Moreover, if $(\Delta m, q) \in \mathscr{P}$ derives from a process suffered by a body of mass M and if α is a positive number, one might expect $\alpha(\Delta m, q)$ to be an element of $\mathscr{P}$ as well – an element that derives from a "scaled copy" of the first process, one suffered by a body of mass αM. With considerations like these in mind, we shall assume that $\mathscr{P} \subset \mathscr{M}^\circ(\Sigma) \oplus \mathscr{M}(\Sigma)$ is a convex cone.[9]

Because the set $\mathscr{P}$ for a particular theory carries all the information we shall require about the nature of processes the theory contains, we shall find it convenient to speak of "the process $(\Delta m, q) \in \mathscr{P}$". In fact, we shall go a step further. In classical thermodynamics one attaches significance not only to

---

[7] The description of a process in terms of what we call the *condition* of a body was suggested by *Noll* [3.4]. The use of what we call *heating measures* in process descriptions and in statements of the Second Law seems to have originated with *Serrin* [3.5].

[8] By the weak-star topology we mean the coarsest topology on $\mathscr{M}(\Sigma)$ such that, for every continuous $\phi: \Sigma \to \mathbb{R}$, the function

$$v \in \mathscr{M}(\Sigma) \to \int_\Sigma \phi \, dv$$

is continuous.

[9] In fact, we shall only require that the topological closure of $\mathscr{P}$ be a convex cone. (If $\mathscr{P}$ is a convex cone, if follows that $\mathrm{cl}(\mathscr{P})$ is a convex cone.) Our discussion here is merely intended to make *plausible* the idea that $\mathrm{cl}(\mathscr{P})$ should have a convex conical structure. Arguments presented in the second of our articles cited earlier are somewhat different and, we think, more satisfactory.

actual processes the bodies under consideration admit but also to idealized "processes" which, while perhaps not among the actual processes, are in some sense approximated by them. Indeed, a substantial part of classical methodology is predicated on the tacit supposition that thermodynamical laws restricting actual processes should govern such idealized "processes" as well. With this in mind, we shall designate as *processes* not only those elements of $\mathscr{M}^\circ(\Sigma) \oplus \mathscr{M}(\Sigma)$ that lie in $\mathscr{P}$ but also those elements that lie in the closure of $\mathscr{P}$ (denoted $\mathrm{cl}(\mathscr{P})$). When we wish to discuss elements of $\mathscr{P}$ in particular, we shall refer to these as *basic processes.*

Definition 1. *A thermodynamical theory* $(\Sigma, \mathscr{P})$ is specified by

i) a non-empty set Σ endowed with a compact Hausdorff topology. Elements of Σ are called *states* (of material points).
ii) a non-empty set $\mathscr{P} \subset \mathscr{M}^\circ(\Sigma) \oplus \mathscr{M}(\Sigma)$ such that $\mathrm{cl}(\mathscr{P})$ is a convex cone. Elements of $\mathrm{cl}(\mathscr{P})$ are called *processes*; in particular, elements of $\mathscr{P}$ are called *basic processes.* If $p = (\Delta m, q)$ is a process, then $\Delta m \in \mathscr{M}^\circ(\Sigma)$ is the *change of condition* for the process p, and $q \in \mathscr{M}(\Sigma)$ is the *heating measure* for the process p.

By a Kelvin-Planck theory we shall mean a thermodynamical theory for which the processes comply with a version of the Kelvin-Planck Second Law. In rough terms this amounts to an assertion that *for no cyclic process can the body suffering the process absorb heat from its exterior without emitting heat as well.* We begin by defining what we mean by a cyclic process.

Definition 2. A *cyclic process* in a thermodynamical theory $(\Sigma, \mathscr{P})$ is a process for which the change of condition is the zero measure. We denote by $\mathscr{C}$ the set of elements of $\mathscr{M}(\Sigma)$ which are heating measures of cyclic processes. That is,

$$\mathscr{C} := \{q \in \mathscr{M}(\Sigma) \mid (0, q) \in \mathrm{cl}(\mathscr{P})\}. \tag{3.8}$$

Elements of $\mathscr{C}$ are called the *cyclic heating measures* for the theory.

Note that for a process to be cyclic we require only that the initial and final *condition* of the *body* suffering the process be identical. We do not impose the stronger requirement that the state of *each material point* be the same at the beginning and end of the process.

A Kelvin-Planck theory is a thermodynamical theory in which no (nonzero) cyclic heating measure is a positive measure. This is to say that if, for a Kelvin-Planck theory, q is a heating measure for a cyclic process and there exists a Borel set $\Lambda \subset \Sigma$ such that $q(\Lambda)$ is positive (corresponding to heat absorption), then there must exist a Borel set $\Lambda' \subset \Sigma$ such that $q(\Lambda')$ is negative (corresponding to heat emission). We shall denote by $\mathscr{M}_+(\Sigma)$ the set of positive Borel measures on Σ – that is, the set of those elements of $\mathscr{M}(\Sigma)$ (including the zero measure) which are non-negative on *every* Borel set. Thus, we have[10]

[10] (3.9) is similar in spirit if not in detail to a version of the Second Law invoked by *Serrin* in the 1978 article cited earlier [3.5].

Definition 3. *A Kelvin-Planck theory* is a thermodynamical theory $(\Sigma, \mathscr{P})$ for which

$$\mathscr{C} \cap \mathscr{M}_+(\Sigma) = \{0\}. \tag{3.9}$$

We are now in position to supply answers to questions posed earlier.

3.3 The Existence of Specific Entropy Functions and Thermodynamic Temperature Scales

The following theorem asserts that, among all thermodynamical theories, the Kelvin-Planck theories are precisely those that admit (continuous) specific entropy functions and thermodynamic temperature scales which are compatible with the Clausius-Duhem inequality in the sense described earlier. Hereafter $C(\Sigma, \mathbb{R})$ will denote the set of real-valued continuous functions on Σ, and $C(\Sigma, \mathbb{P})$ will denote the set of functions in $C(\Sigma, \mathbb{R})$ that take strictly positive values.

Theorem 1. *Let $(\Sigma, \mathscr{P})$ be a thermodynamical theory. The following are equivalent:*

i) *$(\Sigma, \mathscr{P})$ is a Kelvin-Planck theory.*
ii) *There exist functions $\eta \in C(\Sigma, \mathbb{R})$ and $\theta \in C(\Sigma, \mathbb{P})$ such that*

$$\int_\Sigma \eta \, d(\Delta m) \geqq \int_\Sigma \frac{dq}{\theta}, \qquad \forall (\Delta m, q) \in \mathrm{cl}(\mathscr{P}). \tag{3.10}$$

Remark 1. It is worth noting that if, for a particular process, m_i and m_f are the initial and final condition of the body suffering the process, then $\Delta m = m_\mathrm{f} - m_\mathrm{i}$ and

$$\int_\Sigma \eta \, d(\Delta m) = \int_\Sigma \eta \, dm_\mathrm{f} - \int_\Sigma \eta \, dm_\mathrm{i}. \tag{3.11}$$

We shall of course regard (3.10) as a Clausius-Duhem inequality for a (Kelvin-Planck) theory, with $\eta(\cdot)$ playing the role of a specific entropy function and $\theta(\cdot)$ playing the role of a temperature scale. Thus, the terms on the right of (3.11) become the final and initial total entropy of the body suffering the process. It is also worth noting that if, for a particular process, the pair $(\Delta m, q)$ derives from the data specified in Example 1, Sect. 3.2 then, for that process, the inequality appearing in (3.10) can be "pulled back" to a somewhat more traditional expression of the Clausius-Duhem inequality:

$$\int_B \eta(\hat{\sigma}_\mathrm{f}(X)) \, d\mu(X) - \int_B \eta(\hat{\sigma}_\mathrm{i}(X)) \, d\mu(X) \geqq \int_{B \times I} \frac{dh(X,t)}{\theta(\hat{\sigma}(X,t))}. \tag{3.12}$$

Proof of the implication that (ii) implies (i) in Theorem 1 of this section is straightforward, and we shall not give it. We do, however, think it important to give a brief sketch of the proof that (i) implies (ii) if only to demonstrate that *it relies not at all on tacit assumptions about equilibrium states, reversible pro-*

cesses, Carnot cycles, or any of the other hypothetical physical apparatus built into classical arguments for the existence of entropy functions and thermodynamic temperature scales.

Rather, the proof relies exclusively on ideas in functional analysis which, while now standard, were unavailable to the pioneers of thermodynamics. We shall need a few preliminary facts: First, $\mathcal{M}(\Sigma)$ is a locally convex Hausdorff topological vector space, as is $\mathcal{M}^\circ(\Sigma) \oplus \mathcal{M}(\Sigma)$. Second, the compactness of Σ ensures that

$$\mathcal{M}_+^1(\Sigma) := \{ v \in \mathcal{M}_+(\Sigma) \mid v(\Sigma) = 1 \}$$

is a compact subset of $\mathcal{M}(\Sigma)$. Finally, if $f: \mathcal{M}^\circ(\Sigma) \oplus \mathcal{M}(\Sigma) \to \mathbb{R}$ is a continuous linear functional, then there exist in $C(\Sigma, \mathbb{R})$ functions $\alpha(\cdot)$ and $\beta(\cdot)$ such that, for every $(v, w) \in \mathcal{M}^\circ(\Sigma) \oplus \mathcal{M}(\Sigma)$,

$$f(v, w) = \int_\Sigma \alpha\, dv + \int_\Sigma \beta\, dw. \tag{3.13}$$

Sketch of Proof, (i) *implies* (ii). In a Kelvin-Planck theory no process is of the form $(0, w)$ with $w \in \mathcal{M}_+^1(\Sigma)$. Thus, in $\mathcal{M}^\circ(\Sigma) \oplus \mathcal{M}(\Sigma)$ the closed convex cone $\mathrm{cl}(\mathcal{P})$ is disjoint from the compact convex set $\{0\} \times \mathcal{M}_+^1(\Sigma)$. The Hahn-Banach theorem[11] therefore ensures the existence of a continuous linear functional $f: \mathcal{M}^\circ(\Sigma) \oplus \mathcal{M}(\Sigma) \to \mathbb{R}$ such that

$$f(\Delta m, q) \leqq 0\,, \qquad \forall (\Delta m, q) \in \mathrm{cl}(\mathcal{P}) \tag{3.14}$$

and

$$f(0, w) > 0\,, \qquad \forall (0, w) \in \{0\} \times \mathcal{M}_+^1(\Sigma)\,. \tag{3.15}$$

Moreover, there exist in $C(\Sigma, \mathbb{R})$ functions $-\eta(\cdot)$ and $\beta(\cdot)$ such that $f(\cdot, \cdot)$ has a representation

$$f(v, w) = \int_\Sigma (-\eta)\, dv + \int_\Sigma \beta\, dw, \qquad \forall (v, w) \in \mathcal{M}^\circ(\Sigma) \oplus \mathcal{M}(\Sigma)\,. \tag{3.16}$$

Since, for each $\sigma \in \Sigma$, the Dirac measure δ_σ is a member of $\mathcal{M}_+^1(\Sigma)$ it follows from (3.15) and (3.16) that $\beta(\cdot)$ takes positive values. Taking $\theta(\cdot) = 1/\beta(\cdot)$, we obtain (3.10) from (3.14) and (3.16).

Definition 1. Let $(\Sigma, \mathcal{P})$ be a Kelvin-Planck theory. An element $(\eta, \theta) \in C(\Sigma, \mathbb{R}) \times C(\Sigma, \mathbb{P})$ that satisfies condition (3.10) will be called a *Clausius-Duhem pair* for the theory. A function $\theta \in C(\Sigma, \mathbb{P})$ is a *Clausius-Duhem temperature scale* for the theory if there exists $\eta \in C(\Sigma, \mathbb{R})$ such that (η, θ) is a Clausius-Duhem pair, in which case $\eta(\cdot)$ is a *specific entropy function* (corresponding to the Clausius-Duhem temperature scale $\theta(\cdot)$). We denote by $\mathcal{T}_{\mathrm{CD}}$ the set of all Clausius-Duhem temperature scales for the theory.

---

[11] The version of the Hahn-Banach theorem invoked here is one which asserts that, in a locally convex Hausdorff topological vector space, two disjoint non-empty convex sets – one closed, the other compact – admit separation by a hyperplane. See Theorem 21.12 in [3.6]. In fact, [3.6] is a good source of information about the vector space $\mathcal{M}(\Sigma)$.

Because we shall be interested in, among other things, the uniqueness of specific entropy functions and Clausius-Duhem temperature scales for a given Kelvin-Planck theory, we shall be concerned with the supply of Clausius-Duhem pairs a particular theory admits. Inspection of condition (3.10) suggests that the supply of Clausius-Duhem pairs is intimately connected with the supply of processes the theory contains. In rough terms, the larger the set $\mathscr{P}$ for a Kelvin-Planck theory, the more demanding condition (3.10) becomes and the smaller will be the set of (η, θ) pairs that satisfy its requirements. It is this dual relationship between the supply of processes and the supply of Clausius-Duhem pairs for a Kelvin-Planck theory that we begin to explore next.

3.4 Properties of the Set of Clausius-Duhem Temperatures Scales

This section will conclude with a theorem which gives for a Kelvin-Planck theory a condition which is both necessary and sufficient to ensure that all its Clausius-Duhem temperature scales are identical up to multiplication by a positive constant. Even in the absence of such uniqueness we shall want to draw a connection between temperature and "hotness". That is, we shall want to understand the sense in which the Clausius-Duhem temperatures scales of a Kelvin-Planck theory $(\Sigma, \mathscr{P})$ assign to states in Σ numbers which reflect the relative hotness of material points in those various states. Thus, we shall require that the set Σ carry some "hotness structure" wherein it makes sense to say that two distinct states are "of the same hotness" or that one is "hotter than" another. We take the point of view that such a structure should be imposed by the processes the theory contains. Therefore, without recourse to the existence of temperature scales we define the notions "of the same hotness" and "hotter than" in terms of the existence within the theory of processes having specified properties. Then we ask how the resulting hotness structure on Σ is reflected in the Clausius-Duhem temperature scales the theory admits.

We begin by stating what we mean by a reversible process.

Definition 1. A *reversible process* in a thermodynamical theory $(\Sigma, \mathscr{P})$ is a process $(\Delta m, q) \in \mathrm{cl}(\mathscr{P})$ such that $(-\Delta m, -q)$ is also contained in $\mathrm{cl}(\mathscr{P})$.

Note that a reversible process need not be a member of $\mathscr{P}$, the set of basic processes; it need only be a member of $\mathrm{cl}(\mathscr{P})$. This we believe reflects the usual idea that a reversible process need not be a member of the "actual" processes but should be approximated by them. Note also that our definition of a reversible process is somewhat weaker than the traditional one, in which a process and its reverse trace out "paths" in opposite directions. Here we require no notion of "path".

Definition 2. Let $(\Sigma, \mathscr{P})$ be a thermodynamical theory. Two states $\sigma \in \Sigma$ and $\sigma' \in \Sigma$ *are of the same hotness* (denoted $\sigma' \sim \sigma$) if there exists in $\mathrm{cl}(\mathscr{P})$ a reversible cyclic process with heating measure $\delta_{\sigma'} - \delta_{\sigma}$. The equivalence relation $\sim$ induces a partition of Σ into equivalence classes called *hotness levels*. We denote by $\mathbb{H}$ the set of all hotness levels, and we give $\mathbb{H}$ the quotient topology it inherits from Σ.

In rough terms, the reversible cyclic process described in Definition 2 of this section is such that, for the body suffering the process, heat is absorbed only by material points in state σ', heat is emitted only by material points in state σ, and the amount of heat absorbed is equal to the amount emitted (whereupon the First Law requires that the net amount of work done on the body is zero). Here again it is important to recognize that the process described need not be among the basic (or "actual") processes; it need only be approximated by them.

Theorem 1. *Let $(\Sigma, \mathscr{P})$ be Kelvin-Planck theory. The following are equivalent:*

i) *$\sigma' \in \Sigma$ and $\sigma \in \Sigma$ are of the same hotness.*
ii) $\theta(\sigma') = \theta(\sigma)\,, \quad \forall\, \theta \in \mathscr{T}_{\mathrm{CD}}$.

Theorem 1 of this section asserts not only that two states of the same hotness are assigned identical temperatures by every Clausius-Duhem scale but also that *if two states σ' and σ are not distinguished by any Clausius-Duhem scale for the theory then the theory must contain a process of the kind described in Definition 2 of this section.*

Because two states residing in the same hotness level are assigned identical temperatures by every Clausius-Duhem scale, it makes sense to speak of the "temperature of a hotness level".

Definition 3. For a Kelvin-Planck theory $(\Sigma, \mathscr{P})$ let $\theta\colon \Sigma \to \mathbb{P}$ be a Clausius-Duhem temperature scale. By $\theta^*\colon \mathbb{H} \to \mathbb{P}$ we mean the *Clausius-Duhem temperature scale on $\mathbb{H}$ induced by $\theta(\cdot)$* in the following way: For each $h \in \mathbb{H}$

$$\theta^*(h) = \theta(\sigma)\,,$$

where σ is any element of h. The set of all Clausius-Duhem temperature scales induced on $\mathbb{H}$ by members of $\mathscr{T}_{\mathrm{CD}}$ will be denoted by $\mathscr{T}^*_{\mathrm{CD}}$.

Next we wish to give meaning to the idea that one hotness level is "hotter than" another. In preparation for that definition we recall that the *support* of a measure $\nu \in \mathscr{M}_+(\Sigma)$ is defined to be the complement in Σ of the largest open set in Σ of ν-measure zero. We denote the support of ν by $\operatorname{supp}\nu$. In terms which are hardly precise, $\operatorname{supp}\nu$ is that part of Σ which ν acts on in a nontrivial way. In particular, if $h \subset \Sigma$ is a hotness level, then $\operatorname{supp}\nu \subset h$ implies that $\nu(\Lambda) = 0$ for any Borel set $\Lambda \subset \Sigma$ disjoint from h.

Definition 4. For a Kelvin-Planck theory $(\Sigma, \mathscr{P})$ with hotness levels $\mathbb{H}$ we say that $h' \in \mathbb{H}$ is *hotter than* $h \in \mathbb{H}$ (denoted $h' \succ h$) if $h' \neq h$ and there exists in $\operatorname{cl}(\mathscr{P})$ a cyclic process with heating measure of the form

$$q = \nu' - \nu,$$

where ν' and ν are members of $\mathscr{M}_+(\Sigma)$ such that $\operatorname{supp}\nu \subset h$ and $\nu'(h') > \nu(h)$.

In rough terms, the process described is a cyclic (not necessarily reversible) one wherein heat is emitted only by material points in states within hotness level h, and the amount of heat emitted is less than the amount absorbed by material points in states contained within hotness level h'. Thus, the total amount of heat absorbed by the body suffering the process is positive (whereupon the First Law requires that the body do work).[12]

Remark 1. For a Kelvin-Planck theory the relation $\succ$ is a partial order on the set of hotness levels. It need not be the case that the order is total, for the theory may be insufficiently rich in processes to render every pair of hotness levels $\succ$-comparable.

Theorem 2. *Let h' and h be hotness levels for a Kelvin-Planck theory. The following are equivalent:*

i) *h' is hotter than h.*
ii) *There exists $\varepsilon > 0$ such that*

$$\frac{\theta^*(h')}{\theta^*(h)} > 1 + \varepsilon\,, \qquad \forall\, \theta^* \in \mathscr{T}_{\mathrm{CD}}\,. \tag{3.17}$$

Theorem 2 of this section asserts that if h' is hotter than h it is not only true that $\theta^*(h') > \theta^*(h)$ for each $\theta^* \in \mathscr{T}^*_{\mathrm{CD}}$ but also that the ratio of the two temperatures is bounded away from unity as $\theta^*(\cdot)$ ranges over all Clausius-Duhem scales on $\mathbb{H}$. Conversely, if the set of Clausius-Duhem temperature scales for a Kelvin-Planck theory has this property, *then the theory must contain a process of the kind described in Definition 4* of this section.

We turn next to a corollary of Theorem 2 of this section. Recall that for us the hotness levels for a Kelvin-Planck theory $(\Sigma, \mathscr{P})$ were defined objects, and the set $\mathbb{H}$ of all hotness levels inherited its topology (Definition 2 of this section) from that of Σ even *before* $\mathbb{H}$ was endowed with a "hotter than" relation. In the absence of special assumptions one cannot say much about the topological nature of $\mathbb{H}$ other than that $\mathbb{H}$, like Σ, is compact and Hausdorff. If, however, the Kelvin-Planck theory in question is sufficiently rich in processes as to make $\succ$ a total order on $\mathbb{H}$, then one can say quite a bit: *that $\mathbb{H}$ looks very much like a subset of the real line.*

Corollary 1. *Let $\mathbb{H}$ be the set of hotness levels for a Kelvin-Planck theory $(\Sigma, \mathscr{P})$. If $\mathbb{H}$ is totally ordered by $\succ$, then $\mathbb{H}$ is both homeomorphic and order-similar to a subset of the real line.*[13] *In particular, if Σ is connected then $\mathbb{H}$ is homeomorphic and order-similar to a bounded closed interval of the real line.*

---

[12] A similar, though somewhat more special, notion of *hotter than* was used by *Truesdell* [3.7]. Definition 4 of this section corresponds to the relation ${}_4\!\succ$ in the first of our articles cited earlier.

[13] Corollary 1 of this section gives nontrivial information not only about the topological structure of $\mathbb{H}$ but also about its order structure. It is not *generally* the case that a totally ordered set is order-similar to a subset of the real line. For a counterexample see [3.3, Sect. 5].

In questions concerning the essential uniqueness of Clausius-Duhem temperature scales we shall find that Carnot processes play an inexorable role.

Definition 5. A *Carnot process* in a thermodynamical theory $(\Sigma, \mathscr{P})$ is a reversible cyclic process $(0, q) \in \mathrm{cl}(\mathscr{P})$ with the following property: There exist in Σ hotness levels h' and h such that the heating measure q has a representation

$$q = v' - v,$$

where v' and v are non-zero elements of $\mathscr{M}_+(\Sigma)$ with $\operatorname{supp} v' \subset h'$ and $\operatorname{supp} v \subset h$. We say that the Carnot process *operates between hotness levels h' and h.*

In rough terms, the process described in Definition 5 of this section is a reversible cyclic one wherein heat is absorbed by material points in states contained entirely in one hotness level and heat is emitted by material points in states contained entirely in another hotness level.

Theorem 3. *Let $(\Sigma, \mathscr{P})$ be a Kelvin-Planck theory. The following are equivalent:*

i) *All Clausius-Duhem temperature scales for $(\Sigma, \mathscr{P})$ are identical up to multiplication by a positive constant.*
ii) *For each pair of hotness levels h' and h there exists in* $\mathrm{cl}(\mathscr{P})$ *a Carnot process operating between h' and h.*
iii) *For each pair of states $\sigma' \in \Sigma$ and $\sigma \in \Sigma$ there exists in* $\mathrm{cl}(\mathscr{P})$ *a Carnot process with heating measure $c'\,\delta_{\sigma'} - c\,\delta_\sigma$, where c' and c are positive numbers.*[14]

The implications (iii) → (ii) → (i) of Theorem 3 of this section are routine. Of real interest here are the implications (i) → (ii) → (iii). In rough terms these assert that, in order for a Kelvin-Planck theory $(\Sigma, \mathscr{P})$ to have an essentially *unique* Clausius-Duhem temperature scale, it is *necessary* that the theory contain a rich supply of Carnot processes; moreover, *every* state in Σ *must* manifest itself in a *reversible* (Carnot) process.

3.5 Properties of the Set of Specific Entropy Functions

Here we examine connections between the set of specific entropy functions for a Kelvin-Planck theory and the set of processes the theory contains. Let $(\Sigma, \mathscr{P})$ be a Kelvin-Planck theory for which $\mathscr{T}_{\mathrm{CD}}$ is the set of Clausius-Duhem temperature scales. For each $\theta \in \mathscr{T}_{\mathrm{CD}}$ we denote by $\mathscr{S}_\theta$ the set of specific entropy functions that correspond to θ:

$$\mathscr{S}_\theta := \{\eta \in C(\Sigma, \mathbb{R}) \mid (\eta, \theta) \text{ is a Clausius-Duhem pair for } (\Sigma, \mathscr{P})\}.$$

---

[14] In rough terms, the Carnot process described in (iii) is one for which there is net heat absorption only by material points in state σ' and net heat emission only from material points in state σ. It should be remembered that such a process need only lie in the *closure* of $\mathscr{P}$. In terms of the classical picture it is instructive to think about a sequence of Carnot cycles with isothermal parts of decreasing length.

Moreover, we denote by $\mathscr{S}$ the set of all specific entropy functions the theory admits; that is, $\mathscr{S}$ is the union of all $\mathscr{S}_\theta$ as θ ranges over $\mathscr{T}_{CD}$.

Definition 1. An *adiabatic process* in a thermodynamical theory $(\Sigma, \mathscr{P})$ is an element of $\mathrm{cl}(\mathscr{P})$ for which the heating measure is the zero measure.[15]

Definition 2. Let $(\Sigma, \mathscr{P})$ be a thermodynamical theory. Two states $\sigma' \in \Sigma$ and $\sigma \in \Sigma$ are *adiabatically related* if there exists in $\mathrm{cl}(\mathscr{P})$ a reversible adiabatic process for which $\delta_{\sigma'} - \delta_\sigma$ is the change of condition.[16]

The process described in Definition 2 of this section can be regarded as a reversible adiabatic one in which the body (of unit mass) suffering the process has initial condition δ_σ and final condition $\delta_{\sigma'}$. That is, the body begins in a homogeneous condition wherein all material points are in state σ and ends in a homogeneous condition wherein all material points are in state σ'.

Theorem 1. *Let $(\Sigma, \mathscr{P})$ be a Kelvin-Planck theory. The following are equivalent:*

i) *$\sigma' \in \Sigma$ and $\sigma \in \Sigma$ are adiabatically related.*
ii) $\eta(\sigma') = \eta(\sigma)\,, \quad \forall \eta \in \mathscr{S}\,.$

Theorem 1 of this section asserts not only that two adiabatically related states are assigned identical values by every specific entropy function for the theory but also that *if no specific entropy functions distinguishes σ' from σ then the theory must contain a process of the kind described in Definition 2 of this section.*

Our next theorem addresses the following question: In a Kelvin-Planck theory for which all Clausius-Duhem temperature scales are essentially identical, what is required – beyond a rich supply of Carnot processes – in order to ensure that the specific entropy functions corresponding to each temperature scale are also essentially identical?

Theorem 2. *Let $(\Sigma, \mathscr{P})$ be a Kelvin-Planck theory for which all Clausius-Duhem temperature scales are identical up to multiplication by a positive number. The following are equivalent:*

i) *For each $\theta \in \mathscr{T}_{CD}$ the elements of $\mathscr{S}_\theta$ are identical up to an additive constant.*[17]
ii) *For each pair of states $\sigma' \in \Sigma$ and $\sigma \in \Sigma$ there exists in* $\mathrm{cl}(\mathscr{P})$ *a reversible process for which the change of condition is $\delta_{\sigma'} - \delta_\sigma$.*

As in the discussion following Definition 2 of this section, the process described in (ii) can be regarded to be a reversible one in which the body suffering

---

[15] The definition of an adiabatic process given here is somewhat weaker than the usual one. It does not necessarily imply that during the course of the process there is no heat supplied or removed but rather that if there is heat supplied and removed these compensate each other in such a way as to make $\varphi(\Lambda) = 0$ for *every* Borel set $\Lambda \subset \Sigma$.

[16] The equivalence relation on Σ given in Definition 2 of this section serves to partition Σ into *adiabats*. In ways we shall not pursue here, these play for specific entropy functions the role that hotness levels play for Clausius-Duhem temperature scales.

[17] In the context of the theorem it is not difficult to deduce from (ii) that if $\eta(\cdot)$ and $\hat{\eta}(\cdot)$ are any specific entropy functions for the theory, not necessarily corresponding to the same temperature scale, there exist $c \in \mathbb{P}$ and $c' \in \mathbb{R}$ such that $\hat{\eta}(\cdot) = c\eta(\cdot) + c'$.

the process begins with all its material points in state σ and ends with all its material points in state σ'. Here, however, the process need not be adiabatic.

That (ii) implies (i) is well known. That (i) implies (ii) is more interesting. In rough terms this last implication requires that, for a Kelvin-Planck theory $(\Sigma, \mathscr{P})$ to have not only an essentially unique Clausius-Duhem temperature scale but also an essentially unique specific entropy function, it is *necessary* that $\mathscr{P}$ be so rich that, for *each* pair of states in Σ, there exists in $\mathrm{cl}(\mathscr{P})$ a *reversible* process connecting them. Here, as in Theorem 3 of Sect. 3.4 *uniqueness* of the specific entropy function *requires* that each state in Σ manifests itself in some reversible process.

3.6 Concluding Remark

Beginning with a statement of the Second Law *and* the tacit presumption of a rich supply of reversible processes, classical arguments deduce simultaneously *both* the *existence* and *uniqueness* of requisite functions of state. Questions of existence and uniqueness are of course very different, and it should come as no surprise if conditions for one are largely irrelevant to the other. Theorem 1 of Sect. 3.3 ensures the *existence* of a Clausius-Duhem temperature scale and a specific entropy functions for *any* Kelvin-Planck theory – roughly, for any thermodynamical theory that carries a Kelvin-Planck Second Law – *regardless* of the nature of the individual "states" the theory purports to take into account. On the other hand, if these functions are to be essentially *unique* for a particular theory, Theorems 3 of Sect. 3.4 and Theorem 2 of Sect. 3.5 *require* that all the states be of the kind that can be exhibited during the course of a reversible process. Thus, there is little in these theorems to support the position of those who would deny the *existence* of crucial state functions in modern thermodynamical theories. On the contrary, Theorem 1 of Sect. 3.3 calls such a denial into serious question. If, however, *uniqueness* of these functions is at issue, then the more conservative thermodynamicists would be seem to have the weight of Theorem 3 of Sect. 3.4 and Theorem 2 of Sect. 3.5 on their side.

References

3.1 C. Truesdell: *Rational Thermodynamics* (McGraw-Hill, New York 1969) Chap. 2
3.2 R. Feynman: *The Character of Physical Law* (MIT Press, Cambridge, MA 1965) Lecture 2
3.3 M. Feinberg, R. Lavine: Thermodynamics based on the Hahn-Banach Theorem: the Clausius inequality. Arch. Ration. Mech. Anal. **82**, 203–293 (1983)
3.4 W. Noll: On certain convex sets of measures and on phases of reacting mixtures. Arch. Ration. Mech. Anal. **38**, 1–12 (1970)
3.5 J. Serrin: "The Concepts of Thermodynamics", in *Contemporary Developments in Continuum Mechnics and Partial Differential Equations*, ed. by G. de la Penha, L.A.J. Madeiros (North-Holland, Amsterdam 1978)
3.6 G. Choquet: *Lectures on Analysis*, ed. by J. Marsden, T. Lance, S. Gelbart (Benjamin, New York 1969)
3.7 C. Truesdell: Absolute temperatures as a consequence of Carnot's general axiom. Arch. History Exact Sci. **20**, 357–380 (1979)

Chapter 4
Recent Research on the Foundations of Thermodynamics

B. D. Coleman and D. R. Owen*

The principles of thermodynamics have found application in many branches of science. These principles have been employed to understand the efficiency of heat engines, the electromotive force of galvanic cells and thermal junctions, the dependence of chemical equilibrium on temperature and pressure, the properties of phase transitions, and the subject emphasized in this book: the thermomechanics of continuous bodies.

Although thermodynamics is the science of heat and temperature, its principles are often usefully applied to experiments in which heat is not flowing (e.g., those involving poor thermal conductors or insulated reaction chambers) or others in which temperature is not changing (because, say, the object under study is a good thermal conductor in contact with an isothermal environment). One recognizes a thermodynamical argument by its reference to consequences of either the first or the second law. Every student of physics or chemistry has been taught that the first law is an assertion about the balance of heat and work, and that the second law is an assertion about the rate of increase of entropy that, in some sense, is equivalent to a denial of the existence of certain perpetual motion machines, or to a denial of the existence of cycles in which heat is absorbed at some temperatures without emission at others, or to an assumption about the sign of the sum over a cycle of the ratio of the heat absorbed to the temperature at which it is absorbed[1].

We here describe some recent work toward a precise formulation of the second law as a general principle whose implications can be derived with rigor[2]. We do not believe that the results of this work can be dismissed as "mere axiomatics". The development in the 1960s of the thermodynamics of materials with memory raised questions whose resolution required a careful examination of the mathematical foundations of thermodynamics. In its original presentation [1964, 1, 2], the theory of the thermodynamics of materials with fading memory rested

---

* This article appeared as Appendix G1 in C. Truesdell's *Rational Thermodynamics* (Springer, Berlin, Heidelberg, New York, Tokyo 1984).

1 In their explanation of the laws of thermodynamics, particularly the second law, the texts on the subject tend to be obscure, not because the principles sought fail to have the generality claimed, but because of the absence of a *mathematical language* that permits expression of the principles at that level of generality.

2 This brief survey is confined to the study of certain forms of the second law. No attempt is made to present a complete set of axioms for all of thermodynamics. The first law and the concept of work are not discussed. For accounts of the early development of the science of thermodynamics and the discovery of the first and second laws, see the books of *Truesdell* [1980, 2] and *Truesdell* and *Bharatha* [1977, 1].

on the Clausius-Duhem inequality, i.e., on the assumption that for each substance there is a function of state[3], called the entropy, whose difference at two states dominates the ratio of the heat absorbed to the absolute temperature along each process taking one state of the other. The question was raised: Does each substance have such a function of state with the properties of regularity needed to derive the now known consequences of the Clausius-Duhem inequality? Of course, the question is meaningful only if one has a statement of the second law that does not presuppose the presence of entropy as a function of state. A statement of this type can be obtained by making mathematical the ideas behind the familiar assertion that *the sum along a cycle of the ratio of the heat gained to the absolute temperature at which it is gained cannot be positive.* However, to be useful for materials with gradually fading memory and for other substances with few non-trivial cycles, the statement must be formulated in such a way that it has meaning for "approximate cycles". We have obtained such a formulation of the second law and have used it to study various questions, including the existence, uniqueness, and regularity of entropy functions.

In this recent work [1974, 1; 1975, 1], a careful distinction is made between the general structure of thermodynamical systems and the equations defining special classes of systems[4]. The concept of a *system* employed is one in which a system is a pair (Σ, Π) of sets with the following mathematical structure: Σ is a topological space whose elements are the *states*; Π is the set of *processes*; associated with each process is a continuous function ϱ_P mapping a non-empty open subset $\mathscr{D}(P)$ of Σ onto a subset $\mathscr{R}(P)$ of Σ; ϱ_P is called the *transformation induced by* P and its value at a state σ in $\mathscr{D}(P)$ is denoted by $\varrho_P\sigma$; to each pair (P'', P') of processes for which $\mathscr{R}(P')$ intersects $\mathscr{D}(P'')$ is assigned a process $P''P'$ called the *process resulting from the successive application of* (first) P' *and* (then) P''. It is assumed that:

I) for each σ in Σ, the set of states *accessible* from σ, i.e., the set of states of the form $\varrho_P\sigma$ with P in Π, is dense in Σ; and
II) if $P''P'$ is the result of the successive application of P' and P'', then the transformation $\varrho_{P''P'}$ induced by $P''P'$ is the composition of $\varrho_{P'}$ and $\varrho_{P''}$, i.e., is the function defined on $\mathscr{D}(P''P') = \varrho_{P'}^{-1}(\mathscr{D}(P''))$ by the equation $\varrho_{P''P'}\sigma = \varrho_{P''}\varrho_{P'}\sigma$.

The mathematical concept that renders precise and general the idea of a "sum along a process" is that of an *action.* An action is a function that assigns a number $a(P, \sigma)$ to each pair (P, σ) with P in Π and σ in $\mathscr{D}(P)$; $a(P, \sigma)$ is called *the supply of* a *on going from* σ *to* $\varrho_P\sigma$ *via the process* P. Two properties are required of an action:

i) *additivity* in the sense that if P is the result of the successive application of P' and P'', then for each σ in $\mathscr{D}(P)$ the supply of a obtained by going from

3 For a material with fading memory, a *state* can be identified with an appropriate *history*.

4 It is common for the older literature to obscure such a distinction or to employ two languages: the language of mathematics for the treatment of special systems and examples, and a non-mathematical language, reminiscent of metaphysics, for discussion of the general principles of thermodynamics.

σ to $\varrho_P\sigma$ via P is the sum of the supplies of α obtained by going from σ to $\varrho_{P'}\sigma$ via P' and from $\varrho_{P'}\sigma$ to $\varrho_P\sigma = \varrho_{P''}\varrho_{P'}\sigma$ via P'', i.e.,

$$\alpha(P,\sigma) = \alpha(P',\sigma) + \alpha(P'',\varrho_{P'}\sigma) , \tag{4.1}$$

and

ii) *continuity* in the sense that for each P in Π, the function $\alpha(P,\cdot)$ is continuous on $\mathscr{D}(P)$.

A process P and a state σ° are said to form a *cycle* (P,σ°) if σ° is in $\mathscr{D}(P)$ and $\varrho_P\sigma^\circ = \sigma^\circ$. One may consider taking the second law to be the assertion that an appropriate action $\mathscr{s}$ (which, of course, must be be specified) is not positive when its argument is a cycle, i.e., is such that

$$\varrho_P\sigma^\circ = \sigma^\circ \Rightarrow \mathscr{s}(P,\sigma^\circ) \leqq 0 . \tag{4.2}$$

For materials with gradually fading memory, the class of cycles (P,σ°) is too small for (4.2) to have the full implications expected of the second law. To obtain an extension of (4.2) to "approximate cycles", we have employed the following concept: We say that $\mathscr{s}$ has the *Clausius property* at a state σ° if, for each $\varepsilon > 0$, σ° has a neighborhood $\mathscr{O}_\varepsilon(\sigma^\circ)$ for which

$$\varrho_P\sigma^\circ \in \mathscr{O}_\varepsilon(\sigma^\circ) \Rightarrow \mathscr{s}(P,\sigma^\circ) < \varepsilon . \tag{4.3}$$

It is clear that if (4.3) holds and $\varrho_P\sigma^\circ = \sigma^\circ$, we have $\mathscr{s}(P,\sigma^\circ) < \varepsilon$ for every $\varepsilon > 0$, and hence (4.2) holds; i.e., if $\mathscr{s}$ has the Clausius property at σ°, then $\mathscr{s}(P,\sigma^\circ)$ is not positive when (P,σ°) is a cycle.

If an action has the Clausius property at a state σ°, then it has the property at each state in a set Σ° that is dense in Σ and contains all states σ that are accessible from σ°;[5] this set Σ° may be defined as follows. For each state σ° in Σ let $\mathbb{S}(\sigma^\circ,\sigma)$ be the collection of all the open subsets $\mathscr{O}$ of Σ that contain σ and are such that the sets $\mathscr{s}\{\sigma^\circ \to \mathscr{O}\}$, defined by

$$\mathscr{s}\{\sigma^\circ \to \mathscr{O}\} = \{\mathscr{s}(P,\sigma^\circ) \mid P \in \Pi,\ \varrho_P\sigma^\circ \in \mathscr{O}\} , \tag{4.4}$$

are individually bounded above, i.e., have

$$\sup \mathscr{s}\{\sigma^\circ \to \mathscr{O}\} < \infty ; \tag{4.5}$$

Σ° is the set of states for which

$$m(\sigma^\circ,\sigma) := \inf_{\mathscr{O} \in \mathbb{S}(\sigma^\circ,\sigma)} \sup \mathscr{s}\{\sigma^\circ \to \mathscr{O}\} \tag{4.6}$$

is finite, i.e.,

$$\Sigma^\circ = \{\sigma \mid \sigma \in \Sigma,\ m(\sigma^\circ,\sigma) > -\infty\} . \tag{4.7}$$

In [1974, 1] we interpreted the second law as the statement that *a particular action $\mathscr{s}$ has the Clausius property at least at one state.* To prove that such a statement implies the existence of an entropy function that enters a relation with

---

[5] [1974, 1] Thm. 3.1.

the form of the Clausius-Duhem inequality, we there introduced the concept of an upper potential.

A real valued function S on a dense subset $\mathcal{S}$ of Σ is called an *upper potential* for an action $\mathfrak{s}$ if for each pair of states σ_1, σ_2 in $\mathcal{S}$ and each $\varepsilon > 0$ there is a neighborhood $\mathcal{O}_\varepsilon(\sigma_1, \sigma_2)$ of σ_2 such that whenever $\varrho_P \sigma_1$ is in $\mathcal{O}_\varepsilon(\sigma_1, \sigma_2)$,

$$S(\sigma_2) - S(\sigma_1) > \mathfrak{s}(P, \sigma_1) - \varepsilon\,. \tag{4.8}$$

In the special case in which σ_2 is accessible from σ_1, i.e., in which σ_2 has the form $\sigma_2 = \varrho_P \sigma_1$, this relation holds for all $\varepsilon > 0$ and hence implies that the supply of $\mathfrak{s}$ on going from σ_1 to σ_2 is dominated by the difference $S(\sigma_2) - S(\sigma_1)$:

$$S(\varrho_P \sigma_1) - S(\sigma_1) \geqq \mathfrak{s}(P, \sigma_1)\,. \tag{4.9}$$

It is easily seen that an action that has an upper potential has the Clausius property at each state in the domain of the upper potential. Although far less trivial to show, the converse is also true: *the assumption that there are states at which $\mathfrak{s}$ has the Clausius property implies that $\mathfrak{s}$ has an upper potential, in fact, one that is upper semicontinuous*[6]. If we identify $\mathfrak{s}(P, \sigma)$ with the sum of the heat added divided by the temperature at which it is added as the system is taken from the state σ to the state $\varrho_P \sigma$ by the process P, then in (4.9) the upper potential S is playing the role played by entropy in the Clausius-Duhem inequality. Thus, the existence of an upper potential for $\mathfrak{s}$ is tantamount to the existence of entropy as a function of state.

Our construction of an entropy function employs the observation that *if $\mathfrak{s}$ has the Clausius property at σ° and if S° is defined on Σ° by*

$$S^\circ(\sigma) = m(\sigma^\circ, \sigma)\,, \tag{4.10}$$

then not only is the domain Σ° of S° dense in Σ, but S° is an upper potential for $\mathfrak{s}$ and is upper semicontinuous on Σ°.

Under the assumptions stated up to this point, we can say no more about the regularity of upper potentials for $\mathfrak{s}$ than that there is one that is semicontinuous. This should not be surprising, for we have so far assumed very little about systems and actions. At this level of generality, the collection Σ of states has a topology but not the vector-space or manifold structure required to make meaningful the concept of a differentiable function on Σ. When more is assumed about the system (Σ, Π) and the action $\mathfrak{s}$, one expects to be able to prove more about entropy as a function of state.

We have not yet assumed any special properties for the state σ° with which we start when we construct Σ° as shown in equations (4.4) – (4.7). We would like to take this state as a "standard state" and be able to normalize entropy functions S on Σ° so that

$$S(\sigma^\circ) = 0\,. \tag{4.11}$$

---

[6] [1974, 1] Thm. 3.3.

Although the hypotheses we have made so far imply that Σ° is dense in Σ, they are not strong enough to imply that σ° is in Σ°. It does suffice, however, to assume that this selected state σ° is *equilibrated with respect to* $\mathfrak{s}$ in the sense that there is at least one process P° in Π for which

$$\varrho_{P^\circ}\sigma^\circ = \sigma^\circ \quad \text{and} \quad \mathfrak{s}(P^\circ, \sigma^\circ) = 0 \,. \tag{4.12}$$

In fact, we have the following theorem[7]: *Suppose that* σ° *is equilibrated with respect to* $\mathfrak{s}$ *and is a state at which* $\mathfrak{s}$ *has the Clausius property. Then* (1) σ° *is in* Σ°; (2)

$$m(\sigma^\circ, \sigma^\circ) = 0 \,, \tag{4.13}$$

and hence the upper potential S° *defined in* (4.10) *vanishes at* σ°, i.e.,

$$S^\circ(\sigma^\circ) = 0 \,; \tag{4.14}$$

moreover (3) S° *is the smallest entropy function that is normalized in this way: if* S *is an upper potential for* $\mathfrak{s}$ *that is defined on* Σ° *and obeys* (4.11), *then for each state* σ *in* Σ°,

$$S^\circ(\sigma) \leqq S(\sigma) \,. \tag{4.15}$$

If, in addition, $m(\sigma, \sigma^\circ)$ [*defined by interchanging the roles of* σ *and* σ° *in the relations* (4.4) – (4.6)] *is finite for each* σ *in* Σ°, *then the function* $S_\circ$ *defined on* Σ° *by*

$$S_\circ(\sigma) = -m(\sigma, \sigma^\circ) \tag{4.16}$$

also is an upper potential for $\mathfrak{s}$ *and not only obeys the normalization* (4.11), *but is the largest entropy function that does; i.e., each upper potential for* $\mathfrak{s}$ *that is defined on* Σ° *and obeys* (4.11) *has the bounds:*

$$S^\circ(\sigma) \leqq S(\sigma) \leqq S_\circ(\sigma) \,. \tag{4.17}$$

The set of entropy functions on Σ° normalized according to (4.11) is convex: if S_1 and S_2 are two such entropy functions, then so also is each function of the form $\alpha S_1 + (1-\alpha) S_2$, $0 < \alpha < 1$. We have just observed that if $\mathfrak{s}$ has the Clausius property at σ°, if σ° is equilibrated with respect to $\mathfrak{s}$ at σ°, and if $m(\sigma, \sigma^\circ)$ is finite whenever $m(\sigma^\circ, \sigma)$ is, then S° is the minimal and $S_\circ$ is the maximal element of this convex set of normalized entropy functions. Clearly, then, this set reduces to a singleton if and only if $S^\circ = S_\circ$. That is, under these hypotheses about the referene state σ°, in *order that there be only one entropy function* S *on* Σ° *obeying* (4.11) *it is necessary and sufficient that for all* σ *in* Σ°

$$m(\sigma^\circ, \sigma) + m(\sigma, \sigma^\circ) = 0 \,. \tag{4.18}$$

This condition, although met by elastic materials and viscous materials, is not met in general. Among the exceptions are certain elastic-plastic materials,

---

7 [1975, 1] § 3. In [1974, 1] § 7 we show that a weaker hypothesis, namely that σ° be a relaxed state, suffices for the theorem.

materials with fading memory, and certain materials with internal state variables.

We have seen that if one takes the second law to be the assertion that an action $\mathscr{a}$ has the Clausius property, one can deduce the existence of entropy as a function of state and obtain information about the regularity and uniqueness of entropy. We have left open the questions:

i) Which of the many actions one can formulate for a system should be assumed to have the Clausius property?
ii) Can information about the form of $\mathscr{a}$ be deduced from a statement of the second law that makes precise an assertion to the effect that there can be no cycles in which heat is only absorbed?
iii) In what sense is "absolute temperature" a distinguished measure of hotness[8]?

Recent papers of *Serrin* have shed light on these questions. Basic to his theory [1979, 2, 3] is the concept of the *hotness manifold* $\mathscr{H}$, introduced by *Mach* [1896, 1] and assumed by *Serrin* to be a continuous, oriented, one-dimensional manifold whose points L are called *levels of hotness*, or, for short, *hotnesses*. It is assumed that the orientation of $\mathscr{H}$ induces a total strict order "$\prec$" on hotnesses, with "$L_1 \prec L_2$" read "L_1 is a lower level of hotness than L_2", or "L_1 is below L_2", or "L_2 is above L_1". *Serrin's* theory [1979, 2, 3] does not rest on a concept of "state", but does refer to objects that we may here identify with *cycles*, i.e., with pairs (P, σ) in $\Pi \times \Sigma$ with σ in $\mathscr{D}(P)$ and $\varrho_P \sigma = \sigma$. Let us define a *classical thermodynamical system* to be a set $\mathbb{P}_c$ of cycles and a real-valued function Q on $\mathbb{P}_c \times \mathscr{H}$, called the *accumulation function*; the value $Q(\mathscr{P}, L)$ of Q at a point $(\mathscr{P}, L)$ in $\mathbb{P}_c \times \mathscr{H}$ is called the *net heat absorbed by the system at levels of hotness at or below L in the cycle* $\mathscr{P}$. It is assumed that $Q(\mathscr{P}, L)$ varies only over a bounded interval in $\mathscr{H}$ in the sense that for each cycle $\mathscr{P}$ there are levels of hotness $L^l = L^l(\mathscr{P})$, $L^u = L^u(\mathscr{P})$, with $L^l \prec L^u$, such that

$$\left.\begin{aligned} Q(\mathscr{P}, L) &= 0 && \text{for} \quad L \prec L^l\,, \\ Q(\mathscr{P}, L) &= Q(\mathscr{P}, L^u) && \text{for} \quad L^u \preceqq L\,. \end{aligned}\right\} \tag{4.19}$$

For each $\mathscr{P}$, the function $Q(\mathscr{P}, \cdot)$ generates a finitely additive set function $q_{\mathscr{P}}$ for $\mathscr{H}$ whose value $q_{\mathscr{P}}(I) = Q(\mathscr{P}, L_2) - Q(\mathscr{P}, L_1)$ on the set $I = \{L \mid L_1 \prec L \preceqq L_2\}$ is the *net heat absorbed by the system at levels of hotness in I*, (i.e., *above* L_1 *and at or below* L_2); (4.19) implies that this set function has compact support. The number $Q(\mathscr{P}, L^u)$ is called the *overall net gain of heat* (by the system) *in the cycle* $\mathscr{P}$.

---

[8] There is a growing literature devoted to questions related to these. We mention here a few recent contributions: *Boyling* [1972, 1] has discussed the construction of entropy and absolute temperature from axioms of the form proposed by *Carathéodory*. *Truesdell* and *Bharatha* [1977, 1], and *Truesdell* [1979, 4] have clarified and extended *Carnot's, Reech's,* and *Kelvin's* studies of ideal (reversible) systems. *Serrin's* research [1979, 2, 3] and our own research done with him [1981, 1] extend and render mathematical ideas expressed by *Kelvin* and *Clausius*. Recent papers expressing a similar point of view and showing points of contact with the research described below are those of *Šilhavý* [1980, 1], [1982, 3], *Feinberg* and *Lavine* [1982, 1], and *Owen* [1982, 2].

Serrin develops a language for discussing the effect of the operation of two or more systems or the repeated operation of a single system: if $\mathscr{S}' = (\mathbb{P}_c', Q')$ and $\mathscr{S}'' = (\mathbb{P}_c'', Q'')$ are two thermodynamical systems (which may or may not be the same), their *union* $\mathscr{S}' \oplus \mathscr{S}''$ is the thermodynamical system $\mathscr{S} = (\mathbb{P}_c, Q)$ with

$$\mathbb{P}_c = \mathbb{P}_c' \times \mathbb{P}_c'' \tag{4.20}$$

and

$$Q((\mathscr{P}', \mathscr{P}''), L) = Q'(\mathscr{P}', L) + Q''(\mathscr{P}'', L) \tag{4.21}$$

for each pair $(\mathscr{P}', \mathscr{P}'')$ in $\mathbb{P}_c$ and each L in $\mathscr{H}$.

Serrin assumes that a collection $\mathbb{U}$ of classical thermodynamical systems, closed under the union operation, has been given. His statement of the second law is: *If $\mathscr{P}$ is a cycle of a system* $(\mathbb{P}_c, Q)$ *in* $\mathbb{U}$ *with* $Q(\mathscr{P}, L) \geqq 0$ *for every* L, *then* $Q(\mathscr{P}, L) = 0$ *for every* L. In other words, in a cycle for which the net heat absorbed at or below each hotness level is not negative, the accumulation function is identically zero and, in particular, the overall net gain of heat in that cycle is zero.

To show that this statement of the second law permits the construction for each system of a function $\mathfrak{s}$ obeying (4.2), *Serrin* assumes that the collection $\mathbb{U}$ of classical thermodynamical systems contains at least one special system that is the mathematical embodiment of an elastic or viscous substance. His proofs take their simplest form if the distinguished systems are *ideal gases*; these are systems for which each cycle $\mathscr{P}$ can be represented as a closed (oriented) curve $c_{\mathscr{P}}$ in the first quadrant of a coordinate plane, with one coordinate, V, interpreted as the volume of the gas, and the other, θ, a coordinate indicating the level of hotness L in the gas. The number θ is related to L by a strictly increasing, positive-valued, continuous function φ on the manifold $\mathscr{H}$. (Use of the coordinate system φ on $\mathscr{H}$ corresponds to measurement of hotness with an "ideal gas thermometer".) For each process $\mathscr{P}$ of an ideal gas and each level $\bar{L}$ of hotness, $Q(\mathscr{P}, \bar{L})$ equals the integral of a differential form,

$$c(V, \theta)\, d\theta + p(V, \theta)\, dV,$$

over the portion $c_{\mathscr{P}}(\bar{L})$ of $c_{\mathscr{P}}$ on which $L \leq \bar{L}$, i.e., on which the coordinate $\theta = \varphi(L)$ is equal to or less than $\bar{\theta} = \varphi(\bar{L})$; p is the pressure in the gas and is given by the formula,

$$p(V, \theta) = \frac{r\theta}{V}, \tag{4.22}$$

with r a constant; c is the heat capacity of the gas and is given by a function of θ alone:

$$c(V, \theta) = \tilde{c}(\theta). \tag{4.23}$$

(The relations (4.22) and (4.23) distinguish an ideal gas form other homogeneous fluid bodies.) Thus, for an ideal gas,

$$Q(\mathscr{P}, \bar{L}) = \int_{c_{\mathscr{P}}(\bar{L})} \left(\tilde{c}(\theta)\, d\theta + \frac{r\theta}{V}\, dV \right), \tag{4.24}$$

i.e.,

$$Q(\mathscr{P},\varphi^{-1}(\bar{\theta})) = \int_{c_{\mathscr{P}}(\varphi^{-1}(\bar{\theta}))} \left(\tilde{c}(\theta)\,d\theta + \frac{r\theta}{V}\,dV\right). \tag{4.25}$$

Serrin shows that his form of the second law implies that the functions φ corresponding to two distinct ideal gases must be proportional, and hence that ideal gases determine, to within a constant factor, a distinguished coordinate system on $\mathscr{H}$. In [1979, 2] it is shown that the presence in $\mathbb{U}$ of elastic, or even viscous, substances far more general than ideal gases determines the same coordinatization of $\mathscr{H}$. This coordinatization, which is unique if, as in practice, one preassigns the value of the difference in the coordinates of the hotness levels of two phase transitions at a standard pressure, such as the freezing and boiling points of water at one atmosphere, is called the *absolute temperature scale.*

Serrin's principal result is that his statement of the second law is equivalent to asserting that *for every cycle $\mathscr{P}$ of each system in* $\mathbb{U}$

$$\int_0^\infty \frac{Q(\mathscr{P},\varphi^{-1}(\theta))}{\theta^2}\,d\theta \leqq 0\,. \tag{4.26}$$

The importance of this relation, called the *accumulation inequality*, lies in its generality: it refers only to the absolute temperature scale φ and the "accumulation" $Q(\mathscr{P},\cdot)$ of the (countably additive) heat measure $q_{\mathscr{P}}$ on the hotness manifold $\mathscr{H}$; it is independent of the "space-time structure" or the concepts of "body" and "force" used in the specific physical theory to which the thermodynamical concepts of heat and hotness may be applied. When the function $\hat{Q}_{\mathscr{P}}$, defined by $\hat{Q}_{\mathscr{P}}(\theta) = Q(\mathscr{P},\varphi^{-1}(\theta))$, is of bounded variation, for each small $\delta > 0$, an integration by parts yields, in view of (4.19),

$$\begin{aligned}
\int_0^\infty \frac{Q(\mathscr{P},\varphi^{-1}(\theta))}{\theta^2}\,d\theta &= \int_{\varphi(L^l)-\delta}^{\varphi(L^u)} \frac{\hat{Q}_{\mathscr{P}}(\theta)}{\theta^2}\,d\theta + \int_{\varphi(L^u)}^{\infty} \frac{\hat{Q}_{\mathscr{P}}(\varphi(L^u))}{\theta^2}\,d\theta \\
&= \int_{\varphi(L^l)-\delta}^{\varphi(L^u)} \frac{d\hat{Q}_{\mathscr{P}}(\theta)}{\theta} - \left.\frac{\hat{Q}_{\mathscr{P}}(\theta)}{\theta}\right|_{\varphi(L^l)-\delta}^{\varphi(L^u)} + \frac{\hat{Q}_{\mathscr{P}}(\varphi(L^u))}{\varphi(L^u)} \\
&= \int_{\varphi(L^l)-\delta}^{\varphi(L^u)} \frac{d\hat{Q}_{\mathscr{P}}(\theta)}{\theta} = \int_0^\infty \frac{d\hat{Q}_{\mathscr{P}}(\theta)}{\theta}\,.
\end{aligned} \tag{4.27}$$

Thus the accumulation inequality does give a mathematical form to the assertion that "the sum along a cycle of the ratio of heat gained to the absolute temperature at which it is gained cannot be positive".

The integral in the accumulation inequality plays the role of the action $\mathscr{I}$ in (4.2). It is clear that *Serrin's* form of the second law and his derivation of the accumulation inequality go a long way toward the resolution of problems (i), (ii), and (iii). In our recently published joint research with *Serrin* [1981, 1], we have extended *Serrin's* form of the second law and the accumulation inequality so that they are meaningful for "approximate cycles". In this research, by combining definitions and methods of the papers [1974, 1; 1975, 1; 1979, 2, 3], we answer the questions (i – iii) in a way that supplies not only identification of the action $\mathscr{I}$

in (4.2) but also a derivation of the implication (4.3). We consider thermodynamical systems (Σ, Π, Q) that are systems (Σ, Π) in the sense explained above[9] and possess an accumulation function Q that assigns a number $Q(P, \sigma, L)$ to each triple (P, σ, L) with P in Π, σ in $\mathscr{D}(P)$, and L in the hotness manifold $\mathscr{H}$; $Q(P, \sigma, L)$ is called *the net heat absorbed by the system at levels of hotness at or below L in the process P starting at the state σ*. Let $\mathbb{P}$ be the set of pairs (P, σ) with P in Π and σ in $\mathscr{D}(P)$. In addition to a mild regularity condition[10] for Q, it is assumed that, for each (P, σ) in $\mathbb{P}$ [even if (P, σ) is not a cycle, i.e., does not have $\varrho_P \sigma = \sigma$] there are hotness levels $L^l = L^l(P, \sigma)$ and $L^u = L^u(P, \sigma)$, with $L^l \prec L^u$, such that, in analogy with (4.19),

$$\left.\begin{aligned} Q(P, \sigma, L) &= 0 && \text{for} \quad L \prec L^l , \\ Q(P, \sigma, L) &= Q(P, \sigma, L^u) && \text{for} \quad L^u \prec L . \end{aligned}\right\} \tag{4.28}$$

We say that a pair (P, σ) in $\mathbb{P}$ is *absorptive* if $Q(P, \sigma, L) \geqq 0$ for all L in $\mathscr{H}$; i.e., if the heat absorbed by the system at or below each level of hotness is not negative[11].

The *union* $\mathscr{S}' \oplus \mathscr{S}''$ of two thermodynamical systems $\mathscr{S}' = (\Sigma', \Pi', Q')$ and $\mathscr{S}'' = (\Sigma'', \Pi'', Q'')$ is taken to be the system $\mathscr{S} = (\Sigma, \Pi, Q)$ with $\Sigma = \Sigma' \times \Sigma''$, $\Pi = \Pi' \times \Pi''$, $\mathbb{P} = \mathbb{P}' \times \mathbb{P}''$, and with

$$\varrho_{(P', P'')}(\sigma', \sigma'') = (\varrho_{P'} \sigma', \varrho_{P''} \sigma'') \tag{4.29}$$

and

$$Q((P', P''), (\sigma', \sigma''), L) = Q'(P', \sigma', L) + Q''(P'', \sigma'', L) , \tag{4.30}$$

for each $((P', P''), (\sigma', \sigma''))$ in $\mathbb{P} = \mathbb{P}' \times \mathbb{P}''$ and each L in $\mathscr{H}$. As in [1979, 2, 3], it is assumed that a collection $\mathbb{U}$ of thermodynamical systems, closed under the union operation, is given and that $\mathbb{U}$ contains at least one special system that corresponds to an elastic or viscous substance. Again, the discussion of the second law takes its simplest form if the distinguished systems are ideal gases. Each state σ of an ideal gas is represented as a point (V, θ) in the first quadrant of a coordinate plane; each process P_t of the gas is a piecewise continuous function on an interval $[0, t)$ with values $(\dot{V}(\tau), \dot{\theta}(\tau))$ that, for each τ in $[0, t)$, can be interpreted as the rates of change of V and θ at time τ; a pair (P_t, σ°), with $\sigma^\circ = (V^\circ, \theta^\circ)$, is in $\mathbb{P}$ if, for each s in $[0, t]$, $(V(s), \theta(s))$, with

$$\begin{aligned} V(s) &= V^\circ + \int_0^s \dot{V}(\tau)\, d\tau , \\ \theta(s) &= \theta^\circ + \int_0^s \dot{\theta}(\tau)\, d\tau , \end{aligned} \tag{4.31}$$

---

9 I.e., in the sense in which the word *system* is used in [1974, 1; 1975, 1]. It is observed in [1981, 1] that many of the results given there hold under a concept of "system" more general than that introduced in [1974, 1; 1975, 1].

10 Namely that for each choice of P and σ the function $Q(P, \sigma, \cdot)$ is bounded and has at most a countable number of points of discontinuity.

11 When the processes correspond to functions of time, a pair (P, σ) may be absorptive and yet such that heat is emitted by the system during an interval of time; in such a case there will be (for the same process P and initial state σ) other intervals of time during which the system absorbs a compensating amount of heat.

is in Σ, i.e., has $V(s)>0$ and $\theta(s)>0$; in such a case

$$\varrho_{P_t}\sigma^\circ = (V(t), \theta(t)) \,. \tag{4.32}$$

It is again part of the definition of an ideal gas that θ is given by a coordinate system φ on $\mathscr{H}$ and no more is assumed about φ than that it is a strictly increasing, positive-valued, continuous function. For an ideal gas the function Q has the form

$$Q(P_t, \sigma^\circ, \bar{L}) = \int_{M(P_t, \sigma, \bar{L})} \left(\tilde{c}(\theta(s))\,\dot{\theta}(s) + \frac{r\,\theta(s)}{V(s)}\,\dot{V}(s) \right) ds \,, \tag{4.33}$$

where r is a number, $\tilde{c}$ is a function characteristic of the gas, and

$$M(P_t, \sigma, \bar{L}) = \{s \,|\, 0 \leqq s < t, \theta(s) \leqq \varphi(\bar{L})\} \,. \tag{4.34}$$

When the curve with the parametrization (4.31) on $[0, t]$ is a closed curve, and hence $V(t) = V^\circ$, $\theta(t) = \theta^\circ$, (4.32) yields $\varrho_{P_t}\sigma^\circ = \sigma^\circ$, and the pair (P_t, σ°) is a cycle; in such a case, if we write $\mathscr{P}$ for (P_t, σ°), the equations (4.33) and (4.24) become the same. The equation (4.33), which can be written in the line-integral notation used in equations (4.24) and (4.25), is an extension of these equations from pairs (P_t, σ°) that are cycles to pairs (P_t, σ°) with σ° in $\mathscr{D}(P_t)$, i.e., from $\mathbb{P}_c$ to $\mathbb{P}$.

We take the second law to be the following statement that pertains to each system $\mathscr{S} = (\Sigma, \Pi, Q)$ in $\mathbb{U}$: *For each level $\bar{L}$ of hotness and each $\varepsilon > 0$, each state σ has a neighborhood $\mathscr{O}_\varepsilon(\sigma, L)$ in Σ for which*

$$\begin{aligned} \varrho_P\sigma \in \mathscr{O}_\varepsilon(\sigma, \bar{L}), \quad (P, \sigma) \text{ absorptive}, \quad L^{\mathrm{u}}(P, \sigma) \prec \bar{L} \\ \Rightarrow\; 0 \leqq Q(P, \sigma, L^{\mathrm{u}}(P, \sigma)) < \varepsilon \,. \end{aligned} \tag{4.35}$$

In terms more suggestive but less precise: The overall net gain of heat is small in an approximate cycle that is absorptive and operates at or below a fixed level $\bar{L}$ of hotness.

In (4.35), the relation $\varrho_P\sigma \in \mathscr{O}_\varepsilon(\sigma, \bar{L})$ indicates an "approximate cycle" and the relation $L^{\mathrm{u}} = L^{\mathrm{u}}(P, \sigma) \prec \bar{L}$ is the assertion that a pair (P, σ) "operates at or below the fixed level $\bar{L}$ of hotness". The relation $0 \leqq Q(P, \sigma, L^{\mathrm{u}})$ is true for any absorptive pair (P, σ). The relation $Q(P, \sigma, L^{\mathrm{u}}) < \varepsilon$, however, is the assertion that the "overall net gain of heat is small" and is the important conclusion of the implication (4.35).

It is a consequence of this law that the hotness manifold again has a distinguished coordinate system φ that is unique to within a constant multiple, and this coordinate system is that employed in the formula (4.33) for the accumulation function of an ideal gas. The principal results obtained in [1981, 1] are of the following type: *The second law is equivalent to the assertion that for every $\bar{L}$ in $\mathscr{H}$ and every thermodynamical system (Σ, Π, Q) in $\mathbb{U}$, each state σ° has, for each $\varepsilon > 0$, a neighborhood $\mathscr{O}_\varepsilon(\sigma^\circ, \bar{L})$ in Σ for which*

$$\int_0^\infty \frac{Q(P, \sigma^\circ, \varphi^{-1}(\theta))}{\theta^2}\, d\theta < \varepsilon \tag{4.36}$$

whenever (P, σ°) *is in* $\mathbb{P}, L^{\mathrm{u}}(P, \sigma) \prec \bar{L}$, *and*

$$\varrho_P \sigma \in \mathcal{O}_\varepsilon(\sigma, \bar{L}) \,. \tag{4.37}$$

In other words: The second law holds if and only if each system in $\mathbb{U}$ is such that its accumulation integral is approximately negative on approximate cycles. In particular, the second law implies that when

$$\mathscr{A}(P, \sigma^\circ) = \int_0^\infty \frac{Q(P, \sigma^\circ, \varphi^{-1}(\theta))}{\theta^2} d\theta$$

and (P, σ°) operates at or below $\bar{L}$, the implication (4.3), with $\mathcal{O}_\varepsilon(\sigma^\circ) = \mathcal{O}_\varepsilon(\sigma^\circ, \bar{L})$, holds for each system in $\mathbb{U}$, each state σ°, and each $\varepsilon > 0$.

In the second part of this essay we have discussed the problem of characterizing the action $\mathscr{A}$ in a general, essentially context-free, manner in which only the concepts of heat and hotness need be mentioned. Of course, in the thermomechanics of continuous media, formulae for $\mathscr{A}$ have long been known. Of primary interest to researchers in that field will be the ideas and theorems presented in the first part of our discussion, namely in the paragraphs containing the relations (4.1) – (4.18). The concepts set forth there give us an approach to thermodynamics in which the existence and regularity of entropy (and of free energy) as a function of state is to be deduced rather than assumed[12]. In our first paper employing this new approach [1974, 1], we examined the problem of finding the restrictions that the second law places on the constitutive equations of elastic and viscous materials, materials with internal state variables, and materials with fading memory, and we found that the assumption that $\mathscr{A}$ has the Clausius property yields restrictions on the response functions (or functionals) that give such experimentally observable quantities as stress, heat flux, and internal energy (or temperature) agreeing perfectly with restrictions obtained in the treatments that start with a differentiable entropy, or free energy, function and employ the Clausius-Duhem inequality [1963, 1, 2; 1964, 1 – 3; 1967, 1, 2]. For each of these materials more was known about $\mathscr{A}(P, \sigma)$ than its continuity in σ or its general representation as an accumulation integral, and consequently more than semi-continuity was proven about entropy and free energy. In each case it was shown that, *starting with an appropriate expression for* $\mathscr{A}$ *in terms of the response functions or functionals of stress, heat flux, and internal energy, and assuming that* $\mathscr{A}$ *has the Clausius property, one can construct an entropy function (or functional) with the properties of differentiability needed to derive the principal results of the earlier studies.* The failure of the entropy and free energy of a material with fading memory to be unique does not invalidate the earlier studies based on the Clausius-Duhem inequality. The implications of the earlier work for the response functionals for stress, heat flux, and internal energy, required only the existence of an appropriately smooth free-energy functional, not its uniqueness.

It has been found that for some materials one can separate the problem of finding the class of entropy functions from that of deriving the thermodynamical restrictions on response functions relating experimentally accessible quantities. For example, the local thermomechanics of a unidimen-

---

[12] The need for such an approach was brought out in the book of *Day* [1972, 2] which preceded and influenced our paper of 1974. (Another important influence was a paper of *Noll* [1972, 3], which showed the usefulness of the concept of state for a broad class of materials, including materials with memory.) *Day*, starting with a Clausius-type inequality for non-linear materials with fading memory, was able to obtain, albeit under neglect of the contribution to $\mathscr{A}$ of a term involving the inner product of the heat flux vector and the temperature gradients, the existence of entropy and free-energy functionals; however, he assumed, rather than proved, that these functionals have the smoothness needed to proceed further and derive *Coleman's* formula for stress as an instantaneous derivative of free energy. Only the equilibrium response was shown to have the smoothness expected of it. Nevertheless, *Day's* book stimulated the search for general Clausius-type and Kelvin-type formulations of the second law by showing that, even for materials with memory, the existence of entropy can be proved from such starting points.

sional elastic-plastic material (with its elastic behavior linear and its plastic behavior perfect) is described by giving the elastic modulus μ and the yield strain α as functions of the temperature θ, and the heat capacity $\varkappa$, the latent elastic heat Λ_e, and the latent plastic heat Λ_p as functions of the elastic strain λ_e, the plastic strain λ_p, and the temperature:

$$\mu = \mu(\theta) , \qquad \alpha = \alpha(\theta) ,$$
$$\varkappa = \varkappa(\lambda_e, \lambda_p, \theta) , \qquad \Lambda_e = \Lambda_e(\lambda_e, \lambda_p, \theta) , \qquad \Lambda_p = \Lambda_p(\lambda_e, \lambda_p, \theta) .$$

Without mention of entropy or free energy, one can derive relations among the functions μ, σ, $\varkappa$, Λ_e, Λ_p that are necessary and sufficient for compliance with the laws of thermodynamics (see [1976, 2; 1979, 1]). One may separately find conditions on these functions sufficient for the entropy function, normalized as in equation (4.11), to be unique and, in cases where entropy and free energy are not unique, the class of such functions can be precisely described [1975, 1; 1979, 1][13].

Acknowledgment. We are grateful to *James Serrin* for the opportunity to work with him on the theory of the accumulation inequality.

The preparation of this essay was supported in part by the U.S. National Science Foundation and the Italian National Council for Research.

References

1896

[1] E. Mach, *Die Prinzipien der Wärmelehre, Historisch-kritisch entwickelt.* Leipzig, Barth.

1963

[1] B. D. Coleman & V. J. Mizel, "Thermodynamics and departures from Fourier's law of heat conduction", *Archive for Rational Mechanics and Analysis* **13**:245–261.

[2] B. D. Coleman & W. Noll, "The thermodynamics of elastic materials with heat conduction and viscosity", *Archive for Rational Mechanics and Analysis* **13**:167–178.

1964

[1] B. D. Coleman, "Thermodynamics of materials with memory", *Archive for Rational Mechanics and Analysis* **17**:1–46.

[2] B. D. Coleman, "Thermodynamics, strain impulses, and viscoelasticity", *Archive for Rational Mechanics and Analysis* **17**:230–254.

[3] B. D. Coleman & V. J. Mizel, "Existence of caloric equations of state in thermodynamics", *Journal of Chemical Physics* **40**:1116–1125.

1967

[1] B. D. Coleman & M. E. Gurtin, "Equipresence and constitutive equations for rigid heat conductors", *Zeitschrift für Angewandte Mathematik und Physik* **18**:199–208.

[2] B. D. Coleman & M. E. Gurtin, "Thermodynamics with internal state variables", *Journal of Chemical Physics* **47**:597–613.

1972

[1] J. B. Boyling, "An axiomatic approach to classical thermodynamics", *Proceedings of the Royal Society* (London) A **329**:35–70.

[2] W. A. Day, *The Thermodynamics of Simple Materials with Fading Memory*, Springer Tracts in Natural Philosophy, Volume 22, Berlin etc., Springer-Verlag.

[3] W. Noll, "A new mathematical theory of simple materials", *Archive for Rational Mechanics and Analysis* **48**:1–50.

---

[13] We discuss corresponding problems for hypo-elastic materials in [1976, 1] and there show that each hypo-elastic material has a unique normalized free-energy function.

1974

[1] B. D. Coleman & D. R. Owen, "A mathematical foundation for thermodynamics", *Archive for Rational Mechanics and Analysis* **54**:1 – 104.

1975

[1] B. D. Coleman & D. R. Owen, "On thermodynamics and elastic-plastic materials", *Archive for Rational Mechanics and Analysis* **59**:25 – 51.

1976

[1] B. D. Coleman & D. R. Owen, "On thermodynamics and intrinsically equilibrated materials", *Annali di Matematica pura e applicata* (IV) **108**:189 – 199.

[2] B. D. Coleman & D. R. Owen, "Thermodynamics of elastic-plastic materials", *Accademia Nazionale dei Lincei, Rendiconti della Classe di Scienze fisiche, matematiche e naturali* (VIII) **61**:77 – 81.

1977

[1] C. Truesdell & S. Bharatha, *The Concepts and Logic of Classical Thermodynamics as a Theory of Heat Engines, Rigorously Developed upon the Foundation Laid by S. Carnot and R. Reech,* New York etc., Springer-Verlag.

1979

[1] B. D. Coleman & D. R. Owen, "On the thermodynamics of elastic-plastic materials with temperature-dependent moduli and yield stresses", *Archive for Rational Mechanics and Analysis* **70**:339 – 354.

[2] J. Serrin, *Lectures on Thermodynamics,* University of Naples, multiplied typescript.

[3] J. Serrin, "Conceptual analysis of the classical second law of thermodynamics", *Archive for Rational Mechanics and Analysis* **70**:355 – 371.

[4] C. Truesdell, "Absolute temperatures as a consequence of Carnot's General Axiom", *Archive for History of Exact Sciences* **20**:357 – 380.

1980

[1] M. Šilhavý, "On measures, convex cones, and foundations of thermodynamics, I. Systems with vector-valued actions; II. Thermodynamic systems", *Czechoslovak Journal of Physics* **B30**:841 – 861, 961 – 991.

[2] C. Truesdell, *The Tragicomical History of Thermodynamics, 1822 – 1854,* New York etc., Springer-Verlag.

1981

[1] B. D. Coleman, D. R. Owen, & J. Serrin, "The second law of thermodynamics for systems with approximate cycles", *Archive for Rational Mechanics and Analysis* **77**:103 – 142.

1982

[1] M. Feinberg & R. Lavine, "Thermodynamics based on the Hahn-Banach Theorem: the Clausius inequality", *Archive for Rational Mechanics and Analysis* **82** (1983):202 – 293.

[2] D. R. Owen, "The second law of thermodynamics for semi-systems with few approximate cycles", *Archive for Rational Mechanics and Analysis* **80**:39 – 55.

[3] M. Šilhavý, "On the Clausius inequality", *Archive for Rational Mechanics and Analysis* **81**:221 – 243.

Chapter 5
A Third Line of Argument in Thermodynamics

C. Truesdell

We all know the power of arguments resting on cycles. Again and again in the papers we have heard, cycles or approximations to them have delivered the goods. We know also the power of classical ideas and the value of classical thermodynamics as an example of what is wanted and what can be got. Every speaker has called upon classical arguments and classical results, not only as an important special instance that must be included in any modern theory but also as a source of inspiration. But classical thermodynamics, while it may be developed by exploiting the properties of cycles, need not be. David Owen and Manuel Ricou have reminded us that thermodynamics can be founded also upon ideas about a class of processes that need not even include non-trivial cycles.

The thermodynamics of possibly irreversible processes in homogeneous systems, as presented by Planck, associates with a given body an *internal energy* E and an *entropy* H, process-dependent functions of time, which satisfy a "First Law" and a "Second Law" expressible as follows in terms of the heating (heat rate) Q:

(I) $$L + \Delta E = JC,$$

(C-P) $$\Delta H \geqslant \int \frac{Q\,dt}{T},$$

in which the integration with respect to time t is taken over a specified closed interval; the *net gain of heat*

$$C := \int Q\,dt;$$

J is a universal positive constant; and the *net work done*

$$L := \int W\,dt,$$

W being the density of net working (work rate), which may include the rate of decrease of kinetic energy. In Clausius' terms, a process is "compensated" if equality holds in (C-P).

The classical thermodynamics of reversible bodies takes Q and W as linear functions of time rates, thus making heat and work reversible always. E and H are assumed to be (or proved to be) functions of T and a finite-dimensional vector "substate" Υ. "Compensated" and "reversible" are equivalent. If a cycle is defined by the conditions $\Delta T = 0$, $\Delta \Upsilon = 0$, then in a cycle $\Delta E = 0$ and $\Delta H = 0$, and for cycles (C-P) reduces to Clausius' Second Law.

In a more general thermodynamics, the "substate" may be the history of an infinite-dimensional vector, and cycles may be of several different kinds: $\Delta E = 0$

characterizes a cycle in energy; $\Delta H = 0$ a cycle in entropy; *etc.*; and the kinds are generally not co-extensive. Likewise, "compensated" and "reversible" need not be equivalent.

We consider now (I) and (C-P) as they stand. If the *heat absorbed* is

$$C^+ := \int_{Q>0} Q\, dt ,$$

then we may prove that

$$\text{(I) \& (C-P)} \Rightarrow \text{(W)},$$

(W) being the following *estimate of work done*:

$$L + \Delta E \leqslant J\left(1 - \frac{T_{\min}}{T_{\max}}\right) C^+ + T_{\min} \Delta H .$$

The symbols $T_{\max}$ and $T_{\min}$ stand for the essential supremum and essential infimum of T; the former is assumed finite, the latter, positive. The sign of equality holds only for a compensated process which is isothermal or absorbs no heat or absorbs heat at the temperature $T_{\max}$ only and emits heat at the temperature $T_{\min}$ only. A process of the third of these kinds may be called a "compensated Carnot process".

If we assume that (W) holds on all processes, it holds on the restriction of a given process to any subinterval of the time-interval on which that process is defined. We can then prove a converse to the foregoing estimate and so conclude that

$$\text{(I) \& (C-P)} \Leftrightarrow \text{(I) \& (W)} .$$

Thus all properties of Planck's thermodynamics can be deduced from (I) & (W) alone. In other words balance of energy and the same estimate of maximum work as classical thermodynamics delivers suffice to derive all restrictions on constitutive functions, no appeal to any idea of increase of entropy being needed.

The work inequality (W) suggests two possibilities.

1) Defining a "W-thermodynamics" by a sufficiently general concept of "state", a broad enough class of processes, and some appropriate class of constitutive relations, we might use (W) to demonstrate the existence of and calculate a least possible, constitutively determined function H. That would be a construction of entropy functions belonging to the bodies in the W-thermodynamics.

2) Also in a theory not restricted to homogeneous processes we might use a work-inequality similar to (W) to obviate the Clausius-Duhem inequality altogether in the procedure for determining restrictions on constitutive relations. This possibility has been explored by Chi-Sing Man in a thesis accepted in 1980 but not yet published.

If the sets on which $Q > 0$ and $Q < 0$ have positive measure, then T has a finite essential supremum on the former, a positive essential infimum on the latter. Denoting these by T^+ and T^-, we easily prove from (C-P) that

$$\text{(W}'\text{)} \qquad L + \Delta E \leqslant J\left(1 - \frac{T^-}{T^+}\right) C^+ + T^- \Delta H.$$

This estimate of work done is less likely than (W) to serve as a basis for thermodynamics because it holds for only a subclass of processes. When it does hold, it gives a better bound than (W) if and only if

$$\left(\frac{T^-}{T^+} - \frac{T_{\min}}{T_{\max}}\right) JC^+ > (T^- - T_{\min})\, \Delta H.$$

Mr. Serrin has extended the estimate (W′) to greater generality. His statement and proof follow.

"Assume, within the context of my paper in this volume, that the energy-entropy hypothesis holds, namely that there exist functions

$$E: \Sigma \to \mathbb{R}, \qquad H: \Sigma \to \mathbb{R}$$

such that

$$(*) \qquad \begin{aligned} \Delta E &\leqslant J\bar{Q}(P) - \bar{W}(P) \\ \Delta H &\geqslant J\bar{A}(P) \end{aligned}$$

for all processes $P \in P_\Sigma(S)$. Here $\bar{A}(P)$, the accumulation integral, is defined by

$$\bar{A}(P) = \int_0^\infty \frac{Q(P, L)}{T^2}\, dT,$$

where $Q(P, \cdot)$ is the accumulation function of the process P and L is the hotness associated with the absolute temperature T.

"If it is supposed that $Q(P, \cdot)$ is of *bounded variation*, then this function can be represented uniquely in terms of its positive and negative variations $Q^+(P, \cdot)$ and $Q^-(P, \cdot)$ as follows:

$$Q(P, \cdot) \equiv Q^+(P, \cdot) - Q^-(P, \cdot).$$

Clearly both $Q^+(P, L)$ and $Q^-(P, L)$ are constant for all $L \in \mathscr{H}$ which are sufficiently hot or sufficiently cold (depending of course on the process P). Define

$$T^+ = \inf\{\tilde{T} \in \mathbb{R}^+ : Q^+(P, L) \equiv \text{constant for } T > \tilde{T}\}$$
$$T^- = \sup\{\tilde{T} \in \mathbb{R}^+ : Q^-(P, L) \equiv 0 \quad \text{for} \quad T < \tilde{T}\}$$

(obviously $T_{\min} \leqslant T^-$, $T_{\max} \geqslant T^+$).

"*We assert that if* $Q(P, \cdot)$ *is of bounded variation and* $T^+ > 0$, $T^- < \infty$, *then*

$$(**) \qquad \bar{W}(P) + \Delta E \leqslant J\left(1 - \frac{T^-}{T^+}\right) C^+ + T^- \Delta H,$$

where C^+ *is the ultimate (positive) value of* $Q^+(P, \cdot)$, i.e. *the total absorbed heat.*

Proof. Obviously

$$\int_0^\infty \frac{Q^+(P,L)}{T^2}\,dT \geqslant C^+ \int_{T^+}^\infty \frac{dT}{T^2} = \frac{C^+}{T^+}$$

and

$$\int_0^\infty \frac{Q^-(P,L)}{T^2}\,dT \leqslant C^- \int_{T^-}^\infty \frac{dT}{T^2} = \frac{C^-}{T^-},$$

where C^- is the ultimate value of $Q^-(P, \cdot)$, that is

$$C^+ - C^- = \bar{Q}(P)\,.$$

Combining the above lines leads to

$$(\dagger)\qquad \begin{aligned} \bar{A}(P) &= \int_0^\infty \frac{Q^+(P,L) - Q^-(P,L)}{T^2}\,dT \\ &\geqslant \left(\frac{1}{T^+} - \frac{1}{T^-}\right) C^+ + \frac{\bar{Q}(P)}{T^-}\,. \end{aligned}$$

Now from (*) it follows at once that

$$\bar{W}(P) + \Delta E \leqslant J\bar{Q}(P) + T^-\{\Delta H - J\bar{A}(P)\}\,.$$

Eliminating $\bar{A}(P)$ from this inequality by use of (†), we get the required result.

"Inequality (**) is an estimate for the quantity $\bar{W}(P) + \Delta E$ in terms of T^-, T^+, C^+ and ΔH. Another estimate in terms of the same variables can be obtained from $(*)_1$ and the fact that $\bar{Q}(P) = C^+ - C^-$, namely

$$(***)\qquad \bar{W}(P) + \Delta E \leqslant JC^+\,.$$

Which of (**) or (***) is better clearly depends on the quantity $T^+\Delta H - JC^+$. If this is positive, then (***) is stronger; if it is negative then (**) is stronger; and if it is zero then the two are the same.

"It is easily shown that equality holds in (**) if and only if equality holds in (*) and P is a Carnot process with $T^- < \infty$. On the other hand equality holds in (***) if and only if equality holds in (*) and $C^- = 0$, $T^- = \infty$.

"To see the relation between (W), (**) and (***), observe that the proof of (**) also yields the inequality

$$\Delta E + \bar{W}(P) \leqslant J\left(1 - \frac{\tau^-}{\tau^+}\right) C^+ + \tau^- \Delta H$$

for all $\tau^- \in (0, T^-)$ and $\tau^+ \in (T^+, \infty)$. Here (W) results when $\tau^- = T_{\min}$, $\tau^+ = T_{\max}$, while (**) arises when $\tau^- = T^- < \infty$, $\tau^+ = T^+ \geqslant 0$ (note that (**) can be interpreted as $\bar{W}(P) + \Delta E \leqslant T^- \Delta H$ when $T^- < \infty$, $T^+ = 0$).

We now consider the mutually exclusive cases

$$T^+\Delta H - JC^+ > 0 \quad (\text{or } \Delta H > 0 \quad \text{if} \quad C^+ = 0)$$
$$T^+\Delta H - JC^+ < 0 \quad (\text{or } \Delta H < 0 \quad \text{if} \quad C^+ = 0)$$
$$T^+\Delta H - JC^+ = 0 \quad (\text{or } \Delta H = 0 \quad \text{if} \quad C^+ = 0)\,.$$

In the first case the best choice of τ^-, τ^+ is $\tau^- = 0$, τ^+ arbitrary, which yields the optimal bound (***). Hence in this case both (W) and (**) are less good then (***), though which of (W), (**) is better depends on the relation earlier addressed by Professor Truesdell. In the second case, we see from $(*)_2$ and an argument by contradiction that $C^- > 0$, so that $T^- < \infty$. The best choice of τ^-, τ^+ is then $\tau^- = T^-$, $\tau^+ = T^+$, yielding for this case the optimal bound (**). In the third case (**) and (***) are the same, and both arise from the optimal choice $\tau^- = 0$, $\tau^+ = T^+$.

"To return to Professor Truesdell's considerations, the above analysis shows that if his inequality (W) is stronger than (W′), then (***) in turn is stronger than (W). On the other hand if his inequality (W′) is stronger than (W), then either (W′) itself is best, or else (W′) can in turn be improved to (***).

"We conclude with a simple estimate for the work done by cyclic processes. Indeed in this case, assuming $\bar{W}(P) > 0$,

$$\Delta E = \Delta H = 0\,, \qquad T^+ > 0\,, \qquad T^- < \infty$$

(that $C^+ > 0$, and so $T^+ > 0$, follows from $(*)_1$; that $C^- > 0$ and so $T^- < \infty$ then follows from $(*)_2$). Thus in turn $T^+\,\Delta H < JC^+$ so (**) rather than (***) is the stronger inequality. This yields the efficiency estimate

$$\bar{W}(P) \leqslant J\left(1 - \frac{T^-}{T^+}\right) C^+\,,$$

an improved version of the celebrated Kelvin relation

$$\bar{W}(P) \leqslant J\left(1 - \frac{T_{\min}}{T_{\max}}\right) C^+\,.$$

"From this perspective one can profitably view the inequalities (W), (W′) and (**), (***) as generalizations of Kelvin's efficiency formula, providing estimates for the maximum amount of work which can be obtained from a process P when the initial and final states, the maximum temperature T^+ of heat input, the minimum temperature T^- of heat output, and the total absorbed heat C^+ are given."

List of Sources

1 C. Truesdell, "Improved estimates of the efficiency of irreversible heat engines", *Annali di Matematica Pura ed Applicata* (4) **108** (1976):305 – 323.

2 C. Truesdell, "Proof that my work estimate implies the Clausius-Planck inequality", *Accademia Nazionale dei Lincei, Rendiconti della Classe di Scienze Matematiche, Fisiche e Naturali* (8) **68** (1980):191 – 199.

3 Chi-Sing Man, *"Thermodynamics based on a work inequality"*, Ph. D. Thesis, accepted by the Johns Hopkins University, Baltimore, 1980.

Chapter 6
The Laws of Thermodynamics for Non-Cyclic Processes

M. Ricou

If the laws of thermodynamics are stated without reference to the concepts of energy and entropy, the problem of deriving sufficiently general existence and uniqueness theorems for these functions naturally becomes one of central importance in the theory. With few exceptions (for example [6.1]), when the content of the laws is chosen for this purpose, they are phrased as prohibitions against certain types of (possibly approximate) cycles. The difficulties encountered when trying to prove general existence and uniqueness results for energy and entropy from statements of this type are by now reasonably well known and understood.

In spite of these difficulties, it is nevertheless true that existence theorems for entropy have been proved for a wide variety of thermal systems, for example using the notion of approximate cycles as done by *Coleman, Owen* and *Serrin* [6.2], or via separation lemmas associated with the Hahn-Banach Theorem, as done by *Feinberg* and *Lavine* [6.3] and *Šilhavý* [6.4]. Naturally, there is some disagreement as to whether the results thus obtained cover all cases of "practical" interest. In any event it is worth observing that these approaches requires the laws to be stated in forms which are, of necessity, fairly sophisticated and of difficult (or at best not obvious) physical interpretation. Also, they are applied only to systems satisfying certain precise mathematical specifications. Hence, as a matter of principle, their results cannot be considered completely general. Finally, energy and entropy functions are defined axiomatically, by means of inequalities they are supposed to satisfy (see [6.5], Sect. 6). As a result, non-uniqueness questions inevitably arise, and can be dealt with only for systems satisfying fairly restrictive conditions connected with the notion of reversibility.

The purpose of this paper is to give a brief and informal description of an alternate and different resolution of the energy-entropy problem. To be more specific, I shall proceed here under the assumption that our difficulties do not indicate a lack of mathematical sophistication in the statements of the laws, or in the methods used to exploit them, but point instead to the incompleteness of any theory of thermodynamics based solely on properties of cyclic processes. To try to clarify this idea, let me take the First Law as an example. At an elementary level, this law is supposed to prohibit "perpetual machines of the first kind", i.e., machines capable of delivering work without consumption of other forms of energy. Such a prohibition has no obvious, rigorous mathematical form, in part because some systems are clearly capable of producing work, even if not connected to an external power source (consider the case of a box containing an electric motor connected to a charged battery). The mention of cycles in the First Law is a device to avoid examples like the one just described. After all, and as everyone knows, a process by which the above system does work cannot be repeated indefinitely and as such is not cyclic. From our point of view, however, what is important is the following observation: when the First Law is stated only for cycles we gain rigor but lose vital information about the real world: in this instance, the fact that a system as simple as the one above, whether operating in cycles or not, will never produce arbitrarily large amounts of work unless it is connected to an external power source.

Naturally, if statements of the First Law (and possibly of the Second) are presumed to be incomplete if they refer only to cycles, our immediate task is to find concepts which might conceivably be used to replace cyclic processes in their fundamental role. A good place to start is this question: if cycles are to be *defined*, instead of being taken as primitive notions, in what terms can that definition be phrased? Our previous (and informal) description of a cycle as a process which can be repeated after itself an arbitrary number of times contains a hint. If refers implicitly to an algebraic operation on processes, often confused with a property of state spaces, but in fact deeper than and independent of the notion of state. This operation may be loosely described as follows: If P and P' are processes of a system $\mathscr{S}$, it is sometimes possible to apply P and P' to the system, one immediately after the other. If P' is applied after P, the resulting process can be labelled PP', and the operation we are now discussing is the application $(P, P') \to PP'$. In terms of this operation, a cycle is merely a process P for which the combination PP is possible. It is this operation which allows us to phrase the laws of thermodynamics without mentioning the concept of cycle. Before doing so, we must of course introduce the logical framework of the theory. In this we follow ideas recently introduced by Serrin.

6.1 Basic Definitions

The formal definition of thermal system which we shall use here is a modification of that described by Serrin in his paper in this volume. A *thermal system* $\mathscr{S}$ is a triple $\mathscr{S} = (\mathbb{P}(\mathscr{S}), Q, \bar{W})$, where $\mathbb{P}(\mathscr{S}), Q$ and $\bar{W}$ have their usual meanings. In particular

$$\mathbb{P}(\mathscr{S}) \text{ is the set of all } \textit{processes} \text{ available to } \mathscr{S}\ ,$$

$$Q\colon \mathbb{P}(\mathscr{S}) \times \mathscr{H} \to \mathbb{R} \text{ is the } \textit{accumulation} \text{ function of } \mathscr{S}\ ,$$

$$\bar{W}\colon \mathbb{P}(\mathscr{S}) \to \mathbb{R} \text{ is the } \textit{work} \text{ function}\ .$$

$\mathscr{H}$ is of course the *hotness manifold.* (The total heat $\bar{Q}(P)$ absorbed by $\mathscr{S}$ during a process P can of course be obtained directly from Q, (see [6.5], Sect. 3). The set of all thermal systems is the *universe* $\mathscr{U}$. Instead of the fourth element appearing in Serrin's definition of a system (i.e., the set $\mathbb{P}_c(\mathscr{S})$), let me now describe the algebraic structure of $\mathbb{P}(\mathscr{S})$ hinted at in the previous section. This description is obviously an *axiom*:

> There is an associative binary operation defined in a subset $\mathbb{F}(\mathscr{S})$ of $\mathbb{P}(\mathscr{S}) \times \mathbb{P}(\mathscr{S})$ and indicated by juxtaposition. For all P, P' and P'' in $\mathbb{P}(\mathscr{S})$, $(PP')P''$ and $P(P'P'')$ are defined if and only if PP' and $P'P''$ are both defined. Moreover in this case $(PP')P'' = P(P'P'')$. (6.1)

The set $\mathbb{F}(\mathscr{S})$ is called the *follow relation*. Its physical meaning is clear: $(P,P')\in\mathbb{F}(\mathscr{S})$ if and only if the process P' can be applied to $\mathscr{S}$ immediately after P is completed. The functions Q and $\bar{W}$ have algebraic properties connected to this relation, listed in the next *axiom*:

If $(P,P')\in\mathbb{F}(\mathscr{S})$, then

$$\begin{aligned} Q(PP',\cdot) &= Q(P,\cdot)+Q(P',\cdot) \\ \bar{W}(PP') &= \bar{W}(P)+\bar{W}(P') \,. \end{aligned} \tag{6.2}$$

These properties are again of simple physical interpretation.

If P is in $\mathbb{P}(\mathscr{S})$, we introduce the set of *followers* of P, denoted by $\mathrm{Foll}(P)$, and the set of *predecessors* of P, denoted by $\mathrm{Pred}(P)$, by

$$\begin{aligned} \mathrm{Foll}(P) &= \{P'\in\mathbb{P}(\mathscr{S})\colon (P,P')\in\mathbb{F}(\mathscr{S})\} \\ \mathrm{Pred}(P) &= \{P'\in\mathbb{P}(\mathscr{S})\colon (P',P)\in\mathbb{F}(\mathscr{S})\} \,. \end{aligned} \tag{6.3}$$

$\mathrm{Foll}(P)$ is the set of processes available to $\mathscr{S}$ immediately after P is completed. It is also convenient to define the set $\mathbb{P}_f(\mathscr{S})$, consisting of all followers, and the set $\mathbb{P}_p(\mathscr{S})$, formed by all predecessors, namely

$$\begin{aligned} \mathbb{P}_f(\mathscr{S}) &= \{P\in\mathbb{P}(\mathscr{S})\colon \mathrm{Pred}(P)\neq\phi\} \\ \mathbb{P}_p(\mathscr{S}) &= \{P\in\mathbb{P}(\mathscr{S})\colon \mathrm{Foll}(P)\neq\phi\} \,. \end{aligned} \tag{6.4}$$

We denote by $\mathbb{P}_c(\mathscr{S})$ the set of all *cyclic* processes of $\mathscr{S}$. It is *defined* by

$$\mathbb{P}_c(\mathscr{S}) = \{P\in\mathbb{P}(\mathscr{S})\colon (P,P)\in\mathbb{P}(\mathscr{S})\} \,. \tag{6.5}$$

Serrin's assumption about cycles now becomes a lemma (the proof is omitted).

Lemma 6.1.1. If $P\in\mathbb{P}_c(\mathscr{S})$, there is a sequence $\{P^n\}$ in $\mathrm{Foll}(P)\cap\mathrm{Pred}(P)$ such that $Q(P^n,\cdot)=nQ(P,\cdot)$ and $\bar{W}(P^n)=n\bar{W}(P)$.

Here P^n is of course the process P repeated after itself n times. It will occasionally be necessary to refer to *reversible process*. Let us say that $P\in\mathbb{P}(\mathscr{S})$ is *reversible* if and only if there is another process $P'\in\mathbb{P}(\mathscr{S})$ such that

$$\begin{aligned} &PP' \text{ is a cycle} \,, \\ &Q(P,\cdot) = -Q(P',\cdot) \,, \\ &\bar{W}(P) = -\bar{W}(P') \,. \end{aligned} \tag{6.6}$$

If (6.6) holds we shall say that (P,P') is a *reversible pair*.

Finally, we must adapt the notion of *compatible systems* to our previous definitions. Suppose $\mathscr{S}_1$ and $\mathscr{S}_2$ are systems in $\mathscr{U}$, and for convenience denote the pairs (P,Q) in $\mathbb{P}(\mathscr{S}_1)\times\mathbb{P}(\mathscr{S}_2)$ by $P\oplus Q$. The systems $\mathscr{S}_1$ and $\mathscr{S}_2$ are called *compatible* if and only if there is a third system $\mathscr{S}_1\oplus\mathscr{S}_2$ in $\mathscr{U}$ determined as follows:

i) $\mathbb{P}(\mathscr{S}_1\oplus\mathscr{S}_2)=\mathbb{P}(\mathscr{S}_1)\times\mathbb{P}(\mathscr{S}_2)$,
ii) for all $P_1\oplus P_2$ in $\mathbb{P}(\mathscr{S}_1\oplus\mathscr{S}_2)$ we have

$$Q(P_1 \oplus P_2, \cdot) = Q(P_1, \cdot) + Q(P_2, \cdot)$$
$$\bar{W}(P_1 \oplus P_2) = \bar{W}(P_1) + \bar{W}(P_2) ,$$

iii) for all $P_1, P_1' \in \mathbb{P}(\mathscr{S}_1)$ and $P_2, P_2' \in \mathbb{P}(\mathscr{S}_2)$,

$(P_1 \oplus P_2)(P_1' \oplus P_2')$ is defined if and only if $P_1 P_1'$ and $P_2 P_2'$ are defined, in which case

$$(P_1 \oplus P_2)(P_1' \oplus P_2') = (P_1 P_1') \oplus (P_2 P_2') .$$

It is now possible to proceed with the discussion of the First Law at a more rigorous level.

6.2 The First Law and Energy

As suggested above, our aversion to perpetual machines of the first kind does not seem to be directly related to properties of cycles, but to the following observation: once a machine capable of delivering mechanical work is built, it will be unable to deliver arbitrarily large amounts of work unless it is connected to an external power source. It is fairly easy to phrase this idea in the terminology just discussed. The *construction* of a machine, that is, the setting up of a thermal system $\mathscr{S}$ in some particular initial configuration can be interpreted as the execution of a process P in $\mathbb{P}(\mathscr{S})$. Once P is completed, the elements of Foll(P) correspond to the different possibilities for the *operation* of this machine, and naturally depend on the initial configuration, which is in turn determined by P. What we said about the impossibility of generating arbitrarily large amounts of work without consumption of other forms of energy (i.e., heat) is now, very simply, the statement

$$\begin{array}{l}\text{For any } \mathscr{S} \text{ in } \mathscr{U} \text{ and each } P \in \mathbb{P}(\mathscr{S}) , \\ \bar{W} \text{ is bounded above in the set } \{P' \in \text{Foll}(P) \colon \bar{Q}(P') \leqslant 0\} .\end{array} \tag{6.7}$$

This we could take as the First Law. Another possibility, stronger than (6.7) but possibly more suggestive, is what I shall call here the Weak First Law (WFL):

WFL: If $\{P_n\} \in \text{Foll}(P)$, then $\bar{W}(P_n) \to \infty$ implies $\bar{Q}(P_n) \to +\infty$.

This says literally that, following a *fixed* process P, the generation of arbitrarily large amounts of work requires arbitrarily large amounts of heat. Note that WFL implicitly states the existence of a function $W^*\colon \mathbb{P}_{\text{p}}(\mathscr{S}) \times \mathbb{R} \to \mathbb{R}$ such that

$$\text{If} \quad P' \in \text{Foll}(P) \quad \text{and} \quad \bar{Q}(P') \leqslant q \quad \text{then} \quad \bar{W}(P') \leqslant \bar{W}^*(P, q) . \tag{6.8}$$

In particular, the bound mentioned in (6.7) is obviously just $W^*(P, 0)$.

Before listing any of the consequences of WFL, let us pause to indicate what one might hope to accomplish. At this level of generality, our main objective must be to clarify the delicate interplay between the statement just given and the ideas listed below:

1) The properties of cyclic processes,
2) the existence and properties of energy functions for arbitrary thermal systems,
3) the nature and theoretical role of the principle of conservation of energy.

Let me begin by discussing cycles. In our notation, Šilhavý's Weak and Strong First Laws[1] for cycles (respectively WFLC and SFLC) can be stated as follows:

WFLC: If $P \in \mathbb{P}_c(\mathscr{S})$ then $\bar{Q}(P) \leqslant 0 \Rightarrow \bar{W}(P) \leqslant 0$,

SFLC: If $P \in \mathbb{P}_c(\mathscr{S})$ then $\bar{Q}(P) \leqslant 0 \Leftrightarrow \bar{W}(P) \leqslant 0$.

Much has been said about the symmetry (or lack of it) between the roles of $\bar{Q}$ and $\bar{W}$ in SFLC and WFLC. At this point, there is no general agreement as to which of these two statements realistically describes a property of all cycles. Interestingly enough, the lack of symmetry between the roles of $\bar{Q}$ and $\bar{W}$ in WFL is not a matter for philosophical argument: it is evident that interchanging their roles produces a patently false principle. In any case, it is easy to derive WFLC, but not SFLC, from WFL (in fact, from (6.7)).

Lemma 6.2.1. *Weak First Law for Cycles.* If P is a cycle then

$$\bar{Q}(P) \leqslant 0 \Rightarrow \bar{W}(P) \leqslant 0 .$$

Proof. Consider the processes P^n mentioned in Lemma 6.1.1. It is clear that

$$\bar{Q}(P^n) = n\bar{Q}(P) \leqslant 0 \quad \text{and} \quad \bar{W}(P) = \frac{1}{n}\bar{W}(P^n) \leqslant \frac{1}{n} W^*(P,0) .$$

Letting $n \to \infty$, we obtain $\bar{W}(P) \leqslant 0$.

Our next step is the generalization of the inequality

$$\bar{W}(P) \leqslant \mathscr{J}\bar{Q}(P) \quad \text{for all} \quad P \in \mathbb{P}_c(\mathscr{S}) \tag{6.9}$$

to a statement valid for *all* followers of any fixed process $P \in \mathbb{P}(\mathscr{S})$. A simple modification of an argument used to derive (6.9) from WFLC yields.

Theorem 6.2.2. *Work-Heat Inequality.* There is a constant $\mathscr{J} > 0$ such that

$$\operatorname{Sup}\{W(P') - \mathscr{J}Q(P') \colon P' \in \operatorname{Foll}(P)\} < +\infty \quad \text{for all} \quad P \in \mathbb{P}_p(\mathscr{S}) .$$

The proof is omitted. It should be remarked, however, that it uses an auxiliary system $\mathscr{R}$ which must satisfy two conditions: it must be compatible with $\mathscr{S}$, and $\mathbb{P}_c(\mathscr{R})$ must include two cycles R and R' forming a reversible pair, with $\bar{W}(R) > 0$. (Thus in what follows we restrict our attention to thermal systems compatible with $\mathscr{R}$.)

The previous inequality suggests a very natural definition:

---

[1] *Šilhavý* [6.4]. Is should be noted that Šilhavý did not consider the weak statement, by itself, to be an adequate or complete formulation of the First Law.

If $\mathscr{S}$ is a thermal system, its *energy function* $E\colon \mathbb{P}_p(\mathscr{S}) \to \mathbb{R}$ is given by

$$E(P) = \operatorname{Sup}\{\bar{W}(P') - \mathscr{J}\bar{Q}(P')\colon P' \in \operatorname{Foll}(P)\}\,. \tag{6.10}$$

As a consequence of this definition, E obviously satisfies the relation

$$\text{If} \quad P' \in \operatorname{Foll}(P) \quad \text{then} \quad \bar{W}(P') - \mathscr{J}\bar{Q}(P') \leqslant E(P)\,. \tag{6.11}$$

In addition, and again as an easy consequence of its definition, it satisfies the stronger

Theorem 6.2.3. *Energy Inequality.* If $P' \in \mathbb{P}_p(\mathscr{S}) \cap \operatorname{Foll}(P)$, then

$$\mathscr{J}Q(P') - W(P') \geqslant E(P') - E(P)\,.$$

The proof is again omitted. The connection between this inequality and its more common form (using a state function) becomes clear once we realize that the processes P and P' are used here as labels for their *target* states (the state in which the system is when a given process is completed). Note also that one expects the target state of P to be the same as the *initial* state of P'.

There really is no reason to believe in the possibility of proving a general conservation equation for energy within this theory. In fact, the inequality mentioned in Theorem 6.2.3 agrees nicely with the spirit of WFL: energy is not necessarily *conserved*, it is simply never *created*. However, if we restrict our attention to reversible processes, it is not difficult to obtain a balance equation.

Theorem 6.2.4. *Energy Conservation for Reversible Processes.* If (P, P') is a reversible pair of the system $\mathscr{S}$, then

$$\mathscr{J}Q(P') - W(P') = E(P') - E(P)\,.$$

To summarize what was said before, the main consequences of WFL are the following results:

1) Šilhavý's Weak First Law for cycles,
2) the construction of a well-defined energy function for any system compatible with the special system $\mathscr{R}$,
3) the proof of an energy inequality for all processes, and of a balance equation for reversible processes.

Absent from this list are, of course, the Strong First Law for cycles and a general principle of energy conservation. It is therefore natural to ask whether it is possible to reinforce WFL so as to cause the Strong First Law for cycles to hold, and if in so doing energy conservation becomes valid for all processes. Somewhat surprisingly, the answer to these questions is respectively yes and no.

If we interchange in WFL the roles of $\bar{Q}$ and $\bar{W}$, and simultaneously use predecessors instead of followers, the resulting statement does seem to convey the idea of energy indestructibility:

$$\text{If} \quad \{P_n\} \in \operatorname{Pred}(P) \quad \text{then} \quad \bar{Q}(P_n) \to \infty \quad \text{implies} \quad \bar{W}(P_n) \to \infty\,. \tag{6.12}$$

If this statement holds, it is not hard to derive results which are in a sense mirror images of those contained in Theorems 6.2.1, 2 and 3. In particular, if (6.12) is combined with WFL, then the Strong First Law for cycles *can* be proved as a theorem. In spite of this, one can still show that energy conservation does not necessarily hold in general (i.e., for non-cyclic processes as well). In other words, the Strong First Law for cycles and the principle of conservation of energy are *distinct* ideas, requiring different sets of assumptions to hold. It is indeed possible to prove *another* energy inequality, now of the form

$$\text{If } P \in \mathbb{P}_f(\mathscr{S}) \cap \operatorname{Pred}(P') \text{ then } \mathscr{J}\bar{Q}(P) - \bar{W}(P) \leqslant E^*(P') - E^*(P)\,. \quad (6.13)$$

However, the function E^* appearing above is defined for all P in $\mathbb{P}_f(\mathscr{S})$ by:

$$E^*(P) = \operatorname{Sup}\{\mathscr{J}\bar{Q}(P') - \bar{W}(P'): \text{all } P' \in \operatorname{Pred}(P)\} \quad (6.14)$$

and hence E^* and E are, in general, different. Actually, this is one of the major drawbacks of associating WFL and (6.12): if WFL and (6.12) both hold, we must associate with each thermal system *two* energy functions instead of just one. That is, in a word, ugly.

In addition, the existence of two different energy functions must raise some doubts about the proper content of a general conservation principle for energy (for instance, should we choose a balance equation for E or for E^*?). As I tried to show elsewhere [6.6], when (6.12) holds the most reasonable choice for this principle seems to be

$$\text{If } P' \text{ follows } P \text{ then } E(P) = E^*(P')\,. \quad (6.15)$$

Even if the statement above is accepted as the proper expression of the idea of energy conservation, its status in thermodynamics is by no means clear. At first sight, it is very tempting to consider (6.15) as another axiom. But in a system with a high degree of irreversibility in its behavior the sets $\operatorname{Foll}(P)$ and $\operatorname{Pred}(P')$ are likely to be very different. In such a case, on what ideas are we to base a belief in (6.15)? For my part, I cannot imagine arguments making (6.15) plausible which do not involve accessibility conditions (and even notions of reversibility) of dubious generality. Hence, it may be reasonable to treat (6.15) as just another property of *some* thermal systems, a property which of course happens to be common for the systems we have studied in the past (including of course reversible systems).

The possibility of rejecting (6.12) as a basic axiom should be given some consideration. The reasons for doing so are somewhat subjective, but they do exist. First, there is a natural desire for simplicity and elegance in the theory, and that seems quite incompatible with the notion of associating two different energy functions with each thermal system. Second, it is undeniable that (6.12) is *less* convincing than WFL, and thus including it among the basic axioms of this theory must diminish our confidence in its results. Finally, if (6.12) is rejected as a basic axiom our doubts concerning the proper content of the principle of energy conservation disappear, and the principle can only be stated as the usual conservation equation for the function E.

6.3 The Second Law and Entropy[2]

A look at the Weak First Law stated in the previous section suggests the following question: are there any circumstances under which the amount of heat flowing into a body is limited? The purpose of the Second Law (as it is going to be stated here) is precisely to answer this question. It is often said in connection with discussions about the Second Law that a thermal system placed in contact with a single heat reservoir will eventually reach a state of "thermal equilibrium" with the reservoir, at which point heat transfer between the two systems cease. Admittedly, this idea is vague, imprecise, and in some cases plainly wrong, but it does convey an aspect of our intuitions about the Second Law which does not seem to be covered by statements referring only to cycles. Hence, it may very well be a good starting point for our search.

Consider the case of a gas placed in contact with a heat reservoir at some fixed temperature. If the gas is allowed to occupy arbitrarily large volumes, it will certainly absorb arbitrarily large amounts of heat. As a result, the idea discussed above can only be true if applied to processes leading a thermal system to a preassigned *target* configuration. Once the need to look at processes sharing a common target is recognized, it becomes necessary to inquire also whether we may have to fix the *initial* configuration of the system as well. Naturally, if a system is started frozen solid, after which it is placed in contact with a heat reservoir at some fixed temperature and forced to reach some final configuration, it will surely absorb more heat than if started warm. But there is an experimentally recognized limit as to how frozen the system can be initially (I am of course thinking of the Third Law), and it is therefore reasonable to expect the existence of a bound on absorbed heat depending only on the *final* configuration of the system.

Experience with cyclic processes shows that statements of the Second Law referring only to isothermal processes raise serious technical difficulties for the resulting theory. Furthermore, this experience also shows that, when considering processes exchanging heat at more than one hotness level, it is only reasonable to expect a bound on heat absorbed for processes for which the direction of heat transfer is from cold to hot. The formalization of this idea concerning the direction of heat flow has been done by Serrin in his statement of a Second Law for cycles, and will be applied here without modifications: the processes contemplated in the statement of the Second Law given below have *non-negative* accumulation functions (a process with a non-negative accumulation function is called *absorptive*).

Even when considering absorptive processes, it is obvious that the bounds on absorbed heat we are discussing depend on the temperature of the heat reservoir, and conceivably go to infinity at the hot end of $\mathscr{H}$. If we wish to set such bounds for absorptive processes in general, it is clearly necessary to associate with these processes a characteristic hotness level. The simplest way to procede seems to be the following: let P be an absorptive process, and assume $\bar{Q}(P) > 0$. In this case, it is apparent that $Q(P, \cdot)$ can be written (in more than one way) as a sum of a

---

[2] Some of the material in this section will appear in [6.7].

An alle Mitglieder der
Gesellschaft von Freunden und Förderern der Universität München

Stiftungsfest am 28. Juni 1986

Sehr geehrte Damen und Herren.

Die Ludwig-Maximilians-Universität führt nach langjähriger Unterbrechung jetzt wieder jährlich ein Stiftungsfest durch. Es ist geplant, diese Festveranstaltung jeweils am letzten Samstag des Monats Juni durchzuführen.
Unsere Anregung, hierzu auch die „Freunde" einzuladen, stieß bei Herrn Präsident Steinmann auf großes Verständnis. Die Aula der Universität hat jedoch nur ein begrenztes Fassungsvermögen. Herr Präsident Steinmann bat deshalb, vorweg das Interesse an dem Besuch des Stiftungsfestes festzustellen.
Mit dieser Karte möchten wir Sie um Mitteilung bitten, ob Sie daran interessiert sind, das Stiftungsfest zu besuchen, damit wir Ihnen - soweit aus Raumgründen möglich - Einladungskarten vermitteln können.

Mit freundlichen Grüßen
Ihre
Münchener Universitätsgesellschaft

Absender
(bitte deutlich schreiben)

____

____

____

____

Ich bin am Besuch des Stiftungsfestes der Universität München am Samstag, 28.6.1986, vormittags, interessiert mit

______ Person/en

Münchener
Universitätsgesellschaft e.V.
Königinstraße 107

8000 München 40

positive Heavyside function with a non-negative function. Among all such decompositions, one seems rather natural. Define the *Kelvin point* of P, denoted by $K(P)$, by

$$K(P) = g\,\mathrm{lb}\{L \in \mathscr{H}\colon\ Q(P, L') \geqslant \bar{Q}(P), \quad \text{for all} \quad L' \geq L\}\,. \tag{6.16}$$

A moment's thought shows that $Q(P, \cdot)$ can be seen as a combination of an isothermal transfer of $\bar{Q}(P)$ units of heat at the level $K(P)$, together with a transfer of heat from cold to hot with a zero net gain. It is for this reason that we take the Kelvin point of any absorptive process P with $\bar{Q}(P) < 0$ as its characteristic hotness level.

We now have at our disposal all the ingredients needed for a reasonable statement of the Second Law.

Second Law. If $\{P_n\}$ is a sequence of absorptive processes in $\mathrm{Pred}(P)$, then

$$\bar{Q}(P_n) \to +\infty \quad \text{implies} \quad K(P_n) \to +\infty\,. \tag{6.17}$$

Other possibilities for a statement of this type are discussed in [6.7], and in addition there are clear similarities between this statement and that of Owen in [6.1]. Of course Owen's statement is made in the context of Coleman and Owen's theory of semi-systems, and hence makes explicit use of states, instead of the follower relation. However, like (6.17) Owen's formulation provides a limitation on the amount of heat absorbed by a system undergoing processes with non-negative accumulation functions, if those functions have a bounded characteristic hotness level. Owen's Second Law also places a restriction on the *initial* as well as the final states of the processes under consideration, and hence provides an entropy only when the supply of processes available to the system satisfies certain additional restrictions. His characteristic hotness level is also different from the Kelvin point defined by (6.16): in this case it is the level at which $Q(P, \cdot)$ becomes identically equal to $\bar{Q}(P)$.

A statement equivalent to (6.17), but mentioning explicit bounds, is the following:

$$\begin{aligned} &\text{There is a function} \quad Q^*\colon \mathbb{P}_{\mathrm{f}}(\mathscr{S}) \times \mathscr{H} \to \mathbb{R} \quad \text{such that} \\ &P' \in \mathrm{Pred}(P) \quad \text{with} \quad Q(P, \cdot) \geqslant 0 \quad \text{implies} \quad \bar{Q}(P') \leqslant Q^*(P, K(P'))\,. \end{aligned} \tag{6.18}$$

Note that $Q^*(P, \cdot)\colon \mathscr{H} \to \mathbb{R}$ can be chosen in a natural way to be *increasing* and *positive*. Hence, the Second Law in the form (6.17) is definitely very close to the notion of special non-negative temperature scales on $\mathscr{H}$. I would also like to point out the strange but beautiful symmetry existing between (6.17) and WFL. I wonder if that in itself is not one of our best reasons for rejecting (6.12) as another basic axiom of thermodynamics.

As for the First Law, it is now necessary to investigate the major consequences of (6.17). These are, in the order of their appearance, the following:

1) Serrin's Second Law for cycles,
2) a generalized accumulation inequality,

3) a definition of entropy,
4) a proof of the entropy inequality.

Serrin's Second Law for cycles is equivalent to:

$$\text{If} \quad P \in \mathbb{P}_c(\mathscr{S}) \quad \text{is absorptive, then} \quad \bar{W}(P) \leqslant 0 \,. \tag{6.19}$$

Proof. Suppose $P \in \mathbb{P}_c(\mathscr{S})$ is absorptive, and $\bar{Q}(P) > 0$. It is easy to see that the processes P^n mentioned in Lemma 6.1.1 are predecessors of P satisfying

$$P^n \text{ is absorptive}\,, \quad \bar{Q}(P^n) = n\bar{Q}(P)\,, \quad K(P^n) = K(P)\,.$$

Hence

$$\bar{Q}(P) = \frac{1}{n}\,\bar{Q}(P^n) \leqslant \frac{1}{n}\,Q^*(P, K(P))\,.$$

Letting $n \to +\infty$, we conclude that $\bar{Q}(P) \leqslant 0$, contradicting the assumption $\bar{Q}(P) > 0$. We must therefore have $\bar{Q}(P) \leqslant 0$ for all absorptive cycles P, and by the Weak First Law this implies $\bar{W}(P) \leqslant 0$. This completes the proof.

Once (6.19) is established, the existence of absolute temperature scales follows as in [6.8], with the help of a special ideal system $\mathscr{I}$ (for a description of this system, see [6.6, 8]). If one of these scales is chosen, the accumulation integral $\bar{A}$: $\mathbb{P}(\mathscr{S}) \to \mathbb{R}$ can be defined as usual by

$$\bar{A}(P) = \int_0^\infty \frac{Q(P, \pi^{-1}(\pi))}{T^2}\, dT\,, \tag{6.20}$$

where π: $\mathscr{H} \to R^+$ is the chosen scale. Again with the help of the special system $\mathscr{I}$ it is possible to prove a generalization of Serrin's accumulation inequality, valid for all systems compatible with $\mathscr{I}$.

Theorem 6.3.3. *Accumulation Inequality.* If $P \in \mathbb{P}(\mathscr{S})$, then

$$\operatorname{Sup}\{\bar{A}(P')\colon P' \in \operatorname{Pred}(P)\} < +\infty\,.$$

The proof of this result is rather lengthy, involving as it does a fair number of auxiliary lemmas (see [6.6] for details). If $\operatorname{Pred}(P) \neq \phi$ the supremum above is finite and corresponds to an important quantity associated with the process P. It is what I call the *entropy* of $\mathscr{S}$ at (the beginning of) P. We make the following

Definition. The *entropy* of a thermal system $\mathscr{S}$ is the function S: $\mathbb{P}_f(\mathscr{S}) \to \mathbb{R}$ given by

$$S(P) = \operatorname{Sup}\{\bar{A}(P')\colon P' \in \operatorname{Pred}(P)\}\,.$$

Without difficulty we can now prove (again for systems compatible with $\mathscr{I}$):

Theorem 6.3.4. *Entropy Inequality.* If $P \in \mathbb{P}_f(\mathscr{S}) \cap \operatorname{Pred}(P')$, then

$$\bar{A}(P) \leqslant S(P') - S(P)\,.$$

The above inequality is simply a refinement of the (obvious) inequality $\bar{A}(P) \leqslant S(P')$. What is most interesting is its new status in the theory: it is merely a property of entropy, *not* its defining quality, thereby avoiding the non-uniqueness problem previously mentioned.

The function S and E constructed in this and the previous section are not state functions in the usual sense. Moreover, what I have called the energy inequality and the entropy inequality involve differences between the values of E and S at different *processes*, rather than at the initial and final states associated with a single process. We next see how the commonly made assumption of determinism leads (almost automatically) to state functions satisfying the usual forms of the two inequalities.

6.4 Deterministic State Structures

If $\mathscr{S}$ is a thermal system, a state structure for $\mathscr{S}$ is a triple (Σ, i, t), where Σ is a set (the *state space*) and i, t: $\mathbb{P}(\mathscr{S}) \to \Sigma$ are functions. If $P \in \mathbb{P}(\mathscr{S})$, $i(P)$ and $t(P)$ are respectively the initial and final states of $\mathscr{S}$, when undergoing the process P. The triple (Σ, i, t) is not completely arbitrary, and must agree with the follower relation, namely

$$\text{If} \quad PP' \text{ is defined, then} \quad t(P) = i(P') \; . \tag{6.21}$$

In particular, it follows that

$$\text{If} \quad P \quad \text{is a cycle, then} \quad t(P) = i(P') \; . \tag{6.22}$$

Frequently, one assumes also that the state structure further verifies the condition

$$PP' \quad \text{is defined if and only if} \quad t(P) = i(P') \; . \tag{6.23}$$

When (6.23) holds, the structure (Σ, i, t) is called *deterministic*. Condition (6.23) has been assumed by Coleman and Owen, and also by Šilhavý. In Feinberg and Lavine's work, the only statement (vaguely) of this type is:

$$t(P) = i(P) \quad \text{if} \quad P \in \mathbb{P}_c(\mathscr{S}) \; . \tag{6.24}$$

For a system with a deterministic structure it is easy to see that the concomitant energy and entropy functions can be redefined so as to have as their domains a subset of Σ. To this end, it is enough to observe that if (6.23) holds then

$$\begin{aligned} \operatorname{Pred}(P) &= t^{-1}(i(P)) \\ \operatorname{Foll}(P) &= i^{-1}(t(P)) \; . \end{aligned} \tag{6.25}$$

As a result, if P has followers, then

$$E(P) = \operatorname{Sup}\{\bar{W}(P') - \mathscr{J}\bar{Q}(P') \colon P' \in i^{-1}(t(P))\} = \bar{E}(t(P)) \; ,$$

and if P has predecessors then

$$S(P) = \mathrm{Sup}\{\bar{A}(P'):\ P' \in t^{-1}(i(P))\} = \bar{S}(i(P)) \ .$$

It is therefore obvious that the energy inequality and entropy inequality (indicated before) can be rewritten as

$$\begin{aligned} \mathscr{J}\bar{Q}(P) - \bar{W}(P) &\geqslant \bar{E}(t(P)) - \bar{E}(i(P)) \ , \\ \bar{A}(P) &\leqslant \bar{S}(t(P)) - \bar{S}(i(P)) \end{aligned} \tag{6.26}$$

for all $P \in \mathbb{P}_{\mathrm{f}}(\mathscr{S}) \cap \mathbb{P}_{\mathrm{p}}(\mathscr{S})$. If P is reversible, the preceding inequalities naturally become equalities.

Reversible (or ideal) systems are probably the simplest interesting examples of thermal systems with deterministic structures. The concept is by now familiar (see [6.5, Sect. 8]), so we shall simply introduce the notation which will be needed here.

If $\mathscr{I}$ is an ideal system, its state space Σ can be taken to be an open connected subset of $\mathbb{R}^n$. There are continuous differential forms q and w defined in Σ, and each process $P \in \mathbb{P}(\mathscr{I})$ has associated with it a piecewise smooth path $\gamma_{\mathrm{p}}: I \to \Sigma$ defined in a compact interval I of $\mathbb{R}$, such that

$$\bar{W}(P) = \int_{\gamma_{\mathrm{p}}} w \ , \qquad \bar{Q}(P) = \int_{\gamma_{\mathrm{p}}} q \ .$$

To each such path γ there corresponds at least one process P in $\mathbb{P}(\mathscr{I})$. If $\gamma_{\mathrm{p}}: [a, b] \to \Sigma$, then $i(P) = \gamma_{\mathrm{p}}(a)$ and $t(P) = \gamma_{\mathrm{p}}(b)$. The structure (Σ, i, t) is therefore deterministic, and it is not hard to see that all processes in $\mathbb{P}(\mathscr{I})$ are reversible. Finally, there is a function $T: \Sigma \to \mathbb{R}^+$ with the property that

$$\bar{A}(P) = \int_{\gamma_{\mathrm{p}}} \frac{q}{T} \ .$$

Since for the system $\mathscr{I}$ the inequalities in (6.26) are in fact equalities, one obtains immediately

$$\bar{E} \ \text{and} \ \bar{S} \ \text{are potentials for} \ \mathscr{J}q - w \ \text{and} \ q/T \ \text{respectively} \ . \tag{6.27}$$

$q - w$ and q/T evidently have an infinite number of potentials, differing from each other by constants. However, it is easy to identify in these collections the particular potentials $\bar{E}$ and $\bar{S}$, namely

$$\bar{E} \ \text{and} \ \bar{S} \ \text{are the unique potentials for} \ \mathscr{J}q - w \ \text{and} \ q/T \ \text{with infima equal to zero} \ . \tag{6.28}$$

Proof. We prove this statement just for $\bar{E}$. Assume $e: \Sigma \to \mathbb{R}$ is a potential of $\mathscr{J}q - w$, and choose $x \in \Sigma$ and P in $\mathbb{P}(\mathscr{I})$ with $t(P) = x$. Then

$$\bar{E}(x) = E(P) = \mathrm{Sup}\{\bar{W}(P') - \mathscr{J}\bar{Q}(P'):\ \text{all } P' \text{ with } i(P') = x\} \ .$$

Since e is a potential for $q - w$, it is clear that for all $P' \in \mathbb{P}(\mathscr{I})$ we have

$$\bar{W}(P') - \bar{Q}(P') = e(i(P')) - e(t(P')) .$$

In our case, $i(P') = x$ while $t(P')$ is completely arbitrary. Hence

$$\bar{E}(x) = \mathrm{Sup}\{e(x) - e(y)\colon \text{all } y \in \Sigma\} .$$

It follows that $\bar{E}(x) = e(x) - \inf_{\Sigma} e$, and as result $\inf_{\Sigma} \bar{E} = 0$.

We note that the inequalities $E, S \geqslant 0$ should in fact be quite general. In most cases, one should be able to obtain them by considering processes in either Foll(P) or Pred(P) with an arbitrarily short time duration. Moreover, these inequalities can be *forced* to hold in all cases, simply by creating in $\mathbb{P}(\mathscr{S})$, and for each process P, left- and right-identities with respct to the algebraic operation $\oplus$ existing in $\mathbb{P}(\mathscr{S})$.

In the context of this theory the positivity of energy and entropy therefore loses a bit of its mystery.

Acknowledgements. The results here are contained in my doctoral thesis submitted to the University of Minnesota. I wish to thank my thesis adviser Professor James B. Serrin for his help and encouragement, and especially for convincing me that his "follow relation" had to be of fundamental importance in the theory. It is also a pleasure to thank Professor Jerald Ericksen and Professor Clifford Truesdell for their interest and encouragement.

References

6.1 D. Owen: The second law of thermodynamics for semi-systems with few approximate cycles. Arch. Ration. Mech. Anal. **80**, 39 – 55 (1982)
6.2 B. Coleman, D. Owen, J. Serrin: The second law of thermodynamics for systems with approximate cycles. Arch. Ration. Mech. Anal. **77**, 103 – 142 (1981)
6.3 M. Feinberg, R. Lavine: Thermodynamics based on the Hahn-Banach theorem: the Clausius inequality. Arch. Ration. Mech. Anal. **82**, 203 – 293 (1983)
6.4 M. Šilhavý: On measures, convex cones, and foundations of thermodynamics. I. Systems with vector-valued actions; II. Thermodynamic systems. Czech. J. Phys. B **30**, 841 – 861, 961 – 991 (1980)
6.5 J. Serrin: An outline of thermodynamic structure. This volume
6.6 M. Ricou: "Energy, Entropy and the Laws of Thermodynamics", Ph.D. Thesis, University of Minnesota (1983)
6.7 M. Ricou, J. Serrin: A general second law of thermodynamics (to appear)
6.8 J. Serrin: Lectures on Thermodynamics. University of Naples (1979)

Part II

The Thermodynamics of Gibbs and Carathéodory

Chapter 7

What Did Gibbs and Carathéodory Leave Us About Thermodynamics?

C. Truesdell

Apology

One reason I accepted the invitation to speak on this subject is to correct misunderstandings about my opinion of Gibbs. Gibbs needs no estimate from me, but I must defend myself against the contempt I should justly suffer, were I to be so deficient in science and taste as to place his achievements at any level but the highest.

In the terms I have used and sharpened these thirty years and more, I see Gibbs as the one and only creator of *thermostatics*; in contrast, his writings seem to me to bear little on *thermodynamics*. As I myself have never worked in thermostatics but have directed all my effort toward problems of motion – thermodynamics, that is – I have found little reason to refer to Gibbs. I do not think I deserve blame for citing only infrequently the work of a great man whose efforts and successes, magnificent as they are, point in a direction mainly different from that which I have tried to follow.

Table of Contents

7.1 The Words

There is a difference also in usage of words. Gibbs himself always employed "thermodynamics"; as the word "thermostatics" had not yet been coined, its failure to appear in his writings by no means suggests that he did not distinguish statics from dynamics.

In 1849 Kelvin introduced the word "thermodynamic", which, following him, the *Oxford English Dictionary* (1912) defines as follows:

> ... operating or operated by the transformation of heat into motive power.

Noting the reference to *motion* produced by heat, I prefer this early meaning. In 1854 Kelvin (and perhaps also Rankine) used "thermodynamics" in a sense both less and more specific (*ibid.*):

> The theory of the relations between heat and mechanical energy, and of the conversion of either into the other.

In 1867, still a decade before Gibbs's papers appeared, Kelvin & Trait tried to enlarge the meaning of *dynamics*, which they chose to regard as comprising both *statics*, the theory of equilibrium, and *kinetics*, the theory of motions produced by forces. Just a few years earlier, in 1864, Webster had introduced the word "kinetics", then new to our language, in the sense Kelvin & Tait were to use. Can it be that our cousins from the Isles – distant cousins, they were – had let themselves be directed in their native tongue by a transatlantic, and at that the most tendentious of lexicographers? Be that as it may, the *Oxford English Dictionary* [1] in its definition of "Dynamics" (1897) commented as follows:

> ... in earlier use restricted to the action of force in producing or varying motion, and thus opposed to *Statics* (which treats of equilibrium under the action of forces).

Further, remarked the *Dictionary*, despite the efforts of Kelvin & Tait

> the earlier usage, in which Statics and Dynamics are treated as co-ordinate, is still retained by some physicists, and has largely influenced the popular and transferred applications of the word and its derivates.

The term "kinetics" in Webster's sense is now fallen into desuetude; as one of those who introduced the derivate "thermostatics" (1960), I followed "the popular and transferred applications". Perhaps the new term will apear in the still awaited Volume 4 of the *Supplement*. However that may eventuate, I will use in this lecture the old yet still surviving distinction that surely was clear to Gibbs, whatever his choice of terms.

Tomorrow and later you will hear several lectures on recent extensions of Gibbsian thermostatics, some of them deeply influenced by what Gibbs himself published. What I shall say regarding that will be only by way of making the distinction clear. My own research and historical analysis have concerned thermodynamics.

---

[1] The *Supplements* to the *Oxford English Dictionary* issued in 1933, 1972, and 1976 do not add anything to the original definitions of "dynamics" and "kinetics". For "statics" the same holds for the *Supplement* of 1933. For the words "statics" and "thermostatics" in the second *Supplement* we must await its still unpublished Volume 4.

7.2 Statics and Dynamics: the Catenary

Everyone in this audience will know the difference between statics and dynamics. Nevertheless, because our lectures are to be printed and so may fall into the hands of physicists, chemists, heat-transfer men, *etc.*, I feel compelled to adjoin now a statement of the distinction, and to this end I will use the simplest of examples: the catenary curve.

If we attach the two ends of a real and apparently flexible, heavy cord or fine chain from two pins and then let it fall, it will move about for a while and then settle into rest. To determine its shape in equilibrium through a theory that idealizes it as a massy, heavy line, there are two methods. First, we may set up and attempt to solve equations of motion for such a body. Those determine the shapes that the body will assume as time goes on – the *processes* the body may undergo. *Dynamics* is the theory of processes. Among these are the *rest processes*, those in which no part of the chain ever moves. What other processes the chain may undergo, will depend upon the constitutive properties attributed to the chain and its environment: for example, the chain's internal friction and the resistance offered by the air. If both of these are naught, the chain will never come to rest but will keep on oscillating like an assembly of ideal pendulums. Different constitutive properties define different ideal chains; in general, the processes one chain may undergo, another may not. The rest processes, in contrast, are common to all chains. Even so, the problem of the catenary is only partly solved by determining the rest processes, for there are two of these: one bellied upward, the other bellied downward. To obtain the catenary curve, we must select the latter and reject the former. The dynamics of rest does not suffice for that. Some adscititious condition is needed, typically an imposed requirement of stability.

Statics is a theory designed to *avoid commitment* to equations of motion yet determine directly (and alone) the figure of equilibrium and at the same time conditions of stability for it. The statics of the catenary is given by a variational principle: Among all curves of given length that connect two points, a figure of equilibrium gives the center of gravity its lowest possible position. The vanishing of the first variation and the differential equation satisfied by rest processes are identical, and so again there are two solutions, but only at this first step. Since every interior point of the solution bellied upward is higher than every point of the one bellied downward, the latter has a lower center of gravity, and thus it, the catenary curve, is the one and only solution provided by the variational principle: The variational principle by itself delivers the same outcome as does an adscititious condition of stability for rest processes.

The competitors in the variational condition cannot themselves be figures of equilibrium, for there is only one of those, which is the solution itself; neither do they correspond to processes, for in the variational treatment no equations of motion have been specified. The competitors are subject to no mechanical condition at all: They are simply curves of given length joining given endpoints.

To motivate the variational principle, the physical literature often adjoins assertions of a vaguely dynamic tone. For example, "if the chain is forcibly displaced from its figure of equilibrium, its center of gravity will rise, and when the chain is released, its center of gravity will have to fall lower until it becomes as low as possible." Such a statement is not made specific in terms of actual processes; if we try to make it specific, the constitutive relations we provide may well yield a perpetual oscillation, which does not conform with the dynamical claim. Bare assertion of a trend or attempt is not a part of dynamics; it is at best a heuristic, at worst an orison.

Sometimes in mechanics the term "virtual" is utilized here. The vanishing of the first variation is interpreted as a statement that the virtual work done by a virtual displacement equals the virtual increase of the gravitational potential. Since kinetic energy is neglected, a displacement of this kind may be called "quasistatic". This sort of mumbo-jumbo brings comfort to some, but it solves no problems. The variational principle is sufficient to its purpose, which is to solve a problem of statics; for dynamic ends, it is useless.

These distinctions and remarks carry over to thermodynamics and to the thermostatics of Gibbs.

7.3 The Thermostatics of Gibbs

The basic papers of Gibbs on the phenomenology of heat fill 371 pages in his *Collected Works*. While written in elegant, easy English, they are torve, tense, and terse in largely verbal reasoning. I do not understand all of those pages, and I will make no attempt to detail Gibbs's contribution (1873, 1876/8).

We must not confuse what Gibbs *knew* about thermodynamics with what he *did* in thermostatics.

For the latter, I hold in the main to the judgment I wrote thirteen years ago, much as follows now.

> This shy and exotic newcomer found himself before the altar of a goddess far from virgin, her morals eroded by callous abuse and her territory shrunk by the timorosity of her champions. There were two easy possibilities: To set out on a campaign to rejuvenate the goddess and reconquer her lost dominions, or to make her an honest old lady in a safe and clean cottage. Gibbs chose neither. Perceiving that results like those obtained in the thermodynamics of *processes* in bodies of *especially simple material* ought hold also for *all bodies* in *equilibrium*, he created a pure statics of the effects of temperature and heat. Accepting the restrictions Clausius had imposed one by one in his years of retreat from irreversible processes, Gibbs recognized the subject in its starved and shrunken form as being no longer the theory of motion and heat interacting, no long thermo*dynamics*, but only the beginnings of thermo*statics*. It is Gibbs's singular merit to have seen the essence of this thermostatics: a variational *definition* of equilibrium, including its stabilities and instabilities, in which infinitely many putative equilibria are compared, and variational criteria are imposed to select from this class of fields that which, for a given caloric equation of state, corresponds with actual equilibrium, namely that field of fixed total energy which renders the total entropy a maximum. A competitor is not required to obey an "equation of state" and need not be the outcome of any process undergone by the body in

question. A pure statics results, which Gibbs develops fully and deeply in his second and third papers.

The foregoing description fits the opening remarks of Gibbs himself in the abstract[2] of his long paper:

> It is an inference naturally suggested by the general increase of entropy which accompanies the changes occurring in any isolated material system that when the entropy of the system has reached a maximum, the system will be in a state of equilibrium. Although this principle has by no means escaped the attention of physicists, ... little has been done to develop [it] as a foundation for the general theory of thermodynamic equilibrium.
>
> The principle may be formulated as follows, constituting a criterion of equilibrium –
>
> I) *For the equilibrium of any isolated system it is necessary and sufficient that in all possible variations of the state of the system which do not alter its energy, the variation of its entropy shall either vanish or be negative.*
>
> The following form, which is easily shown to be equivalent to the preceding, is often more convenient in application:
>
> II) *For the equilibrium of any isolated system it is necessary and sufficient that in all possible variations of the state of the system which do not alter its entropy, the variation of its energy shall either vanish or be positive.*

My description fits also the modern, rigorous reconstructions of parts of Gibbs's thermostatics, especially those of Coleman & Noll[3] and of Dunn & Fosdick[4]: it fits also the papers of Messrs. Fosdick, James, and Man that that we are to hear; while it fits most of Gibbs's own writings, there are several passages where he seems to do something different, especially in his great tract on "heterogeneous substances", which is more a collection of essays on related topics than a consecutive treatise. Gibbs always specifies his classes of "possible" variations in words, words expressing an idea which was no doubt clear in his mind but which changing usage makes difficult for a modern reader to apprehend precisely. Moreover, these classes vary with the classes of systems he takes up, one after another[5]. On the whole, the outcomes, though not the variations used to get them,

---

[2] Page 354 (*cf.* also page 56) of Volume 1 of Gibbs's *Scientific Papers* (1906) and *Collected Works* (1928). The pagination of this volume is the same in these two collections.

[3] B. D. Coleman & W. Noll, still unpublished work of 1957 mentioned in § 264 of *The Classical Field Theories*, 1960; "On the thermostatics of continuous media", *Archive for Rational Mechanics and Analysis* **4** (1959): 97 – 128; B. D. Coleman, "On the stability of equilibrium states of general fluids", *ibid.* **36** (1970) 1 – 32.

[4] E. Dunn & R. Fosdick, "The morphology and stability of material phases", *Archive for Rational Mechanics and Analysis* **74** (1980): 1 – 99.

[5] On pages 57 – 61 Gibbs introduces δ for a *"possible"* infinitesimal variation and Δ for a variation "in which infinitesimals of the higher orders are not to be neglected" (presumably analytic functions of some unspecified parameter that may tend to 0). He then states what is excluded to make a variation "possible". Some readers have misinterpreted his reference to cases in which "heat can pass by conduction or radiation from every part of the system to every other ... ." In equilibrium such factors as these have finished their work. The "possible" variations in a theory of equilibrium allow for their *effects* but do not specify their nature or their manifestations during processes by which equilibrium might be in the end attained.

refer to piecewise homogeneous conditions in a finite number of constituent bodies. An essentially trivial exception is the inhomogeneity due to a uniform gravitational field[6]. An exception of a different kind is the long, famous Section 6, on the equilibrium of solids[7], in which the field of displacement is varied and the variations are "entirely independent of any supposition in regard to the homogeneity of the solid." This passage is familiar, at least derivatively, to every student of elasticity because it shows, in effect, that for a thermo-elastic body the internal energetic and the free energetic, respectively, serve as stored-energy functions when the allowed variations are adiabatic for the former, isothermal for the latter. The student of mechanics finds the ideas and the analysis in this part of Gibbs's paper easier to follow than the rest, some of which call upon more knowledge of chemistry and physics than he is likely to have. It is an untypical part in that it presents equations in which possibly non-uniform fields are the outcome, and it does not pursue the variational principle far enough to reach conditions of stability for solids.

My subject is thermodynamics, the effects of heat and hotness upon the actual motions of bodies. My foregoing remarks about Gibbs's thermostatics I do not intend as an accurate and comprehensive analysis; even less are they a summary of his work, for they leave aside his astonishing successes with difficult physical situations, in treating which he displays a capacity for physical thought, more "intuitive" than systematic, which might justly be set level with Huygens' and Newton's and Faraday's.

7.4 Gibbs on Thermodynamics

My foregoing remarks are designed only to defend myself against any charge that I fail to value sufficiently the work of Gibbs. I turn now to my subject, thermodynamics, and consider what Gibbs knew about it and what, if anything, he did for it.

The former is easy. He knew in detail almost everything that had been published regarding thermodynamics. An example of what he knew is provided by his being the first to write that it was Rankine, not Clausius, who discovered

Later (page 63 ff.) Gibbs uses d to denote a differential of a function, and he refers (page 64) to these as variations and states that they need not respect the homogeneity of the mass considered. Then he writes "$D\varepsilon$, $D\eta$, etc., for the energy, entropy, etc., of any infinitesimal part". *Cf.* also pages 184 – 218 on elastic solids, where all variations are reduced to statements about differentials and derivatives of functions. On page 222 he introduces and explains a "reversible" variation, warning the reader not to confuse it with a reversible process in the usual sense. On page 247 he writes, "These varied states of the system are not in general states of equilibrium, and the relations expressed by the fundamental equations may not hold true of them." Here, certainly the variations are of the kind allowed by Coleman & Noll. *Cf.* also page 248. On page 326 Gibbs uses all variations compatible with a specified constraint.

Gibbs's variations are analogues to *virtual* displacements in mechanics: changes which *need not conform with such physical laws* as govern *actual* changes.

[6] Pages 188 – 191, 276 – 287, 319 – 323, 329 – 330, 338 – 339.

[7] Pages 184 – 218.

entropy. It is no easy task to discern that fact in the notoriously inscrutable writings of Rankine: Gibbs mentions it in three footnotes[8]. His attempt to set the record straight was not successful; it is an instance of Truesdell's Law of Attributions: For arbitrary *n*, any false attribution outlasts *n* documented corrections.

To justify having said "almost everything" rather than "everything", I remark that Gibbs[9] attributes characteristic functions to Massieu (1869), while in fact the very quantities and thermodynamic potentials Gibbs himself introduced as his own in 1875 had been defined, analysed, and applied by Reech in a paper[10] published in a major journal of mathematics in 1853, when Gibbs was fourteen. Reech, not Gibbs, deserves the credit for first discovery of the idea Gibbs later called "a fundamental equation" [11].

In his course on thermodynamics[12] Gibbs used the first nine lectures to present the traditional theory, basing it on the papers of Kelvin and Clausius. The record of these lectures does not mention "quasistatic processes" or express any doubts regarding the existence of temperature or entropy. Lecture I concerned "thermometry and calorimetry"; Lecture II, the Carnot cycle. E. B. Wilson, a student and worshipper of Gibbs, told me on 3 September 1953 in a conversation that I straightway wrote down,

> Gibbs began his lectures on thermodynamics with the Carnot cycle, which he always got wrong. After getting thoroughly mixed up he concluded the first lecture with an apology, and in the second lecture he gave it letter perfect. It was in this way he introduced entropy, rather than in the formal way in the "Heterogeneous Substances".

While Gibbs's analysis in his three papers almost always concerns thermo*statics*, to motivate his assumptions and interpret his conclusions he tries to convince the reader that those are inferred from and consistent with thermo*dynamics* as that science then stood. At the head of his long paper he put the famous dicta of Clausius in 1865, then only a decade old:

> The energy of the world is constant.
> The entropy of the world tends toward a maximum.

---

[8] Pages 2, 3, and 52.

[9] Pages 86 – 87.

[10] For a summary of Reech's idea of this matter see § 10C, "Reech introduces and analyses the thermodynamic potentials", in my *Tragicomical History of Thermodynamics, 1822 – 1854*, New York *etc.*, Springer-Verlag, 1980.

[11] Pages 2 and 20 – 28, footnote on page 34, pages 86 – 89.

[12] Here I rely on E. B. Wilson, "Papers I and II as illustrated by Gibbs' lectures on "Thermodynamics", Article C in *A Commentary on the Scientific Works of J. Willard Gibbs*, Volume 1, *Thermodynamics*, edited by E. G. Donnan & Arthur Haas, New Haven, Yale University Press, 1936. There Wilson prints the notes, which are scarcely more than an outline, taken by L. I. Hewes on the lectures of 1899/1900.

In the conversation quoted in the text above, Wilson told me that

> Gibbs always lectured above the heads of his students and always refused to teach undergraduates at all. He knew his students did not follow him but did not alter his style on that account, having a definite idea of how the subject should be presented. He once told me that in all his years of teaching he had had only six students sufficiently prepared in mathematics to follow him … .

Immediately thereafter Gibbs refers to processes[13]:

> As the difference of the values of the energy for any two states represents the combined amount of work and heat received or yielded by the system when it is brought from one state to the other, and the difference of entropy is the limit of all the possible values of the integral $\int dQ/t$, (dQ denoting the element of the heat received from external sources, and t the temperature of the part of the system receiving it), the varying values of the energy and entropy characterize in all that is essential the effects producible by the system in passing from one state to another. For by mechanical and thermodynamic contrivances, supposed theoretically perfect, any supply of work and heat may be transformed into any other which does not differ from it either in the amount of work and heat taken together or in the value of the integral $\int dQ/t$.

The reference to "the part of the system" suggests that Gibbs has in mind the variations of temperature in the interior of a body, while in "the heat received from external sources" he may include the heat that comes into one part of the body from other parts. Also he refers to t as "the temperature of the part receiving it", not the temperatures of the sources of heat. Thus his integral $\int dQ/t$ might be interpreted as the sum of the two terms which are now familiar in the "Clausius-Duhem" inequality. On the other hand, in his reference to "the limit of all the possible values" Gibbs seems to suggest that the entropy might be constructed from the class of values that Clausius' integral takes on when it is applied to different processes connecting two states. Later[14] he asserts that the entropy

> is the value of the integral $\int dQ/t$ for any *reversible* process by which [the change in entropy from the value 0 to the actual value] is effected (dQ denoting an element of the heat communicated to the matter thus treated, and t the temperature of the matter receiving it). ... [I]t is understood that at the close of the process, all bodies which have been used, other than those to which [the integral] relates[s], have been restored to their original state ... .

While this remark about reversible processes might suggest that a body's entropy is to be calculated always from its caloric equation of state in thermostatics, Gibbs certainly had no such idea when he wrote his second paper, in which we read[15]

> When the body is not in a state of thermodynamic equilibrium, its state is not one of those which are represented by our surface. The body, however, as a whole has a certain volume, entropy, and energy, which are equal to the sums of the volumes, etc., of its parts.

A footnote tells us that "the word *energy*" here is used "as *including the vis viva of sensible motions.*" Thus Gibbs knew the general principle of energy, and *he*

---

13 Page 55.

14 Page 85.

15 Page 39. Note also page 59, on which Gibbs makes a statement about the entropy of a body which need not be in equilibrium. Likewise in his statistical mechanics Gibbs sets up the expectation of the index of probability as the analogue of the entropy for *any* density in phase, not merely a density appropriate to statistical equilibrium. *Cf.* his *Collected Works*, Volume 2, Part 1, page 20, and Chapter XI.

attributed an entropy to a body in any state of motion. Insofar as he referred to thermodynamics, he did not limit it to slow changes[16] governed by "equations of state" of the classical kind.

Indeed, there is one (and so far as I know, only one) passage[17] in which Gibbs treats processes, and irreversible ones at that. After discussing "a perfect electro-chemical apparatus", he writes

> ... if we give up the condition of the reversibility of the process ... [yet] ... still suppose, for simplicity, that all parts of the cell have the same temperature, which is *necessarily* the case with a perfect electro-chemical apparatus, we shall have, instead of (692),
>
> $$d\eta > \frac{dQ}{t}, \ldots . \tag{695}$$

We recognize this statement, which refers to finite increments, as the isothermal instance of what is now called "the Clausius-Planck inequality". Gibbs's "if we still suppose, for simplicity" suggests that he could easily have stated and applied that inequality in general, without the restriction. Nowhere in his published works did he do so.

Here is the place to finish my long self-defense in regard to Gibbs. I have no doubt that he could, had he so wished, have developed the thermodynamics of irreversible processes. Had he done so, much of what we now call rational thermodynamics might well have been standing in print for a century. Nonetheless it is a fact that what Gibbs chose to produce was in the main thermostatics, not the thermodynamics of irreversible processes.

7.5 The Thermodynamics of Planck

Taking up the thermodynamics of processes at the point where Clausius had surrendered, Planck introduced systems described by n variables. Many of his for-

---

[16] The word "process" occurs in Gibbs's writings sometimes as a term of ordinary speech, a procedure of some kind, or a merely geometrical passage from one point to another on a curve or surface. Of his 371 pages I have found twenty-seven on which he uses "process" to mean a sequence of changes undergone by a body in the course of time or refers to such changes without using the word. Those pages are 1, 27, 33, 37, 39–42, 56, 57, 59, 60, 61, 85, 89, 90, 92, 120, 144, 145, 159, 167, 196, 222, 285, 338, and 339. The discourse there serves mainly to motivate statements about equilibrium and its stability, to visualize changes consistent with thermostatic equations, and to limit the class of competing variations.

On page 39–42 note the imaginary "envelop" to maintain uniformity in the pressure and the temperature of the body, provided that certain strict inequalities hold "if any part of the body has sensible motion."

Mostly, when Gibbs uses the word "reversible", he attaches it to progressive changes for which all equations of thermodynamics are constantly applicable.

[17] Pages 338–339.

mulae look just like some that appear in Gibbs's thermostatics, and in his autobiography[18] he writes

> Unfortunately, as I found out later, the great American theorist Josiah Willard Gibbs was ahead of me in this matter. He had formulated the same theorems [statements?], even partly in a more general form ... .

The vagueness of the German words "Satz" and "formulieren" makes the meaning of Planck's admission unclear. If he had read Gibbs's work carefully enough to master it, he could not have concluded that it anticipated his own, merely that many of the equations and inequalities were formally alike.

Planck in the same book tells us that in his thesis of 1879 he had taken as his basic principle the following statement:

> In every natural process the sum of the entropies of all bodies involved in the process increases.

Restriction of a natural process to a subinterval of the interval of time upon which it is defined delivers another natural process. Thus Planck does, at least implicitly, consider not merely the outcomes of processes but changes that occur as time proceeds; the *actual* changes undergone by *special* systems. Many material bodies that differ in their dynamics share a common statics. In this sense the thermostatics of Gibbs is of broader application than the thermodynamics of Planck. The distinction is confirmed by Planck's own statement regarding the reception of his thesis:

> The impression [that my dissertation presented to the University of Munich in 1879] made upon the public of physics at that time was naught. Of my teachers at the University, as I know precisely from conversations with them, not one understood its contents. They probably let it pass only because they knew me from my other work, in the "Praktikum" for physics and in the mathematical seminar. Even among physicists more nearly concerned with the subject itself I found no interest, let alone approval. Helmholtz probably did not read this paper at all; Kirchhoff rejected its contents expressly with the remark that the concept of entropy, the magnitude of which was measurable only by means of a reversible process and hence was definable only for such, could not legitimately be applied to irreversible processes.

Even 100 years ago, it seems, thermodynamics was already regarded in Germany as a dead field, insusceptible of broadening or deepening; already it was chained to equilibrium and hence inapplicable to "natural processes", which were what Planck designed to represent. Planck went on to conjecture that he obtained a chair not because of his researches on thermodynamics but in spite of them and through the good offices of a friend of his father's.

---

[18] M. Planck, *Wissenschaftliche Selbstbiographie*, Leipzig, J. A. Barth, 1948. I have not used the translation by F. Gaynor, pages 13 – 51 of *Scientific Autobiography and other Papers*, New York, Philosophical Library, 1949.

Planck obtained and applied what we now call the *Clausius-Planck inequality* [19]:

$$\Delta H \geqq \int \frac{dC}{T}, \qquad \dot{H} \geqq \frac{Q}{T}, \qquad (dC = Q\,dt), \tag{7.1}$$

which provides a lower bound for the increase of entropy accompanying the actual accumulation of heat at various temperatures. Gibbs's equation (695) is the special instance of (7.1) in which $T=$ const., and while doubtless he had (7.1) in his hands, this excursus into the theory of processes is unique in his published writings. On the whole, it seems to me, Gibbs's considerations involve bodies putatively subject to non-uniform fields of temperature, energy, entropy, and specific volume. Gibbs demonstrates from his posited variational principle that throughout a body in equilibrium the temperature and other variables are uniform. I find no such argument in the work of Planck, who always associates with any condition of any one body just one temperature, one entropy, *etc.* Planck may be regarded as the originator of the thermodynamics of *homogeneous processes* undergone by bodies susceptible of irreversible change.

Gibbs's thermostatics and Planck's thermodynamics have one common feature: Their conclusions refer – Gibbs's mainly and through proved theorems, Planck's entirely and by assumption – only to homogeneous conditions. *A fortiori,* neither can be compared except in trivial instances with, say, Maxwell's kinetic theory of gas flows.

7.6 Gibbs's Rational Foundations of Thermodynamics: Gibbsian Statistical Mechanics

Gibbs's thermostatics can easily be regarded, at least in its simpler parts, as axiomatic. While he there made his choice of entropy and absolute temperature as primitive concepts because that lent itself to the most compact, mathematically efficient formulation of special problems as well as of the structure of his theory, leading to the delightful "Gibbs relations"

---

[19] M. Planck, "Über das Prinzip der Vermehrung der Entropie", *Annalen der Physik* (2) **30** (1887): 562–582; **31** (1887) 189–208; **32** (1887): 462–503. Planck does not here write (7.1) explicitly. On page 468 of the third paper we find it expressed as follows for "any process that takes place in nature":

$$\Sigma\delta S + \delta\sigma > 0, \qquad \delta\sigma = -\frac{\delta Q}{\theta}.$$

It follows also, expressed as

$$Q \leqq T\,d\Phi,$$

from Equations (68) and (70) in the 7th edition of Planck's famous textbook, *Vorlesungen über Thermodynamik*, Leipzig, Veit & Co., 1891; 2nd ed., 1905; 3rd ed., 1911; 4th ed., 1913; 5th ed., Berlin, De Gruyter, 1917; 6th ed., 1921; 7th ed., 1922. Translation of the 1st ed. by A. Ogg, *Treatise on Thermodynamics, London etc.*, Longmans Green, 1903; 1917; 1921; of the 7th ed., 1927. Of the ingredient relation $dU - T\,d\Phi < W$, Planck writes "All conclusions with regard to thermodynamic chemical changes, hitherto drawn by different authors in different ways, culminate in this equation." Φ is Planck's later symbol for the entropy, replacing the S he used in his paper of 1887.

$$p = -\frac{\partial \varepsilon}{\partial v}, \qquad \theta = \frac{\partial \varepsilon}{\partial \eta}, \tag{7.2}$$

along with many other compact statements, of course Gibbs knew that entropy was not something obvious, not something that comes spontaneously to the burnt child who is learning to avoid the fire. He saw the need for a "Rational Foundation of Thermodynamics". He chose to seek it in *statistical* treatment of the motions of systems of punctual masses. In 1901 he issued his *Elementary Principles in Statistical Mechanics developed with Especial Reference to the Rational Foundation of Thermodynamics* [20]. There he referred to "the slow progress of rational thermodynamics, as contrasted with the rapid deduction of the consequences of its laws as empirically established." He chose to "pursue statistical inquiries as a branch of rational mechanics" so as to gain "clear apprehension of the relation of thermodynamics to rational mechanics".

Gibbs's statistical mechanics, like his phenomenological thermostatics, within the limits he set himself stands a peerless masterpiece. As would be expected, the rational treatment it presents provides a foundation not for thermodynamics, which Gibbs names, but for thermostatics, which he had treated before with entire success in applications. Again like his phenomenal theory, his statistical mechanics has been misrepresented by the tradition. Its high standard of deductive rigor is passed over in silence; its modest and specific theorems are twisted into pronouncements about trends in time which Gibbs suggested but did not claim to prove; and the restrictions that Gibbs imposed as sufficient to obtain thermostatic relations are fraudulently adduced as being necessary to the validity of a thermodynamics of any kind. That confusion may be the root of another and more deadly misconception: that if we are to "explain" irreversibility, we must resort to some kind of statistics. In fact, as P. G. Bergmann has written [21]

> Is is not very difficult to show that the combination of the reversible laws of mechanics with Gibbsian statistics does not lead to irreversibility, but that the notion of irreversibility must be added as an extra ingredient.

Maxwell's kinetic theory of gases provides a brilliant and successful example of an "added extra ingredient", namely, the collisions operator.

The statistical mechanics of Boltzmann, which preceded the work of Gibbs, certainly had an effect and influence; certainly Boltzmann himself strove and struggled toward a rational treatment, but he achieved it only here and there. This place is not fit for the long, detailed and measured analysis that alone would carry conviction regarding the esteem that Boltzmann's assumptions and discoveries deserve. The community of physicists reinforces by rites of intuition and hocus-pocus with formulae its persuasion that statistical treatment provides the

---

[20] Republished as Part 1 of Volume 2 of *The Collected Works of J. Willard Gibbs*, New Haven, Yale University Press, 1928; not included in Gibbs's *Scientific Papers,* 1906.

[21] P. G. Bergmann, "Foundations research in physics", pages 1 – 14 of *Delaware Seminar in the Foundations of Physics*, edited by M. Bunge, New York, Springer-Verlag, 1967. See page 11.

doctrine of the one true church of thermodynamics. Here I quote the only physicist dissenter from that faith I have so far encountered, Bridgman[22]:

> The insight which the probability interpretation of the second law at first seemed to give turns to ashes like apples of Sodom. The small-scale stuff is only a model, obtained by extrapolation of the large-scale stuff. The tactics of this extrapolation certainly cannot be claimed to display any subtlety – it was what any child might have invented. The only check on the extrapolation is that when worked backward it shall again produce the large-scale stuff. ... An understanding of the attitude of physicists toward thermodynamics and kinetic theory is, I think, to be sought only in the realm of psychology.

While I applaud the last sentence, and while I sympathize with part of the statements preceding it, the hundreds of pages of research I have published on kinetic theory and statistical mechanics make it plain that I do not share Bridgman's view *in toto*. Investigations in statistical mechanics carried out since Bridgman's death contain assertions of great interest and claim to prove them. The mathematical style of most of this work is such as to close it to mathematicians not in its small circle of authors. Such a quality is dangerous. Because clear, explicit statement and rigorous proof belong to the essence of rational thermodynamics, it is too soon to assess what progress in rational development recent researches in statistical mechanics have achieved.

7.7 Bryan's Rational Thermodynamics

Statistical thermodynamics is in principle neither more nor less rational than phenomenological thermodynamics. "Rational" is a quality of treatment, not of the object treated. In a book on the "first principles and their direct applications" Bryan[23] in 1907 pointed to rational mechanics not as something to be applied to thermodynamics but rather to be imitated by it. In his preface he writes

> It cannot be denied that the perfection which the study of ordinary dynamics has attained is largely due to the number of books that have been written on *rational* dynamics in which the consequences of the laws of motion have been studied from a purely deductive stand-point. This method in no way obviates the necessity of having books on experimental mechanics, but it has enabled people to discriminate clearly between results of experiment and the consequences of mathematical reasoning. It is maintained by many people (rightly or wrongly) that in studying any branch of mathematical physics, theoretical and experimental methods should be studied simultaneously. It is

[22] P. N. Bridgman, *The Nature of Thermodynamics,* Cambridge, Harvard University Press, 1941. Reprinted "with no essential change", Harper Torchbooks, 1961. I have used only the latter edition, on pages 8 – 9 of which may be found the passage quoted. Some substantiation of detail may be read on pages 152 – 179.

[23] G. H. Bryan, *Thermodynamics, an Introductory Treatise dealing mainly with First Principles and their Applications*, Leipzig, Teubner, 1907.

however very important that the two different modes of treatment should be kept carefully apart and if possible *studied from different books*, and this is particularly important in a subject like thermodynamics.

Part II of Bryan's book, which contains many references to rational mechanics, is called "The Foundations of Rational Thermodynamics".

Praiseworthy as is Bryan's attempt, like Reech's a half century earlier it is a failure. His "axioms" include undefined terms which are not represented by any mathematical concept, and his presentation, while different from those in other works of his day, is no less muddled. Bryan[24] seems also to have been the first to take internal energy rather than heat as the main concept on which to found thermodynamics. We shall soon see a disastrous use of this idea.

Gibbs's success, Boltzmann's uncertainty, and Bryan's failure did not grow from any inherent superiority in statistics or inherent weakness in phenomenology; they merely reflect the capacities of the men.

7.8 Carathéodory's Axioms

Since the purpose of this lecture is to fill some lacunae in the expositions of thermodynamics that I have published, these thirty years past and more, adopting one or another standpoint but each time attempting to take that standpoint seriously and treat it honestly, I come now to the celebrated axiomatization of Carathéodory[25]. Like most of the thermodynamics of the pioneers, Carathéo-

[24] In § 47 of Bryan's book we read

> The following statements may be regarded in the light partly of a definition of energy, and partly of an enunciation of its properties which are assumed as fundamental.
>
> *There is a certain entity called energy which is characterised by the following properties:*
>
> 1) *In an isolated system the total quantity of this entity always remains constant.*
> 2) *The energy of a system cannot be changed without some real physical changes taking place in the state of the system.*
> 3) *The kinetic and potential energies of dynamics are particular forms of this entity.*
>
> The first statement is the *Principle of Conservation of Energy*, and it leads to the following conclusions.
>
> If the energy of a finite non-isolated system or part of a system changes in amount, then changes of equal but opposite amount must occur somewhere outside the system or part considered, so as to make the total amount unaltered.
>
> According to (2) if the physical state of a system is completely defined by certain variables, the energy is a function of those variables only, and does not depend on the past history of the system previous to attaining the state in question.
>
> On the other hand, if the state of the system is defined so far as certain physical phenomena are concerned by certain variables, and we have evidence, from the existence of irreversible phenomena, or from any other cause, that energy changes have occurred in the system which are independent of the changes of these variables, we infer that the variables originally assumed are not sufficient to completely determine the physical state of the system, but that this state depends on some other variables as well.

[25] C. Carathéodory, "Untersuchungen über die Grundlagen der Thermodynamik", *Mathematische Annalen* **67** (1909): 355 – 386 = pages 131 – 166 of Band II of Carathéodory's *Gesammelte Mathematische Schriften*, Munich, C. H. Beck, 1954. Translation by J. Kestin, "Investigation into the foundations of thermodynamics", pages 229 – 256 of *The Second Law of Thermodynamics*,

dory's deals only with systems described by a finite number of scalars and unable to undergo irreversible changes. Not only for that reason but also because it is mathematically incorrect and physically insufficient to deliver even the classical results of Kelvin in the generality Kelvin himself maintained, Carathéodory's treatment has had scant effect on rational thermodynamics, although of course all students of the foundations have consulted and analysed it. Its birth is described as follows by Born[26]:

> From [my reading of Gibbs] sprang an essential piece of progress in thermodynamics – not by myself, but by my friend Carathéodory. I tried hard to understand the classical foundations of the two theorems, as given by Clausius and Kelvin; they seemed to me wonderful, like a miracle produced by a magician's wand, but I could not find the logical and mathematical root of these marvellous results. A month later I visited Carathéodory ... and told him about my worries. I expressed the conviction that a theorem expressible in mathematical terms, namely the existence of a function of state like entropy, with definite properties, must have a proof using mathematical arguments which for their part are based on physical assumptions or experiences but clearly distinguished from these. Carathéodory saw my point at once and began to study the question. The result was his brilliant paper, published in *Mathematische Annalen*, which I consider the best and clearest presentation of thermodynamics.

Presuming that everyone in this audience is familiar with Carathéodory's assumptions and results, I content myself with a list of some of their failings.

1) Carathéodory wrote

> Finally, in order to handle from the beginning systems with arbitrarily many degrees of freedom, it was necessary to use a theorem from the theory of Pfaffian differential equations ... instead of the Carnot cycle, which was always used earlier but is easy to visualize and master only for systems with two degrees of freedom.

The phrase "Finally ... it was necessary to" (Endlich musste) has been misunderstood. While it seems to mean just the personal explanation "I had to", expressed in the impersonal terms with which the modesty of twentieth-century scientists

---

edited by J. Kestin, Stroudsburg (Pa.), Dowden, Hutchinson & Ross, 1976. *Cf.* also "Über die Bestimmung der Energie und der absoluten Temperatur mit Hilfe von reversiblen Prozessen", *Sitzungsberichte der Preussischen Akademie der Wissenschaften, Physikalisch-Mathematische Klasse* (Berlin) 1925: 39 – 47 = pages 167 – 180 of Band II of Carathéodory's *Gesammelte Mathematische Schriften*. There are also expository papers by Born and others.

26 M. Born, *My Life, Recollections of a Nobel Laureate,* London/New York, Taylor and Francis/Charles Scribner's Sons, 1978. See page 119. This work first appeared in a translation into German, 1975. [In view of my words about Born's baneful influence on thermodynamics perhaps it is not out of place to remark here that his autobiography as well as the reputation he left behind him reflect a fine character as well as charm, learning, letters, and eminent humanity. His encounter with the foundations of thermodynamics seems to be the one instance in his life when his physical sense failed him.]

often compels them to cloak their opinions and emotions, it is the source of two widely diffused assertions[27]:

I) Carnot cycles do not lend themselves to rigorous mathematical treatment.
II) The approach to thermodynamics based on Carnot cycles does not work except for systems described by two independent variables.

Half a century of confusion about not only the pioneers' discoveries but also the basic concepts and logical structure of classical thermodynamics, diffused through a forest of pulp and an ocean of talk, grew from these groundless opinions and other aspects of Carathéodory's work. Bharatha & I[28] proved the former assertion false; Pitteri[29], the latter.

2) Mathematical gaps and errors in Carathéodory's work have been discovered by Whaples[30], Bernstein[31], Boyling[32], Cooper[33], Serrin and others.

---

[27] *E.g.* "can be conveniently and convincingly dealt with in the case of a system of two degrees of freedom only" in the Introduction to Johann Walter's "On the definition of the absolute temperature – a reconciliation of the classical method with that of Carathéodory", *Proceedings of the Royal Society of Edinburgh* **82A** (1978): 87–94.

[28] C. Truesdell & S. Bharatha, *The Concepts and Logic of Classical Thermodynamics as a Theory of Heat Engines, Rigorously Developed upon the Foundation Laid by S. Carnot and F. Reech,* New York, Springer-Verlag, 1977.

[29] M. Pitteri, "Classical thermodynamics of homogeneous systems based upon Carnot's General Axiom", *Archive for Rational Mechanics and Analysis* **80** (1982): 333–385.

[30] G. Whaples, "Carathéodory's temperature equations", *Journal of Rational Mechanics and Analysis* **1** (1952): 302–307.

[31] B. Bernstein, "Proof of Carathéodory's local theorem and its global applications to thermostatics", *Journal of Mathematical Physics* **1** (1960): 222–224.

[32] J. B. Boyling, "An axiomatic approach to classical thermodynamics", *Proceedings of the Royal Society* (London) **A 329** (1972): 35–70. See § 1 for references to further mathematical criticisms.

[33] J. L. B. Cooper, "The foundations of thermodynamics", *Journal of Mathematical Analysis and its Applications* **17** (1967): 172–193.

As Cooper pointed out, Carathéodory failed to prove that his "absolute temperature" deserved the name "temperature", which must be an increasing function of any presumed empirical temperature. Cooper's observation induced both Mr. Serrin and me to include proofs that the absolute scales we constructed were not open to any such objection. On the other hand the counterexample Cooper invoked was in effect the caloric theory of heat, in which the heat form is a perfect differential, and so all constants are integrating factors. Professor Walter has remarked (in effect) that Cooper's counterexample cannot apply to Carathéodory's development because he built the First Law into his axioms through his definition of heat, and the First Law excludes the caloric theory (at least if there is some model material which can accomplish cyclic work); *cf.* Johann Walter, "Bemerkungen zur Verwendung der Pfaffschen Formen bei der Definition der absoluten Temperatur nach Carathéodory", pages 504–512 of *Proceedings of the Fourth Conference on Ordinary and Partial Differential Equations, Dundee, Scotland,* March 30–April 2, 1976 (Springer Lecture Notes in Mathematics No. 564). Nonetheless, careful analysis of conditions on Carathéodory's differential forms that suffice to render his argument rigorous supports the gist of Cooper's complaint.

The preceding remarks are open to misinterpretation because they reflect the prejudice that directs him who seeks an absolute temperature to find integrating factors for some differential form. That was not Kelvin's way when he introduced absolute temperature in the first place. He sought directly a scale of temperature with physically desirable properties. Because he then was working with the caloric theory, integrating factors would have been useless to him. He did find an excellent scale. Later, after he had rejected the caloric theory, he introduced the scale now called after him; he did so, not by seeking integrating factors but again by first laying down a physical desiderandum, then determining a function of his old scale such as to satisfy that requirement.

Some analysts uninterested in thermodynamics are offended if they hear any criticism of Carathéodory. They retort, "But in that paper he proved a beautiful theorem on Pfaffian forms!" On that point some other analysts disagree. Remarking that mathematicians had found incomplete Carathéodory's proof of his theorem (repeated in his later publications and in all expositions by physicists), Bernstein provided a precise proof for a precise statement of Carathéodory's type. His theorem delivers Carathéodory's integrating factor only locally, but, as he remarks, if that integrating factor is to be interpreted as an absolute temperature and the essential block from which the entropy is built, local existence is insufficient. To deserve the name "absolute", a scale must be global. While the theorem by itself cannot do the job, Bernstein brings to bear Carathéodory's further assumptions to prove that in the application to classical thermodynamics the scale is indeed global. Even as patched up by Bernstein, Carathéodory's work employs the unnatural assumption that all the coefficients of the Pfaffian form be infinitely many times differentiable. In our rigorous formulation of Carnot's approach Bharatha & I assume only that certain first derivatives, and not all of them, should exist and be continuous. Thus even from the analytical standpoint Carathéodory's results are weaker than need be. Not too much should be made of that, nevertheless, because stronger theorems of existence for differential equations could be invoked now to get Bernstein's results from assumptions more natural to the physics of heat[33a]. Indeed, Mr. Serrin informs me that all mathematical flaws in Carathéodory's work and even the grave error in physics I shall point out below in No. 5 can be overcome by a thorough reformulation and recasting. No such recasting was available before 1975.

3) While the founders had a fair if informal grasp of the difference between a constitutive relation and a generic principle, Carathéodory does not. While Carnot, Clausius, and Kelvin had considered processes having any speed that might be thought appropriate, Carathéodory comes out for "quasistatic processes". He confuses statements regarding a fairly general class of systems in equilibrium with the severe restrictions on constitutive functions that result from applying thermodynamic axioms to systems undergoing processes. Today we are accustomed to this muddle because we find it in every textbook, but in Carathéodory's day it was a new muddle, not yet a sacred cow of a numerous profession. Now or then, a mathematician who cannot get free of it will add nothing to thermodynamics but illusions of rigor where rigor there is none.

To get absolute scales rigorously from Kelvin's line of argument the following assumptions suffice:

1) Carnot's General Axiom
2) Thermometric Axiom (Boyling): for each hotness h there is some body whose latent heat at that hotness fails to vanish for some volume.

Both of these axioms are *scale-independent*; both are compatible with the First Law and the Second Law *but require neither.* The absolute temperatures so provided may or may not be integrating factors for the heat form. For details see my "Absolute temperatures as a consequence of Carnot's General Axiom", *Archive for History of Exact Sciences* **20** (1979): 357–380.

[33a] *Added in Proof.* Indeed Carathéodory's local theorem holds even if the coefficients are only once continuously differentiable, as had been shown by W.-L. Chow, "Über Systeme von linearen partiellen Differentialgleichungen erster Ordnung", *Mathematische Annalen* **117** (1939): 98–105.

4) There is no basis in sensation or other experience for Carathéodory's celebrated axiom of inaccessibility. It reflects neither the origins of thermodynamics in experiment nor the insights provided by the discoveries of the pioneers. Carathéodory sets it down by fiat and expects his readers to accept his trick because it delivers what they already know and accept. The axiom itself is inaccessible in principle to experiment of any kind, for how can anyone assure himself by or from experiment that in each neighborhood of any given point there is another point that cannot be joined to it by any member of an infinite class of paths? Professor Johann Walter in a letter to me has put this quandary as an episode:

> A student bursts into the study of his professor and calls out: "Dear professor, dear professor. I have discovered a perpetual motion of the second kind!"
>
> The professor scarcely takes his eyes off his book and curtly replies: "Come back when you have found a neighborhood U of a state x_0 of such kind that every $x \in U$ is connected with x_0 by an adiabat."

Contrast this demand with "Show me a cyclic heat-engine that works with an efficiency greater than $1 - T_{min}/T_{max}$", or "Show me a cyclic heat-engine that does positive work yet emits as much heat it absorbs."

5) Carathéodory's axioms reflect an error in physics. They presume that the variables sufficient to define mechanical work suffice also to define any thermodynamic system. In the systems treated by the pioneers, the variables that define work are *pressure* and volume. As Thomsen & Hartka[34] remarked, Carathéodory's axioms, since they are expressed in terms of these same variables, cannot apply to water in the range of its anomalous behavior. Be it noted that Carnot, Kelvin, and Clausius in their basic assumptions had always taken *temperature* and volume as independent variables, upon which the latent and specific heats depend. The difference is not trivial, because the pioneers' choice is general for the fluids they considered, water included; this choice, recent studies of rational thermodynamics adopt and extend.

It is one thing for a textbook to leave anomalous behavior out of account because it is untypical of the circumstances most engineers encounter; it is quite another when a mathematician presents as axioms for the whole science of thermodynamics statements that in fact fail to apply to one of the two most useful substances on earth, essential to the existence of life. (Nevertheless, the writers of textbooks may not be wrong always when they claim that their versions of what they call the "classical" treatment of thermodynamics are equivalent to Carathéodory's, for often they, too, start from assumptions that exclude water, although that fact does not keep them from applying their results to it afterward when (and if) they try to explain its anomalous behavior.)

Returning to mathematicians, I must say that of them their colleagues, students, and successors expect a standard in conceptual analysis unreasonable to

---

[34] J. S. Thomsen & T. J. Hartka, "Strange Carnot cycles", *American Journal of Physics* **30** (1962): 26–33, 388–389.

demand of others. While he who does no more than apply a science need not interest himself in its essential physics or its logical structure, he who would set up axioms for a science must first master its main applications. As Hilbert wrote, "only that architect is fit to dispose securely the foundations of a structure who himself knows in all detail the functions it is to discharge."

6) Carathéodory in effect defines heat as a kind or work. Here he cites the article by Bryan[35] in the mathematical *Encyklopädie*, which starts from the First Law with internal energy taken as a primitive idea. Carathéodory, a skilful mathematician, easily sees that Bryan's vague discussion is unnecessary to the mathematical structure. He writes,

> *The whole theory can be derived without assuming the existence of a physical quantity, namely heat, that diverges from the usual mechanical quantities.*

Even thermodynamicists of the classical kind have pounced upon this blunder. As W. J. Hornix[36] has remarked,

> many have tried to reduce thermodynamics to mechanics, but have failed because the concept of heat defies such reduction. An axiomatic approach has made clear the reason for this, by showing that an adiabatic process is necessarily a truly primitive term in thermodynamics; it cannot be derived from mechanics.

I agree with M. Zemansky[37] when he says:

> To deal with the foundations of thermodynamics *as though* you don't know what temperature, heat and work are is nonsense.

That objection in itself suffices to dismiss Carathéodory's axioms. We recall that from D'Alembert's time onward some axiomatizers have thought that mechanics would do very well without a mathematical concept of force – in a word, that Newton in his Laws of Motion had used a dispensable primitive. In thermodynamics such an approach sneers at the struggles of the founders, who had to *discover* the First Law as well as the Second Law. As Mr. Serrin puts it, "Carathéodory's greatest legacy of disaster is his attempt to define heat in terms of work and energy. Any such attempt must fail, since it can neither come to grips with the idea of heat supply as something corresponding in thermodynamics to force in ordinary mechanics nor distinguish between supplies of heat at different temperatures." In this confused and confusing blunder Carathéodory diverted generations of physicists from the major lesson to be learnt from Carnot and Gibbs.

I think that any axiomatization designed to derive thermodynamics as a whole from ideas that represent nature directly is defective if it takes as an axiom a statement about energy. The approach of Carnot, adopted in the main by Clausius, is based plainly and explicitly upon the experimental properties of heat

[35] G. H. Bryan, "Allgemeine Grundlagen der Thermodynamik" (1903), pages 71 – 160 of Band 5, Erster Teil of *Encyklopädie der Mathematischen Wissenschaften*, 1903/1921. See also § 47 of Bryan's book, cited in Footnote 23.

[36] "Proceedings of the international congress on thermodynamics held at Cardiff, U.K., 1 – 4 April, 1970", *Pure and Applied Chemistry* **22** (1970): 211 – 553. See page 537.

[37] *Ibid.* page 552.

in bodies of the simplest kind, as natural to thermodynamics as the ideal pendulum is to mechanics; resurrecting that approach and replacing Carnot's one unacceptable axiom by a statement idealizing a fact of experiment discovered after Carnot's death, Bharatha & I[38] demonstrated by elementary and rigorous use of ordinary calculus the existence and properties not only of entropy but also of internal energy; and Pitteri extended all our results to the generality of reversible systems described by the product of the hotness manifold and an n-dimensional manifold of substates. Heat is not defined away; it is accepted as fundamental and then *proved* to be interconvertible with work and internal energy, restoring Clausius' finest discovery to its just place. I repeat, for reversible systems the *First Law* and the *Second Law*, both of them, emerge as proved consequences of the properties of gases and frictionless idealized heat engines. No assumption that excludes water, no new physical principles, no unmotivated, flashy, and untriable device, was ever called for. The recent work of Šilhavý[39] deserves particular respect for recognizing the general First Law, not restricted to reversible systems, as something that must be derived from natural and immediate assumptions about heat and work. The same holds for the unpublished researches of Ricou.

7) In any kind of thermostatics or thermodynamics adiabats take a central part. In consequence of his attempt to expel from physics the basic concept of heat, Carathéodory assumed that adiabats existed in kind and abundance sufficient for his needs. Apparently he did not see that such existence is a constitutive property which is to be discovered by bringing the theory of differential equations to bear upon particular conditions derived from the constitutive functions that define the thermodynamic system. It is a body's latent and specific heats, Λ_V and K_V for the bodies considered by the pioneers, that determine its adiabats through the differential equation

$$\Lambda_V dV + K_V d\theta = 0 ,$$

θ being the temperature on a given empirical scale. All pioneers knew as much. Carathéodory did not even mention Λ_V; had he attempted anything specific, he would have had no way to determine adiabats without knowing the energy function of the body to which they belong. That energy function, moreover, would have had to operate on temperature and volume or entropy and volume, while his independent variables were pressure and volume. (He could not have used the entropy function or any derivative with respect to entropy because in order to prove entropy to exist he had presumed the existence of adiabats.) Thus he would have been unable to demonstrate the qualitative properties of the adiabats of water. Carathéodory seems to have had no acquaintance with the specific problems thermodynamics is designed to handle. Here in alleviation of his failure we may notice that his somewhat older friend and later colleague, the physicist Som-

[38] *Concepts and Logic*, cited above in Footnote 28.

[39] M. Šilhavý, "On the Second Law of thermodynamics, I. General framework, II. Inequalities for cyclic processes", *Czechoslovak Journal of Physics* **B32** (1982): 987–1010, 1073–1099.

merfeld[40], fared no better when he faced water, and his editors, eminent physicists of a more recent stamp, published a flagrant error regarding it. From the standpoint of Kelvin the matter is simple, and Kelvin (1854) was the first to state that $\Lambda_V < 0$ in the range of anomalous behavior.

8) In presenting a new theory of absolute temperature Carathéodory implied that there was something unsatisfactory in Kelvin's, and that generality could be increased by defining temperature for systems described by many variables. Indeed, there are two gaps in Kelvin's work: He does not recognize the hotness manifold, and he does not perceive need for a thermometric axiom such as to ensure that what he calls "absolute temperature" can qualify as a temperature of any kind. Carathéodory's system shares these faults. More than that, while the founders had worked not only half a century before Mach was to recognize the concept of hotness but also a decade before Riemann was to introduce the concept of manifold, Carathéodory proposed his axioms more than a decade after Mach's book had appeared. Finally, if after accepting an *a priori* concept of hotness and recognizing temperature as being a chart on the hotness manifold we somehow construct a global chart deserving the title "absolute", in regard to it we gain nothing by laboring with more and more complicated systems. Kelvin's construction of absolute temperature as a means of denoting hotness suffices, once we have filled its mathematical gaps, for all systems compatible with Carnot's General Axiom. We need but use it. Only he can refuse it who rejects the concept of hotness or who proposes axioms insufficient to deliver, when specialized to reversible and discrete systems, the basic assumption of Carnot.

9) If for reversible systems Carathéodory's axioms are incomplete, unnecessarily restrictive, and unphysical, they have no compensating virtue in other regards. Nothing like the statements of Clausius, Gibbs, and Planck regarding irreversible changes of entropy was obtained by Carathéodory. Although he does refer to irreversibility, he never makes his notions precise or derives anything concrete about it. As Leaf[41] remarked, the Pfaffian form on which Carathéodory bases all his analysis does not suffice to represent heating or working of bodies that may undergo irreversible changes. While Gibbs had treated the thermostatics of elastic materials with masterly success, and Duhem had formulated and applied thermodynamics in the field theories of linear viscosity and general elasticity, Carathéodory never touches a field.

Not only did Carathéodory himself fail to extend the domain of thermodynamics; he narrowed it, and in my opinion his approach, even after the mathematical deficiencies in his use of it have been emended, brings us further from a thermodynamics of possibly irreversible processes in general deformations of bodies than we are when we read the papers of Carnot, Kelvin, and Clausius.

[40] A. Sommerfeld, *Thermodynamik und Statistik*, herausgegeben von F. Bopp und J. Meixner, Wiesbaden, Dieterichsche Verlagsbuchhandlung, 1952. Translation by J. Kestin, *Thermodynamics and Statistical Mechanics,* New York & London, Academic Press, 1964. See Übungsaufgabe I.6, pages 322–323 of the German edition, page 359 of the English.

[41] B. Leaf, "The principles of thermodynamics", *Journal of Chemical Physics* **12** (1944): 89–98.

Seen against the background of the now understood successes of the pioneers and beyond the foreground of recent treatments, Carathéodory's product reveals itself as an instance of the rule that the misbegotten are often misshapen.

7.9 Carathéodory's Legacy

The axiomatization of Carathéodory has left a baneful legacy: Born's endorsement, which for a long time propagated through physical circles widespread, complacent belief that thermodynamics had been standing on a general and secure foundation since 1909. Until recently physicists used Carathéodory's name as a mall to smash subsequent fundamental inquiry by the mathematically critical. To this I can speak from an experience of my own a third of a century ago, when I submitted an essay of mine to a nabob who enjoyed positive measure on a Nobel base in theoretical physics. The Great Man replied on July 8, 1948,

> the value of such general investigations ... is certainly rather limited. However, I find an axiomatic treatment often very clarifying, and I certainly have no objection against it. I do *not* think that one can go in this way beyond the well established theories. I am therefore very doubtful whether Murnaghan and your generalizations of the classical mechanics of continua, correspond in any way to the behaviour of real solid or fluids. ...
>
> Second, your exposition of thermodynamics certainly goes against the grain ... . I do not understand, why you dismiss the work of Carathéodory (which is a serious and helpful attempt to axiomatize thermodynamics) in one sentence.

In fact I was not doing axiomatics at all; I was merely trying to make thermodynamic statements clear enough to use in problems about deformable bodies; but the Great Man's remarks did send me to the papers by Carathéodory and his admirers among the physicists, and it was then that I realized there must be something more than superficially wrong with thermodynamics, for otherwise a competent mathematician like Carathéodory would not have made such a mess of it. That notwithstanding, I held my peace for thirty years and more, waiting until I should come to find constructive grounds on which to derogate aloud the sanctity of his axioms.

Those who ridicule not only modern axiomatizations but also attempts to create a thermodynamics new in its systematic clarity, its breadth, and its strength sometimes attribute to Carathéodory's work, which they seem to know only by reputation, qualities absent from its pages in print, such as proof that only quasistatic processes obey the laws of thermodynamics, that only bodies nearly in equilibrium have a temperature. A particularly charming quality of the thermopsychic state of this folk is its simultaneous expertise in "heat transfer", a theory in which differences in that quasistatic thing called temperature propagate at infinite speed. To be able to consider the same phenomenon as being both infinitely slow and infinitely fast is a mark of outstanding physical intuition in facing what are nowadays called "real situations".

In reaction, perhaps, against the uncritical, immoderate flood of early praise by Born and other physicists, recent years have seen a slow but monotonically increasing awareness of Carathéodory's inadequacies and errors, not only among mathematicians but also among physicists and engineers of a new generation. Hence has burst out a second disaster. Everybody who finds a gap or mistake, instead of throwing the whole mess away as being just one of those occasional bad roads to goods results, tries to fix it up. The outcome is further confusion and new errors. Professor Walter[42] has provided an extensive but surely far from complete survey and analysis of this interdisciplinary gusher. Some physicists invoke mathematics they have not taken the trouble to learn, for example the theory of differential forms, the terms and notations of which they palaver offhand, concealing their unstated assumptions and orginal mistakes in a fog of mystic manipulation. Some mathematicians, including the author of an otherwise respectable textbook on analysis who offers proof of Carathéodory's conclusions as a trivial exercise for a good analyst, have failed to learn enough about heat and temperature to phrase correctly in their terms what Carathéodory assumed and what he claimed to prove. Parts of the literature of theoretical engineering have taken affirmative action, adopting without prejudice various mixtures of the sins of both foregoing kinds.

Bad as is the literature Professor Walter has analysed, it is probably no worse than what is typical of any other field. The papers we are to hear in this meeting prove, had there been any doubt, that some mathematicians and some engineers are firmly footed in the physics and the mathematics of classical thermodynamics and can make sound and sober extensions of it. I will take it on faith that there are, somewhere, physicists of similar capacity.

7.10 Colophon

What then, did Carathéodory give us? Misconceptions, regress, lacunary mathematics, confusion in physics. Gibbs? A magnificent phenomenological thermostatics, an elegant if rather special statistical thermostatics – all in all, more than any other one man did for the science of heat and hotness.

---

[42] W. Walter, "On the definition of the absolute temperature – a reconciliation of the classical method with that of Carathéodory", *Proceedings of the Royal Society of Edinburgh* **82A** (1978): 87–94; "On the foundations of thermodynamics", pages 695–702 of *Ordinary and Partial Differential Equations* (Proceedings of the Seventh Conference Held in Dundee, Scotland, March 29–April 2, 1982), Springer Lecture Notes in Mathematics No. 964. In the former work Professor Walter offers a mixture of ideas to get Carathéodory's conclusions economically. He takes internal energy as primitive, adopts an axiom of union to exploit the consequences of adjoining a posited thermometric body to an arbitrary body, and by assuming the existence of adiabats of a suitable kind avoids recourse to Carathéodory's axiom of inaccessibility. Walter's development, like Carathéodory's, makes the Second Law contingent on the First.

Acknowledgment. This lecture is drawn in part from the *Historical Introit* prefixed to the second edition of my *Rational Thermodynamics*, Springer-Verlag, 1984. Most of the material on Carathéodory derives from a portion of a lecture to Euromech Colloquium 100 at Pisa, 17 May 1978:

"Schizzo concettuale della termodinamica per gli studiosi di meccanica", *Bollettino della Unione Matematica Italiana* (5) **16-A** (1979): 1 – 20.

I am grateful to Messrs. Coleman, Feinberg, C.-S. Man, and Serrin for enlightening discussions of thermodynamics and thermostatics, some as much as thirty years ago, and for criticism of draughts of the *Introit*.

Chapter 8
Structure and Dynamical Stability of Gibbsian States

R. L. Fosdick

This paper is dedicated to my parents, Mary Helen and Glen Kenneth Fosdick, who provided an early environment for the exercise of imagination and the application of choice.

8.1 Introduction

It can be said that *Gibbs'* theory of thermostatics [8.1] is one of the great constructs of modern science. One can hardly open any textbook on thermodynamics without being exposed to some aspect of his contribution, if not at the foundational level then on matters applied and practical. Gibbs not only opened an area of great conceptual and scientific interest, but also he proposed applications in materials science that are yet today receiving the concentrated attention of researchers.

In 1980, *Dunn* and *Fosdick* [8.2] published a long article entitled "The Morphology and Stability of Material Phases" in which we developed many theorems regarding the existence, uniqueness, structure and stability of coexistent phases in materials which are, roughly, fluid-like in equilibrium but which may have very general constitutive response properties when in motion. We gave a *field* theoretic framework to Gibbs' concepts in thermo*statics* and showed how one could apply continuum thermo*dynamics* to better understand the question of stability of certain static states which within the Gibbsian structure he *defined* as stable. The present paper is based upon that work, and it is intended to represent a fairly comprehensive summary of what I consider to be the main results of that investigation. Seven theorems will be presented; the first two are concerned with the existence and the structure of what Gibbs called equilibrium and stable equilibrium states in thermostatics. The next two theorems describe conditions under which several extremum problems of thermostatics may be considered equivalent. The final three theorems are dynamical in nature and are aimed at the question of stability of thermostatic states.

In Sect. 8.2 we begin by showing how the basic laws of continuum thermodynamics can be used to motivate the idea that those equilibrium or rest states which are expected to be limits of a certain class of dynamical processes may be determined by well-known extremum principles of Gibbsian thermostatics. While it is not essential to restrict the forms of the constitutive response functions of the material outside of equilibrium, we do assume in this work that rest states are such that their fields of specific entropy, specific volume, and specific internal energy take values which lie on an a priori given smooth surface called the static site manisfold $\mathscr{S}$. This surface is, in general, not convex and may only have a parametric representation.

Based upon the motivation of Sect. 8.2, we identify, in Sect. 8.3, Gibbsian states as those fields of specific entropy, specific volume and specific internal energy that are piecewise continuous, take their values on $\mathscr{S}$, and maximize the total entropy of a body for a given total volume and total internal energy. We then go on, in Sects. 8.3 and 4, to describe the existence and detailed structural properties of *normal* Gibbsian states, i.e., Gibbsian states of non-zero thermostatic temperature. In Theorem 1 of Sect. 8.3 we show, as a necessary condition, that a given normal Gibbsian state can only associate with a certain set of support points of $\mathscr{S}$ that lie in a common support plane. Moreover, it is shown how the orientation of this support plane determines the thermostatic temperature and thermostatic pressure of the state. If $\mathscr{S}$ has neither flat patches nor planar arcs we find that any given normal Gibbsian state is at most a step function. We also give a necessary condition on the prescribed values of the total volume and total internal energy in order that a normal Gibbsian state may exist.

We close Sect. 8.3 by showing how Theorem 1 readily provides certain classical comparison principles which apply to sequences of normal Gibbsian states. The idea here is that if one considers a sequence of normal Gibbsian states, certain changes of state are permissible and others are not. General rules which govern the possible changes are called principles of comparison. We show, for example, how local and global forms of the classical Gibbs equation arise in this connection. In addition, we obtain the inequalities of Van't Hoff and Le Chatelier which restrict, respectively, relative changes of temperature and entropy in constant pressure sequences, and relative changes of pressure and volume in constant temperature sequences. Finally, by applying Theorem 1 to a sequence of normal Gibbsian states which support discontinuities (i.e., coexistent phases), we show how the famous Clausius-Clapeyron equation emerges.

The main result of Sect. 8.4 is Theorem 2 which not only gives a sufficient condition on the prescribed total volume and total internal energy for the existence of normal Gibbsian states, but also completely identifies the possible structure of such states. In certain situations, this structure can be fairly complex, leading to states with a high degree, and even a continuum of multiple coexistent phases, and in general it is non-unique. If the static site manifold is such that any single support plane will touch it coincidently in only a finite number of points then the structure is never more general than a step function. We close Sect. 8.4 by introducing the notion of rearrangements and discussing why a normal Gibbsian state that is not only a step function but also unique up to a rearrangement can have no more than 3 coexistent phases. The requirement of uniqueness up to a rearrangement is essential for obtaining the number 3; otherwise, the Gibbs phase rule is not expected to hold within the classical theory of thermostatics for a single substance.

The literature on thermostatics contains many loosely supported claims regarding the "equivalence" of various classical extremum problems. Often, some unstated assumption concerning the invertibility or convexity of certain thermostatic functions is found to be required in order to make any sense of the results. In Theorems 3 and 4 of Sect. 8.5 we provide necessary and sufficient conditions in order that normal Gibbsian states also solve the following four alternative problems of minimization: (i) minimize the internal energy at fixed volume and entropy, (ii) minimize the Gibbs semi-free energy, (iii) minimize the Helmholtz semi-free energy at fixed volume, and (iv) minimize the semi-enthalpy at fixed entropy. While it is fairly elementary to show, as is done in Theorem 3, that a given normal Gibbsian state of positive thermostatic temperature also solves these problems, it is much more involved to prove the converse, especially in the case of problems (iii) and (iv). After a brief discussion of the specific potentials which characterize problems (ii), (iii) and (iv) we carry out this converse argument within the context of Theorem 4.

Finally, in Sect. 8.6 we return to dynamics and the motivation given in Sect. 8.2 for the study of Gibbsian states, and we show how it can be applied more precisely to study the stability of these states. Here, as a matter of convenience we assume that the static site manifold $\mathscr{S}$ can be described globally by a smooth internal energy function of the specific volume and specific entropy. Also, we suppose[1] that for a broad class of thermodynamic processes the thermodynamic state of specific entropy, specific volume, and specific internal energy is bounded below by $\mathscr{S}$.

In all three theorems of Sect. 8.6, we take as given a thermostatic pressure p^* and a positive thermostatic temperature θ^* which are associated with a normal Gibbsian state. We then suppose, in Theorem 5, that the body is allowed to undergo any neo-classical process[2] that is compatible with this temperature and pressure. This includes, for example, any isolated process[2] for which the total volume and total energy (internal plus kinetic) are fixed at the corresponding values of total volume and internal energy that are associated with the normal Gibbsian state. Then, if the static site manifold $\mathscr{S}$ satisfies a certain structural condition, which requires, roughly, that it contain a sufficiently well-distributed set of thermostatic support points of positive thermostatic temperature, we claim, in Theorem 5, that a definite *set* of points on $\mathscr{S}$ determined by θ^* and p^* is stable in the sense of estimate (8.6.7).

For the most part, Theorem 6 focuses on normal Gibbsian states that are unique up to a rearrangement, and shows that they are L^1-stable relative to isolated processes. Roughly, for "small" initial data we show that during any isolated process the fields of specific volume, and specific internal energy always remain close, in an integrated L^1 sense, to the corresponding piecewise constant values of the given normal Gibbsian state over mass measures of the body that are close, in the sense of absolute value, to the unique mass measures which associate with each of the distinct phases of the normal Gibbsian states.

Finally, in Theorem 7 we note that relative to the class of isolated processes, certain very special unique normal Gibbsian states that are homogeneous throughout the body and that associate with sharp isolated points of $\mathscr{S}$ are L^1-stable. While the hypotheses of this theorem, being more strict, also apply in Theorem 5, the claim here is substantially improved in that the earlier less specific conclusion associated with the stability of a *set* is now made more explicit.

8.2 Preliminaries and Motivation

In the subject of continuum thermomechanics, a body B is identified with the regular region of Euclidean three-space which it occupies in a fixed reference

[1] See footnote 16 on page 149 for a relevant comment.
[2] Neo-classical and isolated processes are defined in Sect. 8.2.

configuration. A positive mass measure $m(\cdot)$ is assigned in the usual way, and a process class $\mathbb{P}(B)$, characterizing the material, is introduced. It is convenient to think of the elements of $\mathbb{P}(B)$ as an ordered 8-tuple

$$\pi(\cdot,\cdot) \equiv \{\eta, \chi, \varepsilon, \theta, T, q, b, r\}(\cdot,\cdot) \in \mathbb{P}(B)$$

defined on $B \times \mathbb{R}$, which for every sub-body $P \subset B$ satisfies the *balance of linear momentum*

$$\frac{d}{dt} \int_P \dot{x}\, dm = \int_{\partial P_t} Tn\, da + \int_P b\, dm\,, \tag{8.1}$$

the *balance of energy*

$$\frac{d}{dt} \int_P \left(\varepsilon + \frac{1}{2}|\dot{x}|^2\right) dm = \int_{\partial P_t} (\dot{x} \cdot Tn - q \cdot n)\, da + \int_P (\dot{x} \cdot b + r)\, dm\,, \tag{8.2}$$

and the *Clausius-Duhem inequality*

$$\frac{d}{dt} \int_P \eta\, dm \geqslant - \int_{\partial P_t} \frac{q \cdot n}{\theta}\, da + \int_P \frac{r}{\theta}\, dm\,. \tag{8.3}$$

The components of the process $\pi(X, t)$ at the particle X and time t have the usual respective interpretations of specific *entropy* per unit mass, the *motion*, the specific *internal energy* per unit mass, the *absolute temperature* (>0), the (symmetric) *Cauchy stress* tensor, the *heat flux* vector, and specific *radiant heating* per unit mass. Further, we have used the notation that $\dot{x} \equiv \partial\chi(X, t)/\partial t$ is the *velocity*, $S_t \equiv \chi(S, t)$ for any $S \subset B$, and $n = n(x, t)$ is the outer unit normal to ∂P_t. We shall let $V_t(P)$ denote the volume of P_t and call the positive valued function $v(\cdot,\cdot)$: $B \times \mathbb{R} \to \mathbb{R}^+$ which is such that

$$V_t(P) = \int_P v\, dm\,, \tag{8.4}$$

the specific *volume* per unit mass.

There are two classes of processes which seem to have attracted special attention in the early developments of thermodynamics. Roughly, these correspond to the idea of (i) *isolation*, i.e., any process for which $dV_t(B)/dt = 0$ and there is null global supply of mechanical working $W(t)$, heat working $Q(t)$, and entropy $M(t)$ to B for all $t \geqslant 0$, and (ii) to the notion that the environment of B is held at a fixed temperature and pressure of $\theta^* > 0$ and p^*, respectively, while the global supply of mechanical working, heat working, and entropy to B for all $t \geqslant 0$ satisfy

$$W(t) = -p^* \frac{d}{dt} V_t(B)\,,$$

and

$$M(t) \geqslant \frac{Q(t)}{\theta^*}\,.$$

The latter notion we call *neo-classical*, and we shall denote the set of all such processes with ambient temperature $\theta^* > 0$ and ambient pressure p^* as $\mathbb{P}(\theta^*, p^*) \subset \mathbb{P}(B)$. The conditions of a neo-classical process can be arranged, for example, by considering processes for which b and r are null so that

$$W(t) = \int_{\partial B_t} \dot{x} \cdot Tn \, da = -p^* \int_{\partial B_t} \dot{x} \cdot n \, da = -p^* \frac{d}{dt} V_t(B) ,$$

$$M(t) = -\int_{\partial B_t} \frac{q \cdot n}{\theta} da ,$$

and

$$Q(t) = -\int_{\partial B_t} q \cdot n \, da ,$$

and by applying the inequality

$$(\theta - \theta^*) q \cdot n \geqslant 0 \quad \text{on} \quad \partial B_t$$

for all $t \geqslant 0$. This inequality clearly holds if $\theta = \theta^*$ and $q \cdot n = 0$ on complementary subsets of ∂B_t, and, more generally, has the interpretation that at boundary points of ∂B_t that are $\left\{ \begin{smallmatrix} \text{hotter} \\ \text{colder} \end{smallmatrix} \right\}$ than the environment, heat cannot flux $\left\{ \begin{smallmatrix} \text{into} \\ \text{out of} \end{smallmatrix} \right\}$ the body.

For the class of neo-classical processes $\mathbb{P}(\theta^*, p^*) \subset \mathbb{P}(B)$ we see from (8.2), (8.3) and the above discussion that

$$\frac{d}{dt} \int_B \left(\varepsilon + p^* v + \frac{1}{2} |\dot{x}|^2 \right) dm = 0 , \tag{8.5a}$$

and

$$\frac{d}{dt} \int_B \eta \, dm \geqslant \frac{Q}{\theta^*} . \tag{8.5b}$$

Thus, by introducing the state vector field at time t

$$s_t(\cdot) = (\eta, v, \varepsilon)(\cdot, t) \colon B \to \mathbb{R}^3 \tag{8.6}$$

and the constant ambient vector

$$g^* \equiv \left(-1, \frac{p^*}{\theta^*}, \frac{1}{\theta^*} \right) , \tag{8.7}$$

it follows from (8.5) that

$$\frac{d}{dt} \int_B \left[s_t \cdot g^* + \frac{1}{2} |\dot{x}|^2 / \theta^* \right] dm \leqslant 0 \tag{8.8}$$

for all $t \geqslant 0$. For any neo-classical process in $\mathbb{P}(\theta^*, p^*)$ the functional in (8.8) is non-increasing and, in fact, it would seem reasonable to expect that for all such processes the body B would eventually come to rest. If so, then because of (8.8) it could be that the rest state vector field

$$s_\infty(\cdot) \equiv \lim_{t \to \infty} s_t(\cdot) \colon B \to \mathbb{R}^3 \tag{8.9}$$

has a close relation to a vector field $p(\cdot) \colon B \to \mathbb{R}^3$ which minimizes

$$\int_B p \cdot g^* dm$$

in some sense. At least, it could be that if a body is allowed to undergo a neo-classical process in $\mathbb{P}(\theta^*, p^*)$ which initially at $t = 0$ starts "near" such a minimizing field, then it will remain near or possibily even approach this field with time. In any case, it is evident that such a problem of minimization warrents study and the stability of its resulting solutions merits investigation.

For the class of isolated processes, the right hand sides of (8.2) and (8.3), which for the whole body B represents the sum $W(t)+Q(t)$ and $M(t)$, respectively, vanish and we see that for all $t \geqslant 0$

$$\int_B \left(\varepsilon + \frac{1}{2}|\dot{x}|^2\right) dm = \text{const.}\,,$$

$$\frac{d}{dt}\int_B \eta\, dm \geqslant 0\,.$$

In addition, since the volume is supposed to be fixed for such processes we have

$$\int_B v\, dm = \text{const.}$$

for all $t \geqslant 0$. In this case, if the body B is to approach a state of rest with increasing time while remaining isolated, the above remarks suggest that the rest state vector field (8.9) may be related to a vector field $p(\cdot) \equiv (n, v, e)(\cdot)\colon B \to \mathbb{R}^3$ which in some sense maximizes

$$\int_B n\, dm \tag{8.10a}$$

subject to the constraints

$$\int_B v\, dm = V^*\,, \qquad \int_B e\, dm = E^*\,, \tag{8.10b}$$

where V^* and E^* are given numbers. This is the primary Gibbs problem of thermostatics which, in fact, he took to define stable equilibrium states and which we shall study in Sects. 8.3 and 4. It is possible that solutions of this maximum problem and the earlier mentioned minimum problem could be closely related. This will be investigated, along with other related extremum problems, in Sect. 8.5.

To be definite, we shall assume throughout the remainder of this work that the range of any *rest* state vector field $s_\infty(\cdot)$ is an a priori given 2-dimensional smooth open manifold, called the *static site manifold*[3] $\mathcal{S} \subset \mathbb{R}^3$. Any point $p \in \mathcal{S}$ will be denoted by the ordered triple $p = (n, v, e)$, and all competing vector fields in the variational problems outlined above will have their range in $\mathcal{S}$. While the manifold $\mathcal{S}$ may be considered a generalization of the idea of an equilibrium equation of state, it need not be connected nor convex and may not be expressible globally as a function $e = \bar{e}(n, v)$. However, on patches of $\mathcal{S}$ where such a function may be written, it is common in thermostatics to call $\theta \equiv \partial\bar{e}(n, v)/\partial n$ the *thermostatic temperature* and $p \equiv -\partial\bar{e}(n, v)/\partial v$ the *thermostatic pressure* associated with the point $p = (n, v, e) \in \mathcal{S}$. If $\theta \neq 0$, then a normal vector to $\mathcal{S}$ at $p = (n, v, e)$ may be written as

[3] See Fig. 8.1 for an example of a possible static site manifold.

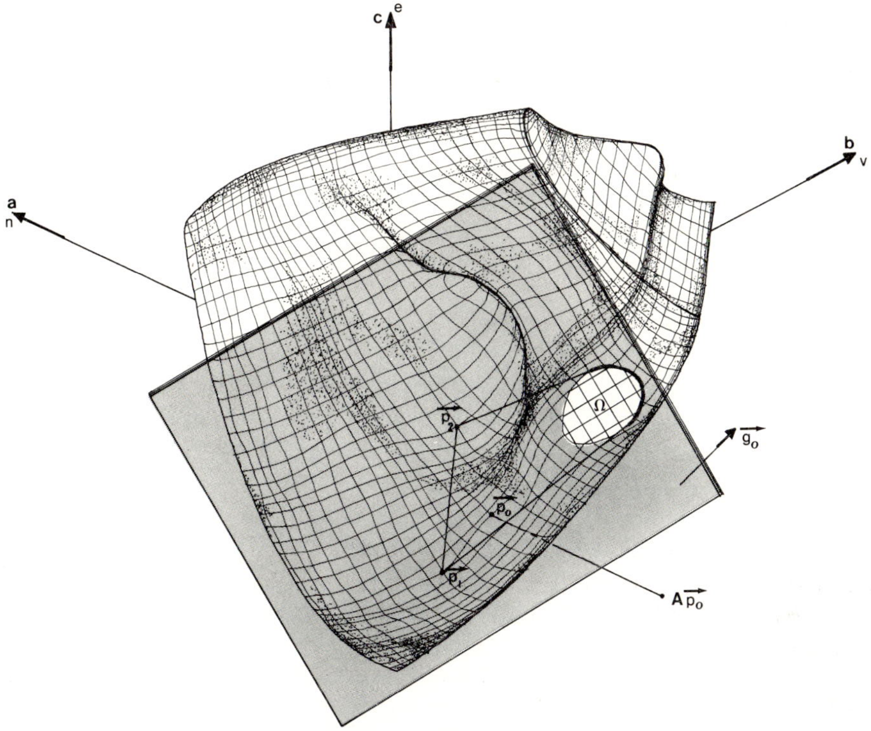

Fig. 8.1. A continuous *static site manifold* $\mathscr{s}$. The points $\boldsymbol{p}_0, \boldsymbol{p}_1, \boldsymbol{p}_2$ and all of the points of the flat patch Ω share a common support (tangent) plane whose normal $\boldsymbol{g}_0$ satisfies $\boldsymbol{g}_0 \cdot \boldsymbol{a} = -1$. Such points are called *thermostatic support points* and the set $\mathscr{s}(\boldsymbol{p}_0)$ of all these points is denoted as the *planar slice at* $\boldsymbol{p}_0$. Points $\boldsymbol{p}_1, \boldsymbol{p}_2$ and those on the darker portion of the boundary of Ω are *sharp thermostatic support points.* $\boldsymbol{A}\boldsymbol{p}_0$ represents the A-projection of $\boldsymbol{p}_0$ into the $v-e$ plane

$$g \equiv \left(-1, \frac{p}{\theta}, \frac{1}{\theta}\right). \tag{8.11}$$

In fact, with a slight generalization, whenever a vector of the form (8.11) is identified as a normal to $\mathscr{s}$ at $\boldsymbol{p} = (n, v, e)$ we shall call θ and p the associated thermostatic temperature and thermostatic pressure at $\boldsymbol{p}$, respectively.

It should be noted that very little has been assumed concerning the general constitutive structure of the material body. Most of the discussion of this section was designed at motivating possible variational problems concerning rest states and the notion of equilibrium. Naturally, the assumption that rest states are to be found on a static site manifold is constitutive and we shall use this in the next section in order to describe the possible states of equilibrium. However, this assumption does not say anything about the dynamic response and, therefore, in particular the state vector $\boldsymbol{s}_t(X)$ for any particle $X \in B$ and time t need not even lie on $\mathscr{s}$. We shall consider the question of stability in Sect. 8.6 wherein a restriction on constitutive response outside of equilibrium will be introduced.

8.3 Gibbsian States: Necessary Conditions and Comparison Principles

Consider now the primary Gibbs problem of thermostatics outlined roughly in (8.10). We shall let C_p denote the set of piecewise continuous functions which map the particles $X \in B$ into points on the static site manifold $\mathscr{S}$, and, for convenience, introduce the fundamental unit triad $\boldsymbol{a} \equiv (1,0,0)$, $\boldsymbol{b} = (0,1,0)$ and $\boldsymbol{c} = (0,0,1)$ of Fig. 8.1.

Definition. *Given two numbers V^* and E^*, let*

$$C_p(V^*, E^*) \equiv \left\{ \boldsymbol{p}(\cdot) = (n, v, e)(\cdot) \in C_p \,|\, \boldsymbol{A} \int_B \boldsymbol{p}\, dm = (0, V^*, E^*) \right\},$$

where the projection tensor[4] $\boldsymbol{A} \equiv \boldsymbol{b} \otimes \boldsymbol{b} + \boldsymbol{c} \otimes \boldsymbol{c}$. *Then,* $\boldsymbol{p}^*(\cdot) \in C_p(V^*, E^*)$ *is a* Gibbsian state *if for all* $\boldsymbol{p}(\cdot) \in C_p(V^*, E^*)$,

$$\boldsymbol{a} \cdot \int_B \boldsymbol{p}^* dm \geqslant \boldsymbol{a} \cdot \int_B \boldsymbol{p}\, dm\,.$$

The primary Gibbs problem is the study of Gibbsian states. Actually, because of (8.11) the thermostatic temperature associated with the value of a Gibbsian state $\boldsymbol{p}^*(X) \in \mathscr{S}$ at a point of continuity $X \in B$ will be zero if a normal vector to $\mathscr{S}$ at $\boldsymbol{p}^*(X)$ has null projection in the direction of $\boldsymbol{a}$. In classical thermostatic it is common, therefore, to consider normal Gibbsian states in order to avoid the possibility of a Gibbsian state associated with zero thermostatic temperature, and we shall do so here. What distinguishes a *normal* Gibbsian state from all other Gibbsian states is that *the range of a* normal Gibbsian state *consists of points on* $\mathscr{S}$ *whose normal vector has a non-zero* $\boldsymbol{a}$*-component.*

We now have

Theorem 1. *Let* $\boldsymbol{p}^*(\cdot) = (n^*, v^*, e^*)(\cdot) \in C_p(V^*, E^*)$ *be a normal Gibbsian state. Then, there exists a unique constant vector* $\boldsymbol{g}^* = (-1, p^*/\theta^*, 1/\theta^*)$, $\theta^* \neq 0$, *such that*

$$[\boldsymbol{p} - \boldsymbol{p}^*(X)] \cdot \boldsymbol{g}^* \geqslant 0 \tag{8.12}$$

for all $\boldsymbol{p} \in \mathscr{S}$ *and all* $X \in B$ *where* $\boldsymbol{p}^*(\cdot)$ *is continuous.*

This theorem, while only a necessary condition for a normal Gibbsian state goes a long way toward solving the primary Gibbs problem. In a sense, it represents both the Euler-Lagrange and the Weierstrass conditions. According to (8.12) a normal Gibbsian state can only associate with *support points* of the static site manifold $\mathscr{S}$. Moreover, if $\boldsymbol{p}^*(X_0)$ is one such support point for some $X_0 \in B$, then $\boldsymbol{g}^*$ must be the normal to $\mathscr{S}$ at this point with $\boldsymbol{g}^* \cdot \boldsymbol{a} = -1$, and if $X \in B$ is any other point of continuity of $\boldsymbol{p}^*(\cdot)$, $\mathscr{S}$ must not only have the same normal vector at $\boldsymbol{p}^*(X)$ but also must lie in the same support plane as does $\boldsymbol{p}^*(X_0)$. The con-

---

[4] The $\boldsymbol{A}$-projection of a point $\boldsymbol{p} = (n, v, e)$ is the image of its perpendicular projection into the $v-e$ plane, i.e., $\boldsymbol{A}\boldsymbol{p} = (0, v, e)$. See Fig. 8.1.

stant vector $\boldsymbol{g}^*$ determines the thermostatic temperature θ^* (non-zero, but possibly negative) and thermostatic pressure p^* of the normal Gibbsian state. It is clear that if the static site manifold $\mathscr{s}$ has neither flat spots nor continuous ridges of constant slope, then, assuming existence, a normal Gibbsian state must generally lie in the class of *step functions* in $C_p(V^*, E^*)$. In this case, the specific structure of a Gibbsian state then identifies the so-called constituent coexistent phases.

The inequality (8.12) has two standard interpretations in classical thermostatics. First, if we let X and Y denote any two distinct point of B where $\boldsymbol{p}^*(\cdot)$ is continuous, then by replacing $\boldsymbol{p}$ in (8.12) by $\boldsymbol{p}^*(Y)$ and by interchanging the roles of X and Y we arrive at the conclusion that

$$[\boldsymbol{p}^*(Y)-\boldsymbol{p}^*(X)]\cdot\boldsymbol{g}^*=0\;. \tag{8.13}$$

Thus, if we let $\boldsymbol{p}^*(\cdot)\equiv(n^*, v^*, e^*)(\cdot)$ it follows that the *chemical potential*

$$\mu^*(X)\equiv e^*(X)-\theta^*n^*(X)+p^*v^*(X)$$

for any normal Gibbsian state is constant at every point $X\in B$ where $\boldsymbol{p}^*(\cdot)$ is continuous. Moreover, if the thermostatic temperature of a Gibbsian state θ^* is positive, (8.12) may be re-written as

$$e^*(X)-\theta^*n^*(X)+p^*v^*(X)-(e-\theta^*n+p^*v)\leqslant 0\;,$$

where we have set $\boldsymbol{p}\equiv(n, v, e)\in\mathscr{s}$. In a terminology suggested by *Keenan* [8.3] this says that at all points $X\in B$ where $\boldsymbol{p}^*(\cdot)$ is continuous the *availability* must be minimized.

Of course, a normal Gibbsian state may not exist, and if one does exist it may not be unique. We shall investigate these matters later in Sect. 8.4 when we establish sufficient conditions for existence and at the same time obtain the complete structural character of any such state. At the moment, however, in addition to the necessary condition of Theorem 1, we record the additional helpful

Remark. As a consequence of an integral mean value theorem for vector-valued functions[5], *if a normal Gibbsian state is to exist it is necessary that the point $(0, V^*, E^*)/m(B)$ belong to the A-projection[6] of the convex hull of a set of support points on $\mathscr{s}$ determined by a plane whose normal $\boldsymbol{g}^*$ satisfies $\boldsymbol{g}^*\cdot\boldsymbol{a}=-1$.*

We turn now to an outline of a *Proof of Theorem 1.* Let $X_0\in B$ be a point where $\boldsymbol{p}^*(\cdot)$ is continuous and set $\boldsymbol{p}_0^*\equiv\boldsymbol{p}^*(X_0)=(n_0^*, v_0^*, e_0^*)$. Let $\boldsymbol{p}=(n, v, e)\in\mathscr{s}$, be any point on the static site manifold and observe that since

$$\lim_{\alpha\to 0}\frac{A(\boldsymbol{p}_0^*-\alpha\boldsymbol{p})}{1-\alpha}=A\boldsymbol{p}_0^*\;,$$

and since $\boldsymbol{p}^*(\cdot)$ is a *normal* Gibbsian state, then for each sufficiently small $\alpha>0$ there exists a point $\boldsymbol{p}(\alpha)\in\mathscr{s}$ such that

---

5 See *Dunn* and *Fosdick* [8.2], p. 90 for a statement and proof of this theorem. Also, Theorem 3 on page 20 of this reference is relevant.

6 See footnote 4 on page 132.

$$\frac{A(p_0^* - \alpha p)}{1-\alpha} = Ap(\alpha)\,. \tag{8.14}$$

Moreover, the one-parameter family $p(\alpha)$, $\alpha > 0$, is a smooth curve on $\mathscr{S}$ such that $p(0) = p_0^*$.

Now, suppose we know that

$$a \cdot [\alpha p + (1-\alpha)p(\alpha)] \leqslant a \cdot p_0^* \tag{8.15}$$

for sufficiently small $\alpha > 0$. Then, by rewriting (8.14) and (8.15) as

$$A[p(\alpha) - p] = A\left[\frac{p(\alpha) - p_0^*}{\alpha}\right],$$

and

$$a \cdot [p(\alpha) - p] \geqslant a \cdot \left[\frac{p(\alpha) - p_0}{\alpha}\right],$$

and by taking the limit $\alpha \to 0$, we get

$$A(p_0^* - p) = A\left.\frac{dp(\alpha)}{d\alpha}\right|_{\alpha=0},$$

$$a \cdot (p_0 - p) \geqslant a \cdot \left.\frac{dp(\alpha)}{d\alpha}\right|_{\alpha=0}.$$

Thus,

$$v_0^* - v = b \cdot p'(0)\,, \qquad e_0^* - e = c \cdot p'(0)\,, \qquad n_0^* - n \geqslant a \cdot p'(0)\,, \tag{8.16}$$

where

$$p'(0) \equiv \left.\frac{dp(\alpha)}{d\alpha}\right|_{\alpha=0}.$$

Since $p'(0)$ must lie in the tangent plane to $\mathscr{S}$ at p_0^*, then if $g^* \equiv (-1, p^*/\theta^*, 1/\theta^*)$ denotes a normal to $\mathscr{S}$ at p_0^*, we have[7]

$$g^* \cdot p'(0) = -a \cdot p'(0) + \frac{p^*}{\theta^*} b \cdot p'(0) + \frac{1}{\theta^*} c \cdot p'(0) = 0\,. \tag{8.17}$$

Whence, (8.16) and (8.17) yield (3.1) at every point $X_0 \in B$ of continuity of $p^*(\cdot)$. It is now a straightforward matter to show by contradiction that g^* in (8.12) is independent of X.

To complete this proof we, thus, need to justify (8.15). As a first step we consider an open ball $B_\delta(X_0)$ of radius δ centered at $X_0 \in B$ and such that $B_\delta(X_0) \subset B$, and note that

$$\lim_{\delta \to 0} \frac{1}{m(B_\delta)} \int_{B_\delta} p^* dm = p_0\,. \tag{8.18}$$

---

[7] Recall that $p^*(\cdot)$ is a *normal* Gibbsian state so that at $p^*(X_0) \in \mathscr{S}$ there exists a normal vector with non-zero a-component.

Thus, with the aid of (8.13) we see that

$$\left\{\frac{1}{m(B_\delta)}A\int_{B_\delta} p^*dm - A\,\alpha p\right\}/(1-\alpha)$$

has the limit $Ap(\alpha)$ as $\delta\to 0$ and therefore lies in any prescribed neighborhood of $Ap(\alpha)$ for sufficiently small δ. This, in turn, implies the existence of a point $p_\delta(\alpha)\in\mathcal{J}$ with the property $p_\delta(\alpha)\to p(\alpha)$ as $\delta\to 0$ and such that

$$\left\{\frac{1}{m(B_\delta)}A\int_{B_\delta} p^*dm - A\,\alpha p\right\}/(1-\alpha) = Ap_\delta(\alpha)\,. \tag{8.19}$$

Now, with $\alpha>0$ sufficiently small, let $P_\delta\subset B_\delta(X_0)$ be a regular sub-part of $B_\delta(X_0)$ with $m(P_\delta)=\alpha m(B_\delta)$ and consider the function

$$p^\delta(X)\equiv\begin{cases} p^*(X)\,, & X\in B-B_\delta\,,\\ p\,, & X\in P_\delta\,,\\ p_\delta(\alpha)\,, & X\in B_\delta-P_\delta\,.\end{cases} \tag{8.20}$$

Clearly $p^\delta(\cdot)\in C_{\mathrm{p}}$, and it is straight forward to see, using (8.18) and (8.19), that

$$\int_B p^\delta dm = (0, V^*, E^*)\,,$$

so that $p^\delta(\cdot)\in C_{\mathrm{p}}(V^*,E^*)$. Whence,

$$a\cdot\int_B p^\delta dm \leqslant a\cdot\int_B p^*dm\,,$$

and by directly integrating (8.20) we obtain

$$a\cdot[\alpha p+(1-\alpha)p_\delta(\alpha)]\leqslant a\cdot\frac{1}{m(B_\delta)}\int_{B_\delta}p^*dm\,.$$

Finally, we reach (8.15) by considering the limit $\delta\to 0$ and by recalling (8.18).

In the remainder of this section, we shall show how the necessary condition (8.12) and its consequence (8.13) lead to certain *comparison principles* for normal Gibbsian states. Consider a sequence of normal Gibbsian states $p_\tau^*(\cdot)\equiv(n_\tau^*, v_\tau^*, e_\tau^*)(\cdot)$ in $C_{\mathrm{p}}(V_\tau^*, E_\tau^*)$ corresponding to the smooth data (V_τ^*, E_τ^*) for $\tau\in\mathcal{I}$, where $\mathcal{I}$ is an open interval. Then, for each $\tau\in\mathcal{I}$ and at every $X\in B$ which is a point of continuity of $p_{\tau+a}^*(\cdot)$, for sufficiently small $a\geqslant 0$ it follows from (8.12) that

$$[p_{\tau+a}^*(X)-p_\tau^*(X)]\cdot g_\tau\geqslant 0\,, \tag{8.21a}$$

$$[p_\tau^*(X)-p_{\tau+a}^*(X)]\cdot g_{\tau+a}\geqslant 0\,. \tag{8.21b}$$

Thus, by dividing (8.21) by a and taking the limit $a\to 0$, and by adding (8.21a) and (8.21b), dividing the result by a^2 and taking the limit $a\to 0$, we obtain the respective conditions

$$\frac{dp_\tau^*(X)}{d\tau}\cdot g_\tau^*=0\,, \tag{8.22}$$

$$\frac{d\boldsymbol{p}_\tau^*(X)}{d\tau} \cdot \frac{d\boldsymbol{g}_\tau^*}{d\tau} \leqslant 0 \,. \tag{8.23}$$

According to Theorem 1, $\boldsymbol{g}_\tau^*$ has the form

$$\boldsymbol{g}_\tau^* = \left(-1, \frac{p_\tau^*}{\theta_\tau^*}, \frac{1}{\theta_\tau^*}\right),$$

where p_τ^* and θ_τ^* denote the (uniform) thermostatic pressure and (uniform) thermostatic temperature associated with $\boldsymbol{p}_\tau^*(\cdot)$. In terms of components, (8.22) and (8.23) have the respective forms

$$\frac{de_\tau^*(X)}{d\tau} = -p_\tau^* \frac{dv_\tau^*(X)}{d\tau} + \theta_\tau^* \frac{dn_\tau^*(X)}{d\tau} \,, \tag{8.24}$$

and

$$\frac{1}{\theta_\tau^*}\left\{\frac{dp_\tau^*}{d\tau}\frac{dv_\tau^*(X)}{d\tau} - \frac{d\theta_\tau^*}{d\tau}\frac{dn_\tau^*(X)}{d\tau}\right\} \leqslant 0 \,. \tag{8.25}$$

The first of these represents a form of the classical (local) *Gibbs equation* for sequences of normal Gibbsian states, and because of its structure may be interpreted as a sort of sequential local balance of energy at points $X \in B$ where normal Gibbsian states are continuous. This interpretation carries with it the idea of, and, in fact, suggests that one identify the last term on the right with the primative notion of *sequential (local) thermostatic heat.*

The inequality (8.25) has the following partial interpretations which are essentially independent observations of Van't Hoff and Le Chatelier in 1884: *For positive thermostatic temperature,* $\theta_\tau^* > 0$, *we have* (i) *in a constant pressure sequence of normal Gibbsian states which are continuous at* $X \in B$, *if the thermostatic temperature* $\left\{\begin{smallmatrix}\text{increases}\\ \text{decreases}\end{smallmatrix}\right\}$ *then the local thermostatic entropy* $n_\tau^*(X)$ $\left\{\begin{smallmatrix}\text{cannot decrease}\\ \text{cannot increase}\end{smallmatrix}\right\}$, *and* (ii) *in a constant temperature sequence of normal Gibbsian states which are continuous at* $X \in B$, *if the thermostatic pressure* $\left\{\begin{smallmatrix}\text{increases}\\ \text{decreases}\end{smallmatrix}\right\}$ *then the local thermostatic specific volume* $v_\tau^*(X)$ $\left\{\begin{smallmatrix}\text{cannot increase}\\ \text{cannot decrease}\end{smallmatrix}\right\}$.

If the inequalities (8.21) are integrated over the body B, then the same arguments as given above may be applied to the integrated forms. In place of (8.24) and (8.25) we get the following *global* comparison principles:

$$\frac{dE_\tau^*}{d\tau} = -p_\tau^* \frac{dV_\tau^*}{d\tau} + \theta_\tau^* \frac{dH_\tau^*}{d\tau} \,,$$

$$\frac{1}{\theta_\tau^*}\left\{\frac{dp_\tau^*}{d\tau}\frac{dV_\tau^*}{d\tau} - \frac{d\theta_\tau^*}{d\tau}\frac{dH_\tau^*}{d\tau}\right\} \leqslant 0 \,,$$

where H_τ^* represents the global thermostatic entropy,

$$H_\tau^* \equiv \int_B n_\tau^*(X)\, dm \,.$$

We now turn to the question of comparison principles at points $X_0 \in B$ on surfaces across which the sequence of normal Gibbsian states $\boldsymbol{p}_\tau^*(\cdot)$ is not continuous. Suppose, first, that X and Y represent two points of continuity of $p_\tau^*(\cdot)$ in B. Then (8.13) yields

$$[p_\tau^*(X)-p_\tau^*(Y)]\cdot g_\tau^*=0\ ,$$

and because of (8.22) we also have

$$[p_\tau^*(X)-p_\tau^*(Y)]\cdot\frac{dg_\tau^*}{d\tau}=0\ .$$

Now, by letting X and Y limit to the point X_0 from either side we see that

$$[\![p_\tau^*(X_0)]\!]\cdot g_\tau^*=0\ , \tag{8.26}$$

and

$$[\![p_\tau^*(X_0)]\!]\cdot\frac{dg_\tau^*}{d\tau}=0\ , \tag{8.27}$$

where $[\![\cdot]\!]$ denotes the jump computed in either order. In terms of components, (8.26) and (8.27) imply the following results:

$$[\![e_\tau^*(X_0)]\!]=-p_\tau^*[\![v_\tau^*(X_0)]\!]+\theta_\tau^*[\![n_\tau^*(X_0)]\!]\ , \tag{8.28}$$

and

$$[\![v_\tau^*(X_0)]\!]\frac{dp_\tau^*}{d\tau}=[\![n_\tau^*(X_0)]\!]\frac{d\theta_\tau^*}{d\tau}\ . \tag{8.29}$$

The first of these needs no further explanation, having mentioned the condition of constant chemical potential earlier. The second genuine comparison principle represents a parametric version of the classical *Clausius-Clapeyron equation* which holds on surfaces of discontinuity in a sequence of normal Gibbsian states, i.e., at points X_0 on any singular surface in B which separates two coexistent phases and which would continue to do so in a sequence as the parameter is continuously changed. In more physical terms, it relates the jumps in the specific volume and specific entropy across a surface of phase transition which separates two coexistent phases to those mutual changes that could take place in the pressure and temperature of the body and yet have the body continue to support two coexistent phases.

8.4 Gibbsian States: Sufficient Conditions and Structure

In the following, it will be convenient to have in mind two geometric notions associated with the (smooth) static site manifold $\mathscr{s}$. We call $p_0\in\mathscr{s}$ a *thermostatic support point*[8] if

$$(p-p_0)\cdot g_0\geqslant 0\quad\forall\quad p\in\mathscr{s}\ ,$$

where g_0 is that normal vector to $\mathscr{s}$ at p_0 which satisfies $g_0\cdot a=-1$. As we saw in Theorem 1, a normal Gibbsian state can only take on values which correspond to

[8] See Fig. 8.1.

the thermostatic support points of $\mathscr{S}$. The set of all thermostatic support points which share the same normal vector $\boldsymbol{g}_0$ at $\boldsymbol{p}_0 \in \mathscr{S}$ must lie in a common support plane for $\mathscr{S}$, and we call this set the *planar slice*[9] *at* $\boldsymbol{p}_0 \in \mathscr{S}$, i.e., *for each thermostatic support point* $\boldsymbol{p}_0 \in \mathscr{S}$,

$$\mathcal{S}(\boldsymbol{p}_0) \equiv \{\boldsymbol{p} \in \mathscr{S} \,|\, (\boldsymbol{p} - \boldsymbol{p}_0) \cdot \boldsymbol{g}_0 = 0,\ \boldsymbol{g}_0 \cdot \boldsymbol{a} = -1\}$$

denotes the planar slice at $\boldsymbol{p}_0 \in \mathscr{S}$. Those points in $\mathcal{S}(\boldsymbol{p}_0)$ which also lie on the boundary of the convex hull of $\mathcal{S}(\boldsymbol{p}_0)$ are called *sharp*[9] thermostatic support points.

Now, the remark following the statement of Theorem 1 has the following interpretation: *a necessary condition for the existence of a normal Gibbsian state is*

$$(0, V^*, E^*)/m(B) \in A\,\mathscr{H}(\mathcal{S}(\boldsymbol{p}_0)) \,, \tag{8.30}$$

where $\mathscr{H}(\mathcal{S}(\boldsymbol{p}_0))$ *is the convex hull of the planar slice at some thermostatic support point* $\boldsymbol{p}_0 \in \mathscr{S}$, *and* $A\,\mathscr{H}$ *denotes the* A*-projection of this convex hull.* In the present section, we shall see that (8.30) is also sufficient for the existence of a normal Gibbsian state in $C_p(V^*, E^*)$. In addition, the specific structure of normal Gibbsian states will be determined and this structure will be related to the question of multiple coexistent phases and to the so-called Gibbs phase rule.

We now have

Theorem 2. *Suppose* (8.30) *holds for some thermostatic support point* $\boldsymbol{p}_0 \in \mathscr{S}$, *so that either*

i) $(0, V^*, E^*)/m(B) = A\boldsymbol{p}'$,

where $\boldsymbol{p}' \in \mathcal{S}(\boldsymbol{p}_0)$, *or*

ii) $(0, V^*, E^*)/m(B)$ *is* not *the* A*-projection of a thermostatic support point.*

In case (i) *if* $\boldsymbol{p}'$ *is sharp, then* $\boldsymbol{p}(X) = \boldsymbol{p}'\ \forall X \in B$ *is the unique normal Gibbsian state.*

In case (i) *if* $\boldsymbol{p}'$ *is not sharp, and in case* (ii), *the complete class* $\mathscr{C}$ *of normal Gibbsian states is given by*

$$\mathscr{C} \equiv \{\boldsymbol{p}(\cdot) \in C_p(V^*, E^*) \,|\, \boldsymbol{p}(X) \in \mathcal{S}(\boldsymbol{p}_0),\ \text{a.a. } X \in B\} \,.$$

This theorem not only affirms that (8.30) is sufficient for the existence of a normal Gibbsian state, but also it catagorizes the structural properties of any such state. If $(0, V^*, E^*)/m(B)$ is the A-projection of a sharp thermostatic support point, then *the* normal Gibbsian state is uniform throughout B, and B is said to support a single phase. If $(0, V^*, E^*)/m(B)$ is either the A-projection of a non-sharp thermostatic support point (so a fortiori (8.30) holds), or (8.30) holds and $(0, V^*, E^*)/m(B)$ is not the A-projection of any thermostatic support point, then in the proof of this theorem we shall show that, in particular, there exists a normal Gibbsian state having the structure of a step function. In this case B is

---

9 See Fig. 8.1.

said to support a finite number of distinct phases, that number equalling the number of different values in the range of the step function. There is, in general, no claim of uniqueness in this situation and, therefore, one should expect not only a multiplicity of normal Gibbsian states that are step functions, but also the possibility for smoothly varying fields that are in the class $\mathscr{C}$. The latter possibility, however, will require that the planar slice $\mathcal{S}(p_0)$ not be discrete and, in fact, contain sets of points of non-zero length or area measure. This, of course, can happen only if the static site manifold $\mathscr{s}$ has either "flat spots" or "flat arcs".

We turn now to an outline of a

Proof of Theorem 2. Suppose, first, that case (i) holds. Then, from the definition of $\mathcal{S}(p_0)$ we may write

$$(p-p')\cdot g_0 \geqslant 0 \quad \forall \quad p \in \mathscr{s} .$$

Because of the identity

$$p-p' = A(p-p') + a[a\cdot(p-p')] ,$$

and $g_0 \cdot a = -1$, we then have

$$g_0\cdot A(p-p') - a\cdot(p-p') \geqslant 0 \quad \forall \quad p \in \mathscr{s} . \tag{8.31}$$

If we now replace p by any $p(\cdot) \in C_{\mathrm{p}}(V^*, E^*)$, integrate over B, and use the hypothesis of case (i) and the properties of $C_{\mathrm{p}}(V^*, E^*)$, it readily follows that

$$a\cdot m(B)p' \geqslant a \cdot \int_B p\, dm$$

for all $p(\cdot) \in C_{\mathrm{p}}(V^*, E^*)$, which shows that the constant function with value p' is a normal [10] Gibbsian state. Now, tracing through the last three inequalities we see that equality holds in the latter if and only if every point of continuity of $p(\cdot)$ $C_{\mathrm{p}}(V^*, E^*)$ is associated with a member of the planar slice $\mathcal{S}(p_0)$. Thus, any other Gibbsian state $p(\cdot) \in C_{\mathrm{p}}(V^*, E^*)$ must satisfy not only this restriction on its range, so that it must be normal and contained in the class $\mathscr{C}$, but also it must meet

$$\int_B p\, dm = m(B)p' .$$

However, in case (i) if $p' \in \mathcal{S}(p_0)$ is sharp (and therefore on the boundary of the convex hull of $\mathcal{S}(p_0)$), an integral mean value theorem for vector valued functions then readily yields [11] $p(X) = p'$ for all $X \in B$, and we reach the uniqueness claim of the first part of the theorem.

On the other hand, in case (i) if $p' \in \mathcal{S}(p_0)$ is not sharp, the uniqueness argument fails and, as we saw above, any member of the class $\mathscr{C}$ will qualify as a normal Gibbsian state. In this case, it is a fortiori clear that the condition (8.30) holds, and whenever this condition does hold it is an elementary result of convex analysis that $(0, V^*, E^*)/m(B)$ may be written as the A-projection of a convex combination of at least two points in $\mathcal{S}(p_0)$; i.e.,

---

[10] Normal, because $p' \in \mathcal{S}(p_0)$.

[11] See *Dunn* and *Fosdick* [8.2], p. 26.

$$(0, V^*, E^*)/m(B) = A \sum_{i=1}^{N} \alpha_i p_i \,, \tag{8.32}$$

where

$$\sum_{i=1}^{N} \alpha_i = 1 \,, \qquad \alpha_i \in (0,1) \,, \qquad p_i \in \mathcal{S}(p_0) \,, \qquad \text{and} \qquad N \geqslant 2 \,.$$

If we now sub-divide the body B into N sub-parts P_i such that $m(P_i) = \alpha_i m(B)$, and introduce the step function $p_s(\cdot)$: $B \to \mathcal{J}$ such that $p_s(X) = p_i$ for $X \in P_i$, then

$$\int_B p_s \, dm = \sum_{i=1}^{N} m(P_i) p_i = m(B) \sum_{i=1}^{N} \alpha_i p_i \,,$$

so that, in fact, $p_s(\cdot) \in \mathscr{C}$. Hence, this step function serves as a normal Gibbsian state in case (i) if $p' \in \mathcal{S}(p_0)$ is not sharp.

In the remaining case (ii) of this theorem, the hypothesis (8.30) allows us to again write (8.32). Thus,

$$(p - p_i) \cdot g_0 \geqslant 0 \quad \forall \quad p \in \mathcal{J}$$

for $i = 1, 2, \ldots, N \geqslant 2$, where equality holds if and only if $p \in \mathcal{S}(p_0)$. Then, analogous to (8.31) we have

$$g_0 \cdot A\,(p - p_i) - a \cdot (p - p_i) \geqslant 0 \quad \forall \quad p \in \mathcal{J} \,,$$

and if we multiply this by α_i, sum over the index i, replace p by any $p(\cdot) \in C_p(V^*, E^*)$, integrate over B, and use (8.32) and the properties of $C_p(V^*, E^*)$, we readily obtain

$$a \cdot m(B) \sum_{i=1}^{N} \alpha_i p_i \geqslant a \cdot \int_B p \, dm$$

for all $p(\cdot) \in C_p(V^*, E^*)$, where equality holds if and only if every point of continuity of $p(\cdot)$ is associated with some member of $\mathcal{S}(p_0)$. This shows that in case (ii) the class $\mathscr{C}$ is, again, the complete class of normal Gibbsian states. The step function introduced earlier applies to this situation and represents just one example of such a state.

This theorem shows that if $(0, V^*, E^*)/m(B)$ is in the A-projection of the convex hull of a collection of thermostatic support points for $\mathcal{J}$ all of which share a common support plane, then normal Gibbsian states exist in $C_p(V^*, E^*)$; indeed, there are at least as many such states as there are distinct collections of such thermostatic support points. Thus, in general any kind of uniqueness of a normal Gibbsian state is the exception rather than the rule, There is, however, a kind of uniqueness that is closely related to the classic Gibbs phase rule of thermostatics for a body which is composed of a single substance. To describe this situation we shall first say that two step functions $p_s(\cdot), p_s'(\cdot)$: $B \to \mathcal{J}$ are *rearrangements* of one another if (i) the two sets of points of continuity of each of these functions in B cover the same range, and (ii) the subsets of B which correspond to equal values of these step functions have equal masses. Then, since any rearrangement of a normal Gibbsian step function is a normal Gibbsian step function, we shall say that a normal Gibbsian step function $p_s(\cdot) \in C_p(V^*, E^*)$ is *unique to within* (or, *up to*) *a rearrangement* whenever the only other normal Gibbsian state in $C_p(V^*, E^*)$ is a rearrangement of $p_s(\cdot)$. It is now straightforward to verify[12] the following

Remark. *Suppose* (8.30) *holds for some thermostatic support point* $\mathbf{p}_0 \in \mathscr{S}$. *Then, there exists a normal Gibbsian state in* $C_p(V^*, E^*)$. *Moreover, such a state is unique up to a rearrangement and is a step function with no more than* 3 *values, if and only if the planar slice* $\mathscr{S}(\mathbf{p}_0)$ *contains* 3 *non-colinear, or less than* 3 *points.*

In more physical than precise terms, we may now conclude that for a given body if a system of coexistent phases which maximize the total entropy at fixed total volume and energy is unique up to a rearrangement then the system can contain at most 3 distinct phases. Without some qualification as for example the "uniqueness up to a rearrangement", the Gibbs phase rule remains simply a rule and not a theorem within Gibbsian thermostatics. While *Gibbs* [8.1], pp. 96–97 was clear on this point, such is unfortunately not the case in other more modern elementary books on thermodynamics.[13]

8.5 Equivalent Problems of Thermostatics

In addition to the primary Gibbs problem of thermostatics outlined in (8.10), there are several other auxillary problems of a similar kind all of which usually are advertised as being equivalent. The conditions under which this equivalence is to hold often are not considered and one wonders about this whole question: Just what can be shown to be equivalent and how should "equivalent" be interpreted?

The first theorem of this section is elementary and while it does not claim any equivalence it does show that normal Gibbsian states of positive temperature also solve certain other extremum problems of thermostatics. The question of equivalence will be considered subsequently. First, we have

Theorem 3. *Let* $\mathbf{p}^*(\cdot) = (n^*, v^*, e^*)(\cdot) \in C_p(V^*, E^*)$ *be a normal Gibbsian state with thermostatic pressure* p^*, *positive thermostatic temperature* $\theta^* > 0$, *and entropy* $H^* \equiv \int_B n^* dm$. *Then,* $\mathbf{p}^*(\cdot)$ *also solves the following problems:*

i) *minimize internal energy at fixed volume and entropy, i.e.,*

$$\text{minimize} \int_B e\, dm$$

among all $\mathbf{p}(\cdot) = (n, v, e)(\cdot) \in C_p$ *with*

$$\int_B n\, dm = H^*\,, \qquad \int_B v\, dm = V^*\,,$$

[12] See *Dunn* and *Fosdick* [8.2], pp. 35–37.

[13] *Dunn* and *Fosdick* [8.2], pp. 81–88, have discussed this point at some length and have indicated where some of the usual "proofs" of the phase rule go awry. See, also, the more recent work of *Man* [8.4].

ii) *minimize Gibbs semi-free energy, i.e.,*

$$\text{minimize} \int_B (e - \theta^* n + p^* v)\, dm$$

among all $\boldsymbol{p}(\cdot) = (n, v, e)(\cdot) \in C_p$,

iii) *minimize Helmholtz semi-free energy at fixed volume, i.e.,*

$$\text{minimize} \int_B (e - \theta^* n)\, dm$$

among all $\boldsymbol{p}(\cdot) = (n, v, e)(\cdot) \in C_p$ *with*

$$\int_B v\, dm = V^*,$$

iv) *minimize semi-enthalpy at fixed entropy, i.e.,*

$$\text{minimize} \int_B (e + p^* v)\, dm$$

among all $\boldsymbol{p}(\cdot) = (n, v, e)(\cdot) \in C_p$ *with*

$$\int_B n\, dm = H^*.$$

Proof. First, if we replace $\boldsymbol{p}$ in (8.12) by any $\boldsymbol{p}(\cdot) \equiv (n, v, e)(\cdot) \in C_p$, multiply the inequality by θ^*, and integrate over B, it follows that

$$\int_B (e^* - \theta^* n^* + p^* v^*)\, dm \leqslant \int_B (e - \theta^* n + p^* v)\, dm$$

for all $\boldsymbol{p}(\cdot) = (n, v, e)(\cdot) \in C_p$. Clearly, then, the normal Gibbsian state $\boldsymbol{p}^*(\cdot) = (n^*, v^*, e^*)(\cdot)$ referred to in this theorem also solves problems (i), (ii), (iii) and (iv).

In this theorem, the potentials of problems (ii), (iii), and (iv) are not the Gibbs free energy, Helmholtz free energy and enthalpy, respectively. For example, in problem (iii) if we were to restrict the class of competitors further so that the range of all such $\boldsymbol{p}(\cdot)$ contained only those points on the static site manifold $\mathscr{S}$ which have the thermostatic temperature θ^*, and if, in addition, it were then possible to write the integrand of the potential as a single-valued function of specific volume on this iso-therm, we would have the problem of minimum Helmholtz free energy at fixed temperature and volume.

Before leaving this theorem it is perhaps worth noting that *while there may be many solutions to each of the problems* (i) – (iv), *if* $\boldsymbol{p}^\dagger(\cdot) = (n^\dagger, v^\dagger, e^\dagger)(\cdot) \in C_p$ *represents any one of them, then*

$$[\boldsymbol{p} - \boldsymbol{p}^\dagger(X)] \cdot \boldsymbol{g}^* \geqslant 0 \tag{8.33}$$

for all $\boldsymbol{p} \in \mathscr{S}$ *and all* $X \in B$ *at which* $\boldsymbol{p}^\dagger(\cdot)$ *is continuous.* Here, $\boldsymbol{g}^* = (-1, p^*/\theta^*, 1/\theta^*)$, where p^* and $\theta^* > 0$ denote the thermostatic pressure and temperature of a normal Gibbsian state in $C_p(V^*, E^*)$. Thus, whenever such a normal Gibbsian state exists, the range of all possible solutions to problems (i) – (iv), as well as all other normal Gibbsian states in $C_p(V^*, E^*)$, is the same. This result is based upon the assumption that a normal Gibbsian state exists. Roughly, a proof of

(8.33) is readily constructed by assuming that its converse holds at a point of continuity of $\boldsymbol{p}^{\dagger}(\cdot)$, replacing $\boldsymbol{p}$ in this converse inequality by the field $\boldsymbol{p}^*(\cdot)$ of the assumed normal Gibbsian state in $C_{\mathrm{p}}(V^*, E^*)$, and integrating over B. After multiplying by the constant $\theta^* > 0$, the inequality

$$\int_B (e^{\dagger} - \theta^* n^{\dagger} + p^* v^{\dagger})\, dm > \int_B (e^* - \theta^* n^* + p^* v^*)\, dm$$

emerges, which contradicts the hypothesis that $\boldsymbol{p}^{\dagger}(\cdot)$ solves any one of the problems (i) – (iv).

We wish to now consider the more difficult question of when solutions to problems (i) – (iv), considered as primative problems, quality as Gibbsian states. The results of this investigation when coupled with Theorem 3 then provide a framework for the interpretation of "equivalence" in these five problems of thermostatics. By analogy to our earlier definition of Gibbsian states, all of these problems may be cast in the following common and convenient form: *Given a projection tensor* $\boldsymbol{U}$, *a vector* $\boldsymbol{u} \in \mathbb{R}^3$ *in the null space of* $\boldsymbol{U}$, *and a vector* $\boldsymbol{P}^* \in \mathbb{R}^3$ *orthogonal to* $\boldsymbol{u}$,

$$\text{maximize}\, \boldsymbol{u} \cdot \int_B \boldsymbol{p}\, dm \tag{8.34a}$$

among all $\boldsymbol{p}(\cdot) \in C_{\mathrm{p}}(\boldsymbol{P}^*)$, *where*

$$C_{\mathrm{p}}(\boldsymbol{P}^*) \equiv \{\boldsymbol{p}(\cdot) = (n, v, e)(\cdot) \in C_{\mathrm{p}} \,|\, \boldsymbol{U} \int_B \boldsymbol{p}\, dm = \boldsymbol{P}^*\}\,.$$

In the primary Gibbs problem outlined in (8.10) and studied in Sects. 8.3 and 4, we have[14] $\boldsymbol{U} = \boldsymbol{A}$, $\boldsymbol{u} = \boldsymbol{a}$, and $\boldsymbol{P}^* = (0, V^*, E^*)$. The four problems introduced in Theorem 3 may be recovered by setting, respectively,

$$\begin{array}{llll}
\text{i)} & \boldsymbol{U} = \boldsymbol{a} \otimes \boldsymbol{a} + \boldsymbol{b} \otimes \boldsymbol{b}\,, & \boldsymbol{u} = -\boldsymbol{c}\,, & \boldsymbol{P}^* = (H^*, V^*, 0)\,, \\
\text{ii)} & \boldsymbol{U} = \boldsymbol{O}\,, & \boldsymbol{u} = (\theta^*, -p^*, -1)\,, & \boldsymbol{P}^* = (0,0,0)\,, \\
\text{iii)} & \boldsymbol{U} = \boldsymbol{b} \otimes \boldsymbol{b}\,, & \boldsymbol{u} = (\theta^*, 0, -1)\,, & \boldsymbol{P}^* = (0, V^*, 0)\,, \\
\text{iv)} & \boldsymbol{U} = \boldsymbol{a} \otimes \boldsymbol{a}\,, & \boldsymbol{u} = (0, -p^*, -1)\,, & \boldsymbol{P}^* = (H^*, 0, 0)\,.
\end{array} \tag{8.34b}$$

Now, since the problem (8.34) is formally identical with the primary Gibbs problem which we associated with the definition of Gibbsian states, its solution may be considered analogously. The main delicate issue is that earlier we introduced and studied *normal* Gibbsian states, and to be strictly analogous we must now identify the counterpart of "normal" here. In this respect it is helpful to recall that the range of a *normal* Gibbsian state is that set of points on $\mathscr{s}$ whose normal vector has a non-zero component in the $\boldsymbol{a}$-direction. While, physically speaking, this consists of all those points on $\mathscr{s}$ of non-zero thermostatic temperature, more pertinate to the mathematical questions which arose in the construction of, say, "normal solutions" to the constrained maximization problem (i.e., *normal* Gibbsian states) was the condition that at points on $\mathscr{s}$ which associated with a solution, the normal vector to $\mathscr{s}$ must have a non-zero component in the null space of the projection tensor $\boldsymbol{A}$ (i.e., non-zero $\boldsymbol{a}$-component) which essen-

[14] Recall the formal definition of a Gibbsian state in Sect. 8.3.

tially identified the constraints of the problem. *In the four problems of* (8.34), to which we now turn, *we shall seek* normal solutions, *i.e., solutions whose range on the static site manifold* $\mathscr{s}$ *consists of those points whose associated normal vector has a non-zero component in the null space of the projection tensor* $\boldsymbol{U}$.

It is now possible to obtain the following analog of Theorem 1:

Theorem 4. *Let* $\boldsymbol{p}^\dagger(\cdot) = (n^\dagger, v^\dagger, e^\dagger)(\cdot) \in C_\mathrm{p}(\boldsymbol{P}^*)$ *be a normal solution to problem* (8.34) *under any one of the conditions* (i) – (iv). *Then, in each case there exists a unique constant vector* $\boldsymbol{g}^\dagger = (-\theta^*, p^*, 1)$ *such that*

$$[\boldsymbol{p} - \boldsymbol{p}^\dagger(X)] \cdot \boldsymbol{g}^\dagger \geqslant 0 \tag{8.35}$$

for all $\boldsymbol{p} \in \mathscr{s}$ *and all* $X \in B$ *where* $\boldsymbol{p}^\dagger(\cdot)$ *is continuous. A fortiori, if* $\theta^* > 0$ *then* $\boldsymbol{p}^\dagger(\cdot)$ *is a normal Gibbsian state in* $C_\mathrm{p}(A \int_B \boldsymbol{p}^\dagger dm)$.

Clearly, in each case, the condition (8.35) is *sufficient* for the existence of a solution to the corresponding version of problem (8.34). This theorem settles completely the relation between normal Gibbsian states and normal solutions of the four problems in (8.34). Because the static site manifold is smooth, we see from this theorem and Theorem 1 that if $\boldsymbol{p}(\cdot) \in C_\mathrm{p}$ is any one of these five types of states, then there exists a constant vector $\boldsymbol{g} = (-1, p^*/\theta^*, 1/\theta^*)$, $\theta^* \neq 0$, in the case of normal Gibbsian states, and $\boldsymbol{g} = (-\theta^*, p^*, 1)$ otherwise, such that

$$[\boldsymbol{p} - \boldsymbol{p}(X)] \cdot \boldsymbol{g} \geqslant 0$$

for all $\boldsymbol{p} \in \mathscr{s}$ and all $X \in B$ where $\boldsymbol{p}(\cdot)$ is continuous; and, therefore, $\boldsymbol{p}(\cdot)$ is also one of the remaining four types of states *if* the thermostatic temperature θ^* associated with $\boldsymbol{p}(\cdot)$ is positive.

We emphasize that the equivalence established in Theorems 3 and 4 require neither the existence nor the positiveness of a thermostatic temperature at every point of the static site manifold $\mathscr{s}$. Further, in the last three problems of (8.34) where the numbers θ^* and/or p^* are prescribed, it is *not* required that the elements of the respective classes of comparison states (i.e., the members of $C_\mathrm{p}(\boldsymbol{P}^*)$) have these numbers as their respective temperature and pressure. In fact, it is typical that both the thermostatic temperature and pressure associated with any $\boldsymbol{p}(\cdot) \in C_\mathrm{p}(\boldsymbol{P}^*)$ are general fields over the body B. Thus, for example, Theorem 4 as applied to problem (iii) claims, roughly, that among all states $(n, v, e)(\cdot)$ (of any temperature field and pressure field on B) with a fixed total volume, any one which minimizes the Helmholtz semi-free energy $\int_B (e - \theta^* n)\, dm$ has, essentially, not only a uniform thermostatic pressure but also a constant thermostatic temperature equal to θ^*.

We turn now to a

Proof of Theorem 4. First, we note that Theorem 4 for problem (i) is strictly analogous to Theorem 1 to within a trivial reblabeling and new identification of the orthonormal triad $(\boldsymbol{a}, \boldsymbol{b}, \boldsymbol{c})$. Normal Gibbsian states and normal solutions to problem (8.34) under condition (i) are identical notions under this relabeling process, and thus the proof of Theorem 1 applies here as well.

Problem (ii) of this theorem may be handeled with similar ease. Here, since the appropriate maximization problem (8.34) has no constraints, a drastically simplified version of the proof of Theorem 1 holds. Summarizing, in particular, we may start as before with $X_0 \in B$ denoting a point where $\boldsymbol{p}^\dagger(\cdot)$ is continuous and observe, next, that (8.18) holds with superscript * replaced by †. Then, we may turn directly to the construction of a test function $\boldsymbol{p}^\delta(\cdot) \in C_\mathrm{p}$ similar to (8.20) but now of the form

$$\boldsymbol{p}^\delta(X) \equiv \begin{cases} \boldsymbol{p}^\dagger(X)\,, & X \in B - B_\delta\,, \\ \boldsymbol{p}_1 \quad , & X \in P_\delta\,, \\ \boldsymbol{p}_2 \quad , & X \in B_\delta - P_\delta\,, \end{cases}$$

where $\boldsymbol{p}_1$ and $\boldsymbol{p}_2$ are arbitrary points on $\mathscr{s}$, and where B_δ and P_δ have the meanings as before. Then, by integration, use of the hypothesis that $\boldsymbol{p}^\dagger(\cdot)$ is a maximizer in the form

$$\boldsymbol{u} \cdot \int_B \boldsymbol{p}^\dagger dm \geqslant \boldsymbol{u} \cdot \int_B \boldsymbol{p}^\delta dm\,,$$

and by a limiting argument as $\delta \to 0$, we reach

$$\boldsymbol{u} \cdot \boldsymbol{p}_0^\dagger \geqslant \boldsymbol{u} \cdot [\alpha \boldsymbol{p}_1 + (1-\alpha)\boldsymbol{p}_2]\,,$$

where $\boldsymbol{p}_0^\dagger \equiv \boldsymbol{p}_0^\dagger(X_0)$. Thus, upon letting $\alpha \to 1$ and setting $\boldsymbol{g}^\dagger \equiv -\boldsymbol{u} = (-\theta^*, p^*, 1)$, from (ii) of (8.34b) we arrive at (8.35) for problem (8.34) under condition (ii).

We now turn to the proof of this theorem for problem (8.34) under the condition (iii). Since condition (iv) may be considered similarly, its formal proof will not be included here. As in Theorem 1, we let $X_0 \in B$ be a point where $\boldsymbol{p}^\dagger(\cdot)$ is continuous, set $\boldsymbol{p}_0^\dagger \equiv \boldsymbol{p}^\dagger(X_0)$, and let $\boldsymbol{p} = (n, v, e)$ be an arbitrary point on the static site manifold $\mathscr{s}$. Then, as in (8.14), and since $\boldsymbol{p}^\dagger(\cdot)$ is a *normal solution* to problem (8.34) under condition (iii), we see that for all sufficiently small $\alpha > 0$ there exists a smooth curve of points $\boldsymbol{p}(\alpha) \in \mathscr{s}$ such that

$$\frac{U(\boldsymbol{p}_0^\dagger - \alpha \boldsymbol{p})}{1-\alpha} = U\boldsymbol{p}(\alpha)\,, \tag{8.36}$$

and $\boldsymbol{p}(0) = \boldsymbol{p}_0^\dagger$. That is, since, under condition (iii), a normal vector to $\mathscr{s}$ at $\boldsymbol{p}_0^\dagger$ does not have only a $\boldsymbol{b}$-component, and $\mathscr{s}$ is smooth, then this property of a normal not aligning with the $\boldsymbol{b}$-direction persists in a neighborhood of $\boldsymbol{p}_0^\dagger$ on $\mathscr{s}$. This means that for any given number $(v_0^\dagger - \alpha v)/(1-\alpha) \equiv v(\alpha)$ close to $v^\dagger$ ($\alpha > 0$ and sufficiently small), there exists numbers $n(\alpha)$ and $e(\alpha)$ so that $\boldsymbol{p}(\alpha) \equiv (n, v, e)(\alpha)$ is not only on $\mathscr{s}$ and close to $\boldsymbol{p}_0^\dagger$, but also is differentiable. We note, for later use, that

$$\alpha = \frac{v_0^\dagger - y}{v - y}$$

is the inverse of $v(\alpha) = y$, and that by introducing $\hat{\boldsymbol{p}}(y) \equiv \boldsymbol{p}(\alpha(y)) = [\hat{n}(y), y, \hat{e}(y)]$ we have

$$\begin{aligned} \boldsymbol{p}'(0) &= (v_0^\dagger - v)\hat{\boldsymbol{p}}(v_0^\dagger) \\ &= [\hat{n}'(v_0^\dagger), 1, \hat{e}'(v_0^\dagger)](v_0^\dagger - v)\,. \end{aligned} \tag{8.37}$$

Since $\alpha > 0$, it follows that $v < (>)\, v_0^\dagger$ implies $y > (<)\, v_0^\dagger$.

Now suppose, analogous to (8.15), that we have established the inequality

$$\boldsymbol{u}\cdot[\alpha\boldsymbol{p}+(1-\alpha)\boldsymbol{p}(\alpha)]\leqslant\boldsymbol{u}\cdot\boldsymbol{p}_0^\dagger \tag{8.38}$$

for sufficiently small $\alpha>0$, where $\boldsymbol{u}=(\theta^*,0,-1)$ as is noted in (8.34) under condition (iii). Then, as in the proof of Theorem 1 in obtaining (8.16), we see from (8.38) that

$$\theta^*(n_0^\dagger-n)-(e_0^\dagger-e)\geqslant(\theta^*\boldsymbol{a}-\boldsymbol{c})\cdot\boldsymbol{p}'(0)\ ,$$

and, with the aid of (8.37), that

$$\theta^*(n_0^\dagger-n)-(e_0^\dagger-e)\geqslant[\theta^*\hat{n}'(v_0^\dagger)-\hat{e}'(v_0^\dagger)](v_0^\dagger-v)\ .$$

Thus, there exists a number $p^*(X_0)$, say, depending on $\theta^*, \boldsymbol{p}_0^\dagger=\boldsymbol{p}^\dagger(X_0)$, and the structure of $\mathscr{S}$ at $\boldsymbol{p}_0^\dagger$, such that

$$[\boldsymbol{p}-\boldsymbol{p}^\dagger(X_0)]\cdot(-\theta^*,p^*(X_0),1)\geqslant 0 \tag{8.39}$$

for all $\boldsymbol{p}\in\mathscr{S}$. Clearly, from this inequality and the smoothness of $\mathscr{S}$, we may conclude that the vector $(-\theta^*,p^*(X_0),1)$ is proportional to a normal vector to $\mathscr{S}$ at $\boldsymbol{p}_0^\dagger$, and therefore that $p^*(X_0)$ is unique, given θ^*, $X_0\in B$, and $\boldsymbol{p}^\dagger(\cdot)$.

There are now two matters left to be established. First, we remark that the inequality (8.38) can be obtained by an argument similar to that given for (8.15) in Theorem 1. In that argument, which begins with (8.18), we need only replace the superscript * by †, $\boldsymbol{A}$ by $\boldsymbol{U}=\boldsymbol{b}\otimes\boldsymbol{b}$, and $\boldsymbol{a}$ by $\boldsymbol{u}=(\theta^*,0,-1)$. We shall not consider this issue any further here.

Finally, we must show that the number $p^*(X_0)$ in (8.39) is independent of the point $X_0\in B$ at which $\boldsymbol{p}^\dagger(\cdot)$ is continuous, and that, in fact, this number also is not dependent upon the particular normal solution $\boldsymbol{p}^\dagger(\cdot)\in C_{\mathrm{p}}(\boldsymbol{P}^*)$ that one assumes for problem (8.34) under condition (iii). To this end, suppose first that $\boldsymbol{p}^\dagger(\cdot)$ is not constant valued. Then, an integral mean value theorem for vector valued functions[15] asserts that there are $N>1$ numbers $\alpha_i\in(0,1)$ and points $X_i\in B$ at which $\boldsymbol{p}^\dagger(\cdot)$ is continuous such that

$$\sum_{i=1}^N\alpha_i=1\ ,\qquad \boldsymbol{p}^\dagger(X_i)\neq\frac{1}{m(B)}\int_B\boldsymbol{p}^\dagger dm\ ,$$

and

$$\sum_{i=1}^N\alpha_i\boldsymbol{p}^\dagger(X_i)=\frac{1}{m(B)}\int_B\boldsymbol{p}^\dagger dm\ . \tag{8.40}$$

Now, for an arbitrary collection of N vectors $\{\boldsymbol{t}_i\}$ it is clear that

$$\boldsymbol{U}[\boldsymbol{p}^\dagger(X_i)+\tau\boldsymbol{t}_i]\to\boldsymbol{U}\boldsymbol{p}^\dagger(X_i)$$

as $\tau\to 0$, and, since $\boldsymbol{p}^\dagger(\cdot)$ is a normal solution to problem (8.34) under condition

[15] See, e.g., *Dunn* and *Fosdick* [8.2], p. 90.

(iii), we see that for sufficiently small $|\tau|$ there exist N smooth curves of points $\boldsymbol{p}_i(\tau) \in \mathscr{s}$ such that

$$U[\boldsymbol{p}^\dagger(X_i) + \tau \boldsymbol{t}_i] = U\boldsymbol{p}_i(\tau) \,, \tag{8.41}$$

and $\boldsymbol{p}_i(0) = \boldsymbol{p}^\dagger(X_i)$. Thus, by differentiating (8.41) and recalling that $U = \boldsymbol{b} \otimes \boldsymbol{b}$ we get

$$\boldsymbol{b} \cdot \boldsymbol{t}_i = \boldsymbol{b} \cdot \boldsymbol{p}_i'(0) \,. \tag{8.42}$$

In addition, if we multiply (8.41) by α_i, sum over the index, and restrict the collection $\{\boldsymbol{t}_i\}$ so that

$$\sum_{i=1}^{N} \alpha_i \boldsymbol{b} \cdot \boldsymbol{t}_i = 0 \,, \tag{8.43}$$

we reach

$$U \sum \alpha_i \boldsymbol{p}_i(\tau) = U \sum \alpha_i \boldsymbol{p}^\dagger(X_i) \,.$$

It is now helpful to note that with the aid of (8.40), and the fact that $\boldsymbol{p}^\dagger(\cdot) \in C_{\mathrm{p}}(\boldsymbol{P}^*)$, we may write this as

$$U \int_B \boldsymbol{p}_\tau \, dm = U \int_B \boldsymbol{p}^\dagger dm = \boldsymbol{p}^*$$

where, for each τ with $|\tau|$ sufficiently small, the function $\boldsymbol{p}_\tau(\cdot): B \to \mathscr{s}$ is a step function taking values $\boldsymbol{p}_i(\tau)$ on N regular sub-parts of B each having mass measure $\alpha_i m(B)$. Thus, $\boldsymbol{p}_\tau(\cdot) \in C_{\mathrm{p}}(\boldsymbol{P}^*)$ and consequently we have

$$\boldsymbol{u} \cdot \int_B \boldsymbol{p}^\dagger dm \geqslant \boldsymbol{u} \cdot \int_B \boldsymbol{p}_\tau dm = \boldsymbol{u} \cdot m(B) \sum_{i=1}^{N} \alpha_i \boldsymbol{p}_i(\tau) \,.$$

Now, since this inequality becomes an equality and the right hand side is maximized at $\tau = 0$ because of (8.40) and the fact that $\boldsymbol{p}_i(0) = \boldsymbol{p}^\dagger(X_i)$, it follows that

$$\boldsymbol{u} \cdot \sum_{i=1}^{N} \alpha_i \boldsymbol{p}_i'(0) = 0 \,. \tag{8.44}$$

On the other hand, each vector $\boldsymbol{p}_i'(0)$ is tangent to $\mathscr{s}$ at $\boldsymbol{p}^\dagger(X_i)$ and by repeatedly identifying X_0 with X_i, $i = 1, 2, \ldots, N$ in (8.39) we see that the vector $(-\theta^*, p^*(X_i), 1) = -\boldsymbol{u} + p^*(X_i)\boldsymbol{b}$ is normal to $\mathscr{s}$ at $\boldsymbol{p}^\dagger(X_i)$ so that

$$[-\boldsymbol{u} + p^*(X_i)\boldsymbol{b}] \cdot \boldsymbol{p}'(0) = 0 \,.$$

Thus, using (8.42) we see that

$$\boldsymbol{u} \cdot \boldsymbol{p}'(0) = p^*(X_i)\boldsymbol{b} \cdot \boldsymbol{t}_i \,,$$

and by multiplying by α_i and summing over the index we find that (8.44) yields

$$\sum_{i=1}^{N} p^*(X_i)\, \alpha_i \boldsymbol{b} \cdot \boldsymbol{t}_i = 0 \,.$$

Since this must hold for all collections $\{\boldsymbol{t}_i\}$ which satisfy (8.43), it readily follows that $p^*(X_i) \equiv p^*$ is the same number for each X_i, $i = 1, 2, \ldots, N$, and, based upon (8.39), we may write

$$[\boldsymbol{p}-\boldsymbol{p}^{\dagger}(X_i)]\cdot(-\theta^*,p^*,1)\geqslant 0 \tag{8.45}$$

for all $\boldsymbol{p}\in\mathscr{I}$ and for each of the $N>1$ points $X_i\in B$. Thus, the N points $\boldsymbol{p}^{\dagger}(X_i)\in\mathscr{I}$ are support points of $\mathscr{I}$ which share a common support plane. In addition, equality holds in (8.45) if and only if $\boldsymbol{p}\in\mathscr{I}$ lies in this common support plane.

If we now multiply (8.45) by α_i, sum on the index, and use (8.40), it follows that

$$[m(B)\boldsymbol{p}(X)-\int_B \boldsymbol{p}^{\dagger}dm]\cdot(-\theta^*,p^*,1)\geqslant 0\ , \tag{8.46}$$

where X is an arbitrary point in B and $\boldsymbol{p}(\cdot)\in C_{\mathrm{p}}$. Equality holds here only if $\boldsymbol{p}(X)$ lies in the common support plane determined in (8.45). Integrating (8.46) and noting that $(-\theta^*,p^*,1)=-\boldsymbol{u}+p^*\boldsymbol{b}$, we reach

$$\boldsymbol{u}\cdot\int_B \boldsymbol{p}^{\dagger}dm\geqslant\boldsymbol{u}\cdot\int_B \boldsymbol{p}\,dm$$

for all $\boldsymbol{p}(\cdot)\in C_{\mathrm{p}}(\boldsymbol{P}^*)$, where equality holds only if when $X\in B$ is a point of continuity of $\boldsymbol{p}(\cdot)$, $\boldsymbol{p}(X)$ lies in the common support plane determined in (8.45). *A fortiori*, the range of $\boldsymbol{p}^{\dagger}(\cdot)$ at all its points of continuity must lie in this common support plane, and, thus, by returning to (8.39) we see that $p^*(X_0)$ must be independent of $X_0\in B$ as long as X_0 is a point of continuity of $\boldsymbol{p}^{\dagger}(\cdot)$. In addition, since any other normal solution of problem (8.34) under condition (iii) must, at its points of continuity in B, have values which also lie in the common support plane introduced above, we conclude that the constant p^* is common to all normal solutions in $C_{\mathrm{p}}(\boldsymbol{P}^*)$.

Finally, if the normal solution $\boldsymbol{p}^{\dagger}(\cdot)$ is constant valued, much of the argument given above may be by-passed and we need only start with (8.39) and directly apply the reasoning from (8.46) onward.

8.6 Dynamical Stability

We return, in this final section, to the motivation, based on dynamics, given in Sect. 8.2 for the study of Gibbsian states, and show more specifically how it is related to the question of stability of these static states. Our aim is to present and discuss some theorems in this regard relative to neo-classical and isolated processes. For the most part, we shall not give proofs here, but, rather, refer to the original work of *Dunn* and *Fosdick* [8.2] for some of the detailed arguments. For convenience, we shall assume here that the points $\boldsymbol{p}=(n,v,e)$ of the static site manifold $\mathscr{I}$ lie on a connected surface which is expressible globally as a smooth function

$$e=\bar{e}(n,v)\ . \tag{8.47}$$

In general, the state vector field at time t, $s_t(\cdot)=(\eta,\nu,\varepsilon)(\cdot,t)\colon B\to\mathbb{R}^3$, of a process need not associate with points on this surface. However, throughout Sect. 8.6 we shall employ the following

Assumption[16]. *$\bar{e}(\cdot,\cdot)$ is a sub-potential for all neo-classical processes in the sense that the sub-set*

$$\mathscr{D} \equiv \{s = (\eta, v, \varepsilon) \mid \varepsilon \geqslant \bar{e}(\eta, v)\} \subset \mathbb{R}^3 \tag{8.48}$$

is the only part of $\mathbb{R}^3$ that is accessible to the states of any such process.

Consider, now, the set $\mathbb{P}(\theta^*, p^*)$ of all neo-classical processes with ambient temperature $\theta^* > 0$ and ambient pressure p^*, and suppose that $\boldsymbol{g}^* \equiv (-1, p^*/\theta^*, 1/\theta^*)$ determines a planar slice of thermostatic support points $\mathcal{S}^* \subset \mathscr{s}$ in the sense introduced in Sect. 8.4. Since, if $\boldsymbol{p} \in \mathcal{S}^*$, it follows from (8.48) that for any $\boldsymbol{s} \in \mathscr{D}$,

$$(\boldsymbol{s}-\boldsymbol{p}) \cdot \boldsymbol{g}^* \geqslant (\boldsymbol{r}(\boldsymbol{s})-\boldsymbol{p}) \cdot \boldsymbol{g}^* \geqslant 0 \,, \tag{8.49}$$

where

$$\boldsymbol{r}(\boldsymbol{s})\,|_{s=(\eta, v, \varepsilon)} \equiv (\eta, v, \bar{e}(\eta, v)) \in \mathscr{s} \,, \tag{8.50}$$

then, for any neo-classical process in $\mathbb{P}(\theta^*, P^*)$, (8.8) requires

$$d\Psi^*(t)/dt \leqslant 0 \,,$$

where

$$\Psi^*(t) \equiv \int_B [(\boldsymbol{s}_t-\boldsymbol{p}) \cdot \boldsymbol{g}^* + \tfrac{1}{2}\,|\dot{\boldsymbol{x}}\,|^2/\theta^*]\,dm \geqslant 0 \,. \tag{8.51}$$

This, with (8.49), in turn, implies the following mild conclusions regarding stability:

$$\begin{aligned} &\text{i)} \quad 0 \leqslant \int_B \tfrac{1}{2}\,|\dot{\boldsymbol{x}}\,|^2 dm \leqslant \theta^* \Psi^*(t) \,, \\ &\text{ii)} \quad 0 \leqslant \int_B (\boldsymbol{s}_t-\boldsymbol{p}) \cdot \boldsymbol{g}^* dm \leqslant \Psi^*(t) \,, \\ &\text{iii)} \quad 0 \leqslant \Psi^*(t) \leqslant \Psi^*(0) \,. \end{aligned} \tag{8.52}$$

Roughly, whenever B undergoes a neo-classical process in $\mathbb{P}(\theta^*, p^*)$ with small initial data (i.e., $\Psi^*(0)$ small), the kinetic energy of B will remain small for all time, and the state vector field at time t, $\boldsymbol{s}_t(\cdot)$, of the process will remain "close" to the plane of states that contains the planar slice $\mathcal{S}^*$. This latter remark is based upon the observation that the integrand in (8.52, ii) is non-negative and is proportional to the perpendicular distance between $\boldsymbol{s}_t$ and the plane containing $\mathcal{S}^*$.

It is possible to obtain a more explicit result on stability provided the static site manifold $\mathscr{s}$ has a "properly distributed" set of thermostatic support points, each having a positive thermostatic temperature, which "surround" the planar slice $\mathcal{S}^*$. In order to more clearly state the precise result, it is convenient to have available the set theoretic notions of a *star appropriately surrounding* $\mathcal{S}^*$ and a *radiant shell* of the star.

We say that $S \subset \mathscr{s}$ *is a star appropriately surrounding* $\mathcal{S}^* \subset \mathscr{s}$ *if S is compact and if there exists some subset $R_S \subset S$, called a radient shell surrounding $\mathcal{S}^*$, such*

[16] This assumption naturally is valid (with equality) for any classical, linearly viscous, compressible fluid. In addition, as remarked by *Coleman* and *Greenberg* [8.5] and *Coleman* [8.6], it also is valid within the theory of simple fluids with fading memory and for certain fluids with internal state variables. *Dunn* and *Fosdick* [8.7], and *Fosdick* and *Rajagopal* [8.8] have applied a version of this assumption to study the thermodynamics and questions of asymptotic stability of certain Rivlin-Ericksen fluids.

that (i) *no points of* $\mathcal{S}^*$ *are limit points of* R_S, *and* (ii) *for each* $\mathbf{p} \in \mathscr{s} - S$, *among all of the projected straight lines onto the* $n - v$ *plane between every* $\mathbf{p}^* \in \mathcal{S}^*$ *and* $\mathbf{p}$, *at least one of minimum length has a nonempty intersection with the projection of* R_S *onto the* $n - v$ *plane.*

It can be shown that $\mathcal{S}^* \subset S$. Also, as an example of a star S appropriately surrounding $\mathcal{S}^*$ and a radiant shell R_S of the star, one may associate S with a patch of $\mathscr{s}$ that contains all of the planar slice $\mathcal{S}^*$ in its interior, and take for R_S a closed continuous curve on $\mathscr{s}$ but in S such that the projection of $\mathcal{S}^*$ onto the $n - v$ plane is enclosed by the $n - v$ projection of the curve R_S. Other, more general, examples for which R_S is a disconnected set of arcs on $\mathscr{s}$ can be constructed.

We now have the following

Theorem 5. *Let* $\mathbf{g}^* = (-1, p^*/\theta^*, 1/\theta^*)$, $\theta^* > 0$, *determine a planar slice of thermostatic support points* $\mathcal{S}^* \subset \mathscr{s}$. *Suppose* $\mathbf{s}_t(\cdot)$: $B \to \mathscr{D}$ *is the state vector field at time t of a neo-classical process in* $\mathbb{P}(\theta^*, p^*)$. *Let* $\mathscr{s}$ *admit of a star appropriately surrounding* $\mathcal{S}^*$ *such that every point in one of its radiant shells about* $\mathcal{S}^*$ *is a thermostatic support point of* $\mathscr{s}$ *with a positive thermostatic temperature. Then, for every* $\delta > 0$ *there exists* $C(\delta) > 0$, *monotone increasing and unbounded as* $\delta \to 0$, *such that*

$$0 \leqslant \int_B \inf_{\mathbf{p} \in \mathcal{S}^*} |\mathbf{s}_t - \mathbf{p}| \, dm \leqslant m(B)\,\delta + C(\delta)\, \Psi^*(t) \tag{8.53}$$

for all $t \geqslant 0$.

For a proof of this theorem we refer to Theorem 15 and its Corollary in the work of *Dunn* and *Fosdick* [8.2]. The theorem, itself, asserts a type of stability of the *set* $\mathcal{S}^*$ with regard to processes in $\mathbb{P}(\theta^*, p^*)$ in the sense that the distance between $\mathcal{S}^*$ and $\mathbf{s}_t(X)$ is bounded above in mean by a non-increasing function of time. A result of this type was established in the works of *Coleman* and *Greenberg* [8.5] and *Coleman* [8.6] but only in the elementary case when the planar slice $\mathcal{S}^*$ consisted of a single point – the case where the Gibbsian state corresponding to a temperature and pressure of θ^* and p^*, respectively, would be homogeneous throughout B. In the more general situation considered here, the estimate (8.53) establishes the stability of the *set* $\mathcal{S}^*$ and not the stability of any one of the multitude of Gibbsian states, all of which have $\mathcal{S}^*$ as their effective range. Further, this conclusion is the best that can be expected in light of the fact that the theorem admits the entire collection of neo-classical processes in $\mathbb{P}(\theta^*, p^*)$, and, therefore, is not prejudiced by requiring additional specific initial and boundary data.

In the remainder of this paper we shall consider *isolated processes.* Such processes, although more controlled, nevertheless are neo-classical processes in $\mathbb{P}(\theta^*, p^*)$ for *any* $\theta^* \neq 0$ and *any* p^*, and, thus, (8.53) remains valid. Being more controlled in the sense that during any isolated process the volume of B is fixed and the global supply of mechanical working, heat working and entropy is null for all time, one should expect to obtain additional conclusions regarding stability under isolation which reflect these additional constraints. In fact, the following theorem represents two such conclusions which are refinements of (8.53). First, it will be helpful to recall from Sect. 8.2 and the notion of an A-projection[17] that if $s_t(\cdot) = (\eta, v, \varepsilon)(\cdot, t)\colon B \to \mathscr{D}$ is the state vector field at time t of an isolated process, then

$$A \int_B s_t dm + (0, 0, K(t)) = (0, V^*, E^*) \,, \tag{8.54}$$

where

$$K(t) \equiv \int_B \tfrac{1}{2} |\dot{x}|^2 dm \tag{8.55}$$

is the kinetic energy of B_t, and where $(0, V^*, E^*)$ is the prescribed initial value of the left hand side of (8.54) at $t = 0$. As a second prerequisit, suppose that the planar slice $\mathscr{S}^*$ of Theorem 5 is a *finite collection* of $n < \infty$ points, $\mathscr{S}^* = \{p_1, p_2, \ldots, p_n\}$. In this case, the domain $\mathscr{D}$ of (8.48) can be decomposed into a disjoint union of n domains $\mathscr{D} = \bigcup_{i=1}^n \mathscr{D}_i$, each of which, say $\mathscr{D}_i$, has the property that it contains those $s = (\eta, v, \varepsilon) \in \mathscr{D}$ that are *closer* to $p_i = (n_i, v_i, e_i) \in \mathscr{S}^*$ in specific volume v and specific energy ε than to any other member $p_j \in \mathscr{S}^*$ with $j < i$, and *at least as close* to $p_i \in \mathscr{S}^*$, in this sense, than to *all* members $p_j \in \mathscr{S}$. Thus,

$$\mathscr{D}_i \equiv \{s \in \mathscr{D} \,\|\, |A(s - p_i)| \leqslant |A(s - p_j)| \quad \textit{for all} \\ j = 1, 2, \ldots, n; \ \textit{strict inequality for } j < i\} \,,$$

and we note that if $n = 1$, then $\mathscr{D}_1 = \mathscr{D}$.

It is now possible to establish the following

Theorem 6. *Let all the hypotheses of Theorem 5 hold and, further, suppose that the planar slice $\mathscr{S}^*$ is a finite collection of n points, and that $s_t(\cdot)\colon B \to \mathscr{D}$ is the state vector field at time t of an isolated process for which* (8.54) *holds. Then,*

i) *if $(0, V^*, E^*)/m(B)$ does* not *belong to the A-projection of the convex hull of $\mathscr{S}^*$, it is impossible for both*

$$\int_B \inf_{p \in \mathscr{S}^*} |s_t - p| \, dm \qquad \textit{and} \qquad K(t) \,,$$

and, therefore, also $\Psi^*(t)$, to go to zero as $t \to \infty$;

ii) *if $(0, V^*, E^*)/m(B)$ belongs to the A-projection of the convex hull of $\mathscr{S}^*$ and if either $n < 3$, or $n = 3$ and the points of $\mathscr{S}^*$ are not colinear*[18]*, then for every $\delta > 0$ there exists $C(\delta) > 0$, monotone increasing and unbounded as $\delta \to 0$, such that for each $i = 1, \ldots, n$,*

$$|\alpha_i^* - \alpha_i(t)| \leqslant \delta + C(\delta)\, \Psi^*(t) \,, \tag{8.56}$$

where $\alpha_i^*, \ldots, \alpha_n^*$ are the n unique numbers in $[0, 1]$ *which enter the convex combination in the representation*

---

[17] See footnote 4 on page 132.

[18] See the formal Remark and the discussion following this remark at the end of Sect. 8.4, wherein the number 3 is related to the question of uniqueness of Gibbsian states.

$$(0, V^*, E^*)/m(B) = A \sum_{i=1}^{n} \alpha_i^* p_i \,, \qquad \sum_{i=1}^{n} \alpha_i^* = 1 \,, \tag{8.57}$$

and where $\alpha_i(t)$ is the mass fraction at time t of those particles $X \in B$ that have their state $s_t(X)$ in $\mathscr{D}_i$, i.e.,

$$\alpha_i(t) \equiv \frac{m(s_t^{-1}(\mathscr{D}_i))}{m(B)} \,, \qquad \sum_{i=1}^{n} \alpha_i(t) = 1 \,. \tag{8.58}$$

Part (i) of this theorem is not unexpected since the hypothesis violates a necessary condition for the existence of a normal Gibbsian state as given in the formal Remark of Sect. 8.3.

Under the condition of part (ii) of this theorem, we know[19] that a normal Gibbsian state $p^*(\cdot) \in C_p(V^*, E^*)$ is unique up to a rearrangement and that it takes on at most the $n \leqslant 3$ distinct values corresponding to the members of the planar slice $\mathscr{S}^*$. Moreover, the mass fractions of those particles $X \in B$ that take on distinct values $p^*(X)$ in $\mathscr{S}^*$ are given by the numbers α_i^*. Thus, in the case $n = 1$, (8.56) is vacuous since both α_1^* and $\alpha_1(t)$ equal one. However, more generally, since

$$\int_{s_t^{-1}(\mathscr{D}_i)} |A(s_t - p_i)| \, dm \leqslant \int_B \inf_{p \in \mathscr{S}^*} |A(s_t - p)| \, dm \,,$$

and since $|A(s_t - p)| \leqslant |s_t - p|$, we interpret (8.52, iii), (8.53), (8.65) and (8.58) to imply, roughly, that under isolation if $\Psi^*(0)$ is sufficiently small, the specific volume and energy fields associated with $s_t(\cdot)$ will stay close in a L^1 sense to those of $p_i \in \mathscr{S}^*$ over a set of particles $s_t^{-1}(\mathscr{D}_i) \subset B$ whose mass measure is always close in an absolute sense to $\alpha_i^* m(B)$.

We now outline a

Proof of Theorem 6. First, we develope a key inequality that will be used to establish both parts of this theorem. For any state at time t, $s_t(\cdot): B \to \mathscr{D}$, associated with a neo-classical process and for $p_i \in \mathscr{S}^*$, we have

$$\sum_{i=1}^{n} \int_{s_t^{-1}(\mathscr{D}_i)} (s_t - p_i) \, dm = \int_B s_t \, dm - \sum_{i=1}^{n} \alpha_i(t) p_i m(B) \,,$$

where we have used (8.58). Thus, with some rearrangement, it follows with the aid of (8.54) that

$$(0, V^*, E^*)/m(B) - A \sum_{i=1}^{n} \alpha_i(t) p_i$$

$$= \frac{1}{m(B)} \left\{ \sum_{i=1}^{n} \int_{s_t^{-1}(\mathscr{D}_i)} A(s_t - p_i) \, dm + (0, 0, K(t)) \right\} ,$$

and, because of the identity and inequality

[19] See footnote 18 on page 151.

$$|A(s_t(X)-p_i)| = \inf_{p\in\mathcal{S}^*} |A(s_t(X)-p)| \leqslant \inf_{p\in\mathcal{S}^*} |s_t - p|$$

for all $X \in B$ with $s_t(X) \in \mathcal{D}_i$, we have

$$\left|(0, V^*, E^*)/m(B) - A\sum_{i=1}^{n} \alpha_i(t) p_i\right| \leqslant \frac{1}{m(B)}\left\{\int_B \inf_{p\in\mathcal{S}^*} |s_t - p|\, dm + K(t)\right\}. \tag{8.59}$$

If both (8.52i) and (8.53) are applied to this result we get

$$\left|(0, V^*, E^*)/m(B) - A\sum_{i=1}^{n} \alpha_i(t) p_i\right| \leqslant \delta + \frac{1}{m(B)}\,[C(\delta) + \theta^*]\,\Psi^*(t). \tag{8.60}$$

Now, to arrive at part (i) of the theorem, we note that since the summation in (8.59) or (8.60) is a convex combination of the elements of $\mathcal{S}^*$ for any t, its value must always lie in the convex hull of $\mathcal{S}^*$. By performing an A-projection and using the hypothesis, we conclude that the left hand side of the inequalities (8.59) and (8.60) cannot vanish. Thus, the right hand sides of (8.59) and (8.60) cannot vanish and this confirms the claim of this part.

To begin the proof of part (ii), we shall take for granted the uniqueness of the numbers α_i^*, as claimed, since this is an elementary result in convex analysis. The remainder of the proof then rests on an identity which we now develop. Since

$$\boldsymbol{g}^* = A\boldsymbol{g}^* - \boldsymbol{a},$$

and A is symmetric, it follows that for $p_i \in \mathcal{S}^*$ and for any n numbers λ_i we have

$$\boldsymbol{a}\cdot\sum_{i=1}^{n}\lambda_i p_i = \boldsymbol{g}^*\cdot A\sum_{i=1}^{n}\lambda_i p_i - \boldsymbol{g}^*\cdot\sum_{i=1}^{n}\lambda_i p_i.$$

But, $\boldsymbol{g}^*\cdot(p_i - p_j) = 0$ for any points p_i and p_j in $\mathcal{S}^*$, and, thus, if the numbers λ_i are such that $\sum_{i=1}^{n}\lambda_i = 0$ we have

$$\boldsymbol{a}\cdot\sum_{i=1}^{n}\lambda_i p_i = \boldsymbol{g}^*\cdot A\sum_{i=1}^{n}\lambda_i p_i.$$

Now, since

$$\sum_{i=1}^{n}\lambda_i p_i = A\sum_{i=1}^{n}\lambda_i p_i + \boldsymbol{a}\left\{\boldsymbol{a}\cdot\sum_{i=1}^{n}\lambda_i p_i\right\},$$

it follows that for $\sum_{i=1}^{n}\lambda_i = 0$,

$$\sum_{i=1}^{n}\lambda_i p_i = (\boldsymbol{1} + \boldsymbol{a}\otimes\boldsymbol{g}^*)A\sum_{i=1}^{n}\lambda_i p_i. \tag{8.61}$$

Clearly, if $n = 1$ we have $\lambda_1 = 0$, and (8.61) is vacuous. In the two cases when $n \leqslant 3$ and the planar slice $\mathcal{S}^*$ contains either 3 non-colinear or just 2 points, it is straightforward to show that there exists a set of unique vectors $\{\boldsymbol{d}_i\}$, $i = 1, \ldots, n$, in the plane determined by $\mathcal{S}^*$ such that

$$\boldsymbol{d}_i \cdot \sum_{j=1}^{n} \lambda_j \boldsymbol{p}_j = \lambda_i ,$$

whenever $\sum_{i=1}^{n} \lambda_i = 0$. In these cases, (8.61) yields

$$\lambda_i = \boldsymbol{d}_i \cdot (\mathbf{1} + \boldsymbol{a} \otimes \boldsymbol{g}^*) A \sum_{j=1}^{n} \lambda_j \boldsymbol{p}_j . \tag{8.62}$$

Now, if we set $\lambda_i = \alpha_i^* - \alpha_i(t)$ for $i = 1, \ldots, n$, in (8.62), use (8.57), and invoke the hypothesis concerning n and the planar slice $\mathcal{S}^*$, we find that

$$|\alpha_i^* - \alpha_i(t)| \leqslant |\boldsymbol{d}_i| |\mathbf{1} + \boldsymbol{a} \otimes \boldsymbol{g}^*| \left| (0, V^*, E^*)/m(B) - A \sum_{j=1}^{n} \alpha_j(t) \boldsymbol{p}_j \right| .$$

Finally, by using (8.60) and appropriately redefining the numbers δ and $C(\delta)$ we arrive at (8.56).

There is an interesting result on the stability of Gibbsian states relative to isolated processes which can be obtained when the data $(0, V^*, E^*)/m(B)$ corresponds to the A-projection of certain special points in the planar slice $\mathcal{S}^*$, even though the planar slice itself may be quite complex. This is the content of our next and final

Theorem 7. *Let all the hypotheses of Theorem 5 hold and, further, suppose that* $\boldsymbol{p}_0 \in \mathcal{S}^*$ *is both a sharp thermostatic support point*[20] *and an isolated discrete point of the set* $\mathcal{S}^*$. *Let* $s_t(\cdot): B \to \mathcal{D}$ *denote the state vector field at time t of an isolated process for which* (8.54) *holds, and suppose that* $s_t(\cdot)$ *is bounded for all* $t \geqslant 0$ *and* $(0, V^*, E^*)/m(B) = A\boldsymbol{p}_0$.[21] *Then, for every* $\delta > 0$ *there exists* $C(\delta) > 0$, *monotone increasing and unbounded as* $\delta \to 0$, *such that*

$$\int_B |s_t - \boldsymbol{p}_0| \, dm \leqslant m(B)\,\delta + C(\delta)\, \Psi^*(t)$$

for all $t \geqslant 0$.

A proof of this theorem is contained in *Dunn* and *Fosdick* [8.2], Theorem 17 and its Corollary.

References

8.1 J. W. Gibbs: On the equilibrium of heterogeneous substances. (Trans. Conn. Acad. 1875 – 1878). The Scientific Papers, Vol. 1, (Yale Univ. Press, New Haven 1907) pp. 55 – 353

8.2 J. E. Dunn, R. L. Fosdick: The morphology and stability of material phases. Arch. Ration. Mech. Anal. **74**, 1 – 99 (1980)

8.3 J. H. Keenan: *Thermodynamics*, 6th printing (Wiley, New York 1947)

---

[20] Recall Sect. 8.4 where this notion was introduced.

[21] In this case, according to Theorem 2, the normal Gibbsian state corresponding to this data is unique and equal to $\boldsymbol{p}_0$.

8.4 Chi-Sing Man: Material stability, the Gibbs conjecture and the first phase rule for substances. Arch. Ration. Mech. Anal. **91**, 1 – 53 (1985)
8.5 B. D. Coleman, J. M. Greenberg: Thermodynamics and the stability of fluid motion. Arch. Ration. Mech. Anal. **25**, 321 – 341 (1967)
8.6 B. D. Coleman: On the stability of equilibrium states of general fluids. Arch. Ration. Mech. Anal. **36**, 1 – 32 (1970)
8.7 J. E. Dunn, R. L. Fosdick: Thermodynamics, stability and boundedness of fluids of complexity 2 and fluids of second grade. Arch. Ration. Mech. Anal. **56**, 191 – 252 (1974)
8.8 R. L. Fosdick, K. Rajagopal: Thermodynamics and stability of fluids of third grade. Proc. Roy. Soc. London A **339**, 351 – 377 (1980)

Chapter 9

Genericity and Gibbs's Conjecture on the Maximum Number of Coexistent Phases

C.-S. Man

9.1 Gibbs's Conjecture and the First Phase Rule

Consider a body B, homogeneous in "substance" and placed in a medium of constant (absolute) temperature θ° and pressure p°. For such a system, a traditional thermodynamicist would agree without hesitation that the body B, when in equilibrium, can exhibit at most three coexistent phases. His conviction would be founded on what we now call the Gibbs phase rule, a "proof" of which appears in every textbook on thermodynamics.

Gibbs himself, however, was much more cautious than most of his followers. In fact he wrote, "it is entirely improbable that there are four coexistent phases of any simple substance", and in a more general context, "it does not seem probable that r can ever exceed $n+2$" (*Gibbs* [9.1], p. 97); here n denotes the number of "independent variable components", and r the number of "coexistent phases" in equilibrium. For definiteness, we shall henceforth call Gibbs's foregoing assertion *Gibbs's conjecture* on the maximum number of coexistent phases. We call it a conjecture because Gibbs nowhere explained what he meant by the words "entirely improbable" and "does not seem probable". The reader will recognize the statement "$r \leqslant n+2$", which is usually regarded as part of the Gibbs phase rule; here we shall call it *the first phase rule.*

If the paper of *Noll* [9.2] did not produce the immediate impact it deserved, the more recent discussions by *Wightman* [9.3], *Feinberg* [9.4], and *Dunn* and *Fosdick* ([9.5], § 7) should have made clear to thermodynamicists that Gibbs's phase rule may not always hold and that the usual conviction about the phase rule is really founded on erroneous "proofs". (See § 7 of *Dunn* and *Fosdick* [9.5] for a critique of the usual "proofs" of the first phase rule.) That there are counterexamples to the first phase rule, however, does not invalidate Gibbs's original conjecture, for it asserts only that violation of the rule is "entirely improbable". Indeed my objective here is to give, for the special instance of substances, a precise interpretation of the words "entirely improbable" so that Gibbs's conjecture is replaced with exact theorems. For want of space no proofs will be given, and at places the discussion will be sketchy and incomplete. The reader is referred to *Man* [9.6] where he will find more results, proofs of all the theorems asserted below, and a full discussion of various aspects of the issue.

9.2 Counterexamples

Henceforth we shall consider the Gibbs conjecture and the first phase rule only in their restrictions to bodies homogeneous in substance and placed in environments of constant temperature and pressure.

To motivate what we shall do below, let us consider how the first phase rule could be violated. We shall use the field formulation of Gibbsian thermostatics, which Gibbs himself applied in some important analyses in his major memoir [9.1], was abandoned in the traditional main-stream of thermodynamics, and was revived by *Coleman* and *Noll* [9.7] in the fifties. In this formulation the properties of a substance Σ are determined by a two-dimensional manifold S in the $\eta-\upsilon-e$ space; here η denotes the entropy density (per unit mass), υ the specific volume, and e the internal energy density (per unit mass). We call such two-dimensional manifolds Gibbs surfaces.

Now consider a body B of substance Σ placed in a medium of constant temperature θ° and pressure p°. By assumption B is endowed with a positive Borel measure m; $m(P)$ gives the mass of the Borel set P in B. We assume that $m(B) < \infty$. Following *Coleman* [9.8], we call each Borel-measurable map $(\eta(\cdot), \upsilon(\cdot), e(\cdot))$: $B \to S$ a static state of the body B.

For the body B under consideration, the maximum number $r_{\max}$ of phases that can coexist in equilibrium is determined by Gibbs's equilibrium criterion: the "availability" $A \equiv \int_B (e-\theta^\circ \eta + p^\circ \upsilon)\, dm$, as a function over static states, assumes its minimum value at an equilibrium state (*Gibbs* [9.9], p. 43). It follows that we can determine $r_{\max}$ by a simple geometric construction. Let $a \equiv e-\theta^\circ \eta+p^\circ \upsilon$ be the integrand of A. Let $\Pi[d]$ be the plane defined by the equation $e-\theta^\circ \eta+p^\circ \upsilon = d$ or $a = d$. We shall regard the $\Pi[d]$'s as the level-planes of the integrand a. Let d_0 denote the minimum value of d such that $\Pi[d_0]$ is a supporting plane of the Gibbs surface S in the $\eta-\upsilon-e$ space, by which we mean $\Pi[d_0]$ is tangent to S and divides the $\eta-\upsilon-e$ space into two parts so that the Gibbs surface S lies in the part defined by the inequality $e-\theta^\circ \eta+p^\circ \upsilon \geqslant d_0$. For any static state $(\eta(\cdot), \upsilon(\cdot), e(\cdot))$: $B \to S$, and for any bodypoint X in B, $\Pi[d_0]$ is the lowest level-plane that $a(X)$ can reach.

To minimize the "availability" A, an equilibrium state $(\bar{\eta}(\cdot), \bar{\upsilon}(\cdot), \bar{e}(\cdot))$ must satisfy the condition that $(\bar{\eta}(\cdot), \bar{\upsilon}(\cdot), \bar{e}(\cdot))(B) \subset \Pi[d_0] \cap S$, to within a set of (mass) measure zero in B which we always ignore. Conversely, any static state which satisfies the foregoing geometric condition is an equilibrium state of B. Hence $r_{\max}$ is the number of distinct points at which the plane $\Pi[d_0]$ and the Gibbs surface S touch. It follows that $r_{\max}$, depending on the Gibbs surface S, could be any number ranging from 1 to ∞; no postulate of Gibbsian thermostatics is violated. Thus counterexamples which contradict the assertion "$r_{\max} \leqslant 3$" abound.

9.3 Naive Reformulation of the Gibbs Conjecture for Substances

From the above discussion it is clear that Gibbs surfaces could be subdivided into two classes: (I) those which have a supporting plane that touches the Gibbs surface in question at more than three points; (II) those which do not have this property. Should the Gibbs surface of a substance fall into class (I), it would be theoretically possible for us to take a body of that substance and choose an environment of the right temperature and pressure such that the body when placed in the chosen environment could exhibit more than three coexistent phases in equilibrium, thus producing a counterexample to the first phase rule. No counterexample will ensue from a surface of class (II). Accordingly, henceforth Gibbs surfaces of class (I) are simply called the violators, whereas those of class (II) are said to be rule-abiding.

For a Gibbs surface S of class (I), there is at least one supporting plane which touches the surface at more than three points. Let us do a thought-experiment with the surface S: Suppose we deform the surface slightly so that another Gibbs surface S' results from the deformation; will the surface S' fall into class (I) or class (II)? Since three points in the $\eta - \upsilon - e$ space determine a plane, there seems to be something unusual about the surface S that it should possess a tangent plane which contains more than three points of S. By deforming the Gibbs surface S it seems likely that we shall revert the unusual to the commonplace. Thus we will bet that the surface S' will fall into class (II). On the other hand, if we take a surface of class (II) and deform it slightly, by intuition we would guess that the resulting surface will remain in class (II). Our judgement is in fact based on the following supposition: If we pick a Gibbs surface at random, it is "entirely improbable" that the surface should belong to class (I); or in the set of Gibbs surfaces those of class (I) constitute a subset that is very small in some appropriate sense. Remembering that Gibbs surfaces of class (II) obey the first phase rule while those of class (I) are the violators, we dare say that Gibbs himself might have had something similar in mind when he asserted that a violation of the first phase rule was "entirely improbable".

Gibbs's original assertions and the discussion in the preceding paragraph are of course heuristic. They are, however, hopelessly imprecise. We did not say what we really meant by a small deformation, nor did we explicate the meaning of Gibbs's "entirely improbable". To proceed beyond Gibbs, first of all we have to recast the Gibbs conjecture and make it a precise statement susceptible of proof or disproof.

Given a set E, to indicate precisely how small a subset F of E is we could proceed in two ways: (i) Make E a measure space; then show that F is a subset of small measure or a set of measure zero. (ii) Endow E with a topology; show that the complement of F as a subset of E is open and dense, or residual, etc. Motivated by the above discussion on deforming Gibbs surfaces, we will here follow the second approach and tentatively reformulate the Gibbs conjecture as follows: When the set of Gibbs surfaces G is endowed with a suitable topology, those surfaces which obey the first phase rule constitute an open and dense subset. Henceforth we shall refer to the preceding assertion as *the naive reformulation of the Gibbs conjecture*. In the naive reformulation, denseness manifests our expectation that by slightly deforming a violator we shall change it into a rule-abiding citizen, and openness reflects our surmise that one which obeys the rule remains faithful to it after a sufficiently small deformation. The topology on G will give a precise meaning to the words "sufficiently small deformation".

9.4 The Set of Gibbs Surfaces G

Of course, the naive reformulation remains empty until we specify the "suitable topology" we choose. But we cannot define a topology on G unless the set of Gibbs surfaces itself is well-defined, for which we must agree on how smooth a Gibbs surface should be and what adscititious inequalities it should satisfy.

We introduce the following notations: Let M and N be C^∞ manifolds (in this paper every manifold is assumed to be Hausdorff, paracompact and with a countable base, and a manifold may have "corners" if not specified otherwise). For $0 \leqslant r \leqslant \infty$, let $C^r(M,N)$, $\mathrm{Emb}^r(M,N)$ and $\mathrm{Imm}^r(M,N)$ denote the set of C^r mappings, C^r embeddings and C^r immersions from M to N, respectively. Let R be the set of real numbers, and let $R^n \equiv R \times \ldots \times R$ (n times).

Roughly speaking, in this paper we require every Gibbs surface to be of class C^2 and to satisfy at each surface point the adscititious inequalities that the specific heat at constant specific volume and the isothermal compressibility both be positive. More precisely, we introduce the following

Definition. A Gibbs surface is represented by a 4-tuple $(M_\alpha, \eta, \upsilon, e)$, where M_α is the constitutive domain, η, υ and e are the entropy density (per unit mass), the specific volume and the internal energy density (per unit mass), respectively. By assumption the 4-tuple $(M_\alpha, \eta, \upsilon, e)$ satisfies the following postulates:

i) M_α is a Hausdorff, paracompact, second-countable, 2-dimensional C^∞ manifold-without-boundary, which need not be connected.
ii) The constitutive functions η, υ, and e are in $C^2(M_\alpha, R)$; (η, υ, e): $M_\alpha \to R^3$ is a C^2 embedding, and (η, υ): $M_\alpha \to R^2$ is a C^2 immersion.
iii) The functions η, υ and e satisfy the following adscititious constitutive inequalities: $\upsilon > 0, e,_\eta > 0, e,_{\eta\eta} > 0$ and $e,_{\eta\eta}\, e,_{\upsilon\upsilon} - (e,_{\eta\upsilon})^2 > 0$.

In the above definition $e,_\eta$ is defined to be the function in $C^1(M_\alpha, R)$ such that at x in M_α,

$$e,_\eta(x) = \frac{\partial(e, \upsilon)}{\partial(x_1, x_2)}(x) \Big/ \frac{\partial(\eta, \upsilon)}{\partial(x_1, x_2)}(x) \; ;$$

here (x_1, x_2) denotes a local coordinate system at x, and the numerator and denominator are both Jacobian determinants. Since (η, υ) is an immersion, the denominator is nonzero at x. Moreover, the value $e,_\eta(x)$ is independent of the choice of local coordinate system at x. Thus $e,_\eta$ is a well-defined function in $C^1(M_\alpha, R)$. Similarly we define the functions $e,_\upsilon \equiv (\partial e/\partial \upsilon)_\eta$, $e,_{\eta\eta} \equiv (\partial^2 e/\partial \eta^2)_\upsilon$, $e,_{\eta\upsilon} \equiv \partial^2 e/\partial \upsilon \partial \eta$, and $e,_{\upsilon\upsilon} \equiv (\partial^2 e/\partial \upsilon^2)_\eta$ pointwise through local coordinate systems. The function $e,_\upsilon$ is in $C^1(M_\alpha, R)$, and the functions $e,_{\eta\eta}$, $e,_{\eta\upsilon}$, $e,_{\upsilon\upsilon}$ are in $C^0(M_\alpha, R)$.

Let $\theta \equiv e,_\eta$, $p \equiv -e,_\upsilon$. Then for each x in M_α and each local coordinate system (x_1, x_2) at x,

$$\frac{\partial(\theta, \upsilon)}{\partial(x_1, x_2)}(x) = e,_{\eta\eta}(x)\frac{\partial(\eta, \upsilon)}{\partial(x_1, x_2)}(x) \neq 0 .$$

The function $(\partial p/\partial \upsilon)_\theta$ is defined pointwise through local coordinates as follows: For each x in M_α and each local coordinate system (x_1, x_2) at x,

$$\left(\frac{\partial p}{\partial \upsilon}\right)_\theta (x) = -\frac{\partial(p, \theta)}{\partial(x_1, x_2)}(x) \bigg/ \frac{\partial(\theta, \upsilon)}{\partial(x_1, x_2)}(x) ;$$

$(\partial p/\partial \upsilon)_\theta$ is in $C^0(M_\alpha, R)$.

Remark. Since $e,_\eta = \theta$, $e,_{\eta\eta} = \theta/\varkappa_\upsilon$ and $e,_{\eta\eta}\, e,_{\upsilon\upsilon} - (e,_{\eta\upsilon})^2 = -(\theta/\varkappa_\upsilon)(\partial p/\partial \upsilon)_\theta$, where θ is the absolute temperature, p the pressure and $\varkappa_\upsilon$ the specific heat at constant specific volume, postulate (iii) is equivalent to the constitutive assumption that $\upsilon > 0$, $\theta > 0$, $\varkappa_\upsilon > 0$ and $(\partial p/\partial \upsilon)_\theta < 0$. The reader will recognize that the last two inequalities are simply the usual "stability conditions". Gibbs himself, however, preferred not to impose these conditions and he let his "primitive surface" include "states of essential instability" [9.9], p. 47. The C^2 assumption on smoothness would certainly invite criticism. The reader is referred to *Man* [9.6] for more discussion on these questions. Because our Gibbs surface of a substance could be taken as that which results when "states of essential instability" are excised from Gibbs's "primitive surface" of a unit mass of that substance, in postulate (i) we allow the constitutive domain to be disconnected. As a result, what we call a Gibbs surface here need not be connected. This usage of the term "surface" deviates from common practice in the mathematical literature in which "surfaces" are defined to be connected. Since every $C^r (r \geqslant 1)$ differentiable structure on a manifold-without-boundary contains a compatible C^∞ structure (see *Hirsch* [9.10], Theorem 2.2.9), there is no loss in generality to assume that M_α be a C^∞ manifold rather than a C^2 manifold. The reason that we require M_α to be a manifold-without-boundary rather than a manifold-with-corners will be given in Sect. 9.6.

Let G_α be the subset of mappings in $C^2(M_\alpha, R^3)$ which represent Gibbs surfaces and thence satisfy the conditions laid down in the definition above. Let G_β, G_γ, etc., be similarly defined when M_α in the definition above is replaced by the 2-dimensional C^∞ manifolds M_β, M_γ, etc., respectively. Every function in G_α, G_β, etc., represents some Gibbs surface, but two different functions $(\eta, \upsilon, e) \in G_\alpha$ and $(\eta^*, \upsilon^*, e^*) \in G_\beta$ may describe the same surface. Indeed, two functions $(\eta, \upsilon, e) \in G_\alpha$ and $(\eta^*, \upsilon^*, e^*) \in G_\beta$ describe the same Gibbs surface if and only if the embeddings (η, υ, e) and $(\eta^*, \upsilon^*, e^*)$ map M_α and M_β, respectively, onto the same surface in R^3. This motivates the following

Postulate of Equivalence. $(\eta, \upsilon, e) \in G_\alpha$ and $(\eta^*, \upsilon^*, e^*) \in G_\beta$ describe the same Gibbs surface if and only if there is a C^2 diffeomorphism $\zeta_{\beta\alpha}$: $M_\beta \to M_\alpha$ such that $\eta^* = \eta \circ \zeta_{\beta\alpha}$, $\upsilon^* = \upsilon \circ \zeta_{\beta\alpha}$, and $e^* = e \circ \zeta_{\beta\alpha}$.

The set of Gibbs surfaces G is precisely the quotient set that remains when we identify the functions in $\bigcup_\alpha G_\alpha$ that describe the same surface. The set G is obviously non-empty.

9.5 Strong and Weak Topologies on G

After the set of Gibbs surfaces G has been well-defined, we can begin our study of the Gibbs conjecture and its naive reformulation. As mentioned above, our surmise is that the violators constitute a "very small" subset of G, and we will attempt at a topological characterization of its smallness. To this end it is plain that any topology on G which can serve the above purpose is "suitable" and the more "suitable" topologies we put on G, the more different descriptions of smallness we could obtain, all the better.

The naive reformulation, on the other hand, is one specific guess that ensues from our geometrical thought-experiment in deforming Gibbs surfaces. Here a "suitable topology" on G should capture the intuitive notion of "small deformations"; a "sufficiently small deformation" of a Gibbs surface S will then be determined by an appropriate neighbourhood of S. To proceed further we should first analyse the notion of "small deformations" more closely and sharpen it as far as possible. We shall achieve this aim by adding physics into the otherwise purely geometrical notion.

Without violence to our intuition, let us assume that every deformation of a Gibbs surface S establishes a one-to-one correspondence between the points of S and those of the resulting surface S', and that this correspondence is mathematically a diffeomorphism. Let Ξ be a deformation which maps S onto S' and satisfies the following condition at each point x in S for all the constitutive quantities: the value a constitutive quantity assumes at x is so close to the value the same constitutive quantity assumes at the corresponding point $\Xi(x)$ in S' that their difference defies any empirical resolution; here the constitutive quantities include η, υ, e, and those (e.g., the absolute temperature θ, the pressure p, the specific heat at constant specific volume, the isothermal compressibility, etc.) which could be defined through the first and second partial derivatives of the function $e = e(\eta, \upsilon)$, should the surface in question be the graph of such a function. A deformation such as Ξ must be regarded physically as "small". It follows that a "suitable topology" for the naive reformulation should be sufficiently fine to provide each S in G a neighbourhood of surfaces in which every surface could be obtained from S through a deformation "small" in the sense above.

Motivated by the above discussion, we define a topology on G as follows: We endow each $C^2(M_\alpha, R^3)$ with the Whitney C^2 topology (or the strong C^2 topology),[1] and denote the topological space that results by $C^2_S(M_\alpha, R^3)$. The set

[1] For convenience of the reader we describe here a base for the Whitney C^r topology on $C^r(M, R^n)$, where M is a 2-dimensional C^∞ manifold. Let $\Phi \equiv \{(U_i, \phi_i)\}_{i\in\Lambda}$ be a *locally finite* family of charts on M such that each U_i contains a compact set K_i, the union of which covers M, i.e., $\cup_i K_i = M$. Let $\varepsilon \equiv \{\varepsilon_i\}_{i\in\Lambda}$ be a family of positive numbers. Let pr_j: $R^n \to R$ denote the projection to the

G_α is open in $C^2_S(M_\alpha, R^3)$ (*Man* [9.6], Lemma 3.3), and we give G_α the subspace topology. Let $\mathrm{Diff}^2_S M_\alpha$ be the set of C^2 diffeomorphisms on M_α, equipped with the Whitney C^2 topology. $\mathrm{Diff}^2_S M_\alpha$ is a topological group under the operation of composition, and the map $\mathrm{Diff}^2_S M_\alpha \times G_\alpha \to G_\alpha$ defined by $(\zeta, (\eta, \upsilon, e)) \mapsto (\eta, \upsilon, e) \circ \zeta$ is a continuous right action of $\mathrm{Diff}^2_S M_\alpha$ on G_α. Let $G_\alpha/\mathrm{Diff}^2_S M_\alpha$ be the orbit space, equipped with the quotient topology. The postulate of equivalence says that two functions in G_α describe the same Gibbs surface if and only if they lie on the same orbit in $G_\alpha/\mathrm{Diff}^2_S M_\alpha$. Let π_α: $G_\alpha \to G_\alpha/\mathrm{Diff}^2_S M_\alpha$ be the natural surjection; π_α is open and continuous.

Let M_β be a constitutive domain diffeomorphic to M_α. Let $\mathrm{Diff}^2(M_\beta, M_\alpha)$ be the set of C^2 diffeomorphisms from M_β onto M_α. Let $\zeta_{\beta\alpha} \in \mathrm{Diff}^2(M_\beta, M_\alpha)$. The map $h_{\beta\alpha}$: $G_\alpha \to G_\beta$ defined by $h_{\beta\alpha}((\eta, \upsilon, e)) = (\eta, \upsilon, e) \circ \zeta_{\beta\alpha}$ is a homeomorphism. Moreover two functions (η, υ, e) and $(\bar{\eta}, \bar{\upsilon}, \bar{e})$ belong to the same orbit in $G_\alpha/\mathrm{Diff}^2_S M_\alpha$ if and only if $h_{\beta\alpha}((\eta, \upsilon, e))$ and $h_{\beta\alpha}((\bar{\eta}, \bar{\upsilon}, \bar{e}))$ belong to the same orbit in $G_\beta/\mathrm{Diff}^2_S M_\beta$. Then by passing to the quotient, the map $h^*_{\beta\alpha}$: $G_\alpha/\mathrm{Diff}^2_S M_\alpha \to G_\beta/\mathrm{Diff}^2_S M_\beta$ defined by the commutative diagram

$$\begin{array}{ccc} G_\alpha & \xrightarrow{\quad h_{\beta\alpha}\quad} & G_\beta \\ {\scriptstyle \pi_\alpha}\big\downarrow & & \big\downarrow{\scriptstyle \pi_\beta} \\ G_\alpha/\mathrm{Diff}^2_S M_\alpha & \xrightarrow{\quad h^*_{\beta\alpha}\quad} & G_\beta/\mathrm{Diff}^2_S M_\beta \end{array}$$

is a homeomorphism. By the postulate of equivalence, $S_\alpha \in G_\alpha/\mathrm{Diff}^2_S M_\alpha$ and $h^*_{\beta\alpha}(S_\alpha) \in G_\beta/\mathrm{Diff}^2_S M_\beta$ correspond to the same Gibbs surface S. Thus in accord with that postulate we should "paste" or "sew" $G_\alpha/\mathrm{Diff}^2_S M_\alpha$ and $G_\beta/\mathrm{Diff}^2_S M_\beta$ together by means of the homeomorphism $h^*_{\beta\alpha}$. This leads to the following

Definition. The space of Gibbs surfaces G is that which results when we "paste" or "sew" the spaces $G_\alpha/\mathrm{Diff}^2_S M_\alpha$, $G_\beta/\mathrm{Diff}^2_S M_\beta$, etc., together by means of the homeomorphisms $h^*_{\beta\alpha}$.

The topology on G given above is clearly derived from the strong C^2 topolopy on the function spaces $C^2(M_\alpha, R^3)$. Hence we simply say, for short, that the space of Gibbs surfaces is the set G endowed with the strong topology. If to

---

jth coordinate. Let $s = (s_1, s_2)$ be an ordered pair of non-negative integers, and let $|s| \equiv s_1 + s_2$. If $f \in C^r(M, R^n)$, a basic neighbourhood $N^r(f; K, \Phi, \varepsilon)$ is defined to be the set of C^r maps h: $M \to R^n$ such that for each i in Λ, for all $\underset{\sim}{x} \equiv (x_1, x_2)$ in $\phi_i(K_i)$, for $j = 1, 2, \ldots, n$, and $|s| = 0, 1, \ldots, r$,

$$|D^s(\pi_j h \phi_i^{-1})(\underset{\sim}{x}) - D^s(\pi_j f \phi_i^{-1})(\underset{\sim}{x})| < \varepsilon_i;$$

here $D^s = (\partial/\partial x_1)^{s_1}(\partial/\partial x_2)^{s_2}$ and for $k = 1, 2$, $(\partial/\partial x_k)^{s_k}$ is the identity operator when $s_k = 0$. For $0 \leq r < \infty$, the Whitney C^r topology on $C^r(M, R^n)$ has all possible sets of the above form for a base. The Whitney C^∞ topology on $C^\infty(M, R^n)$ is simply the union of the topologies induced by the inclusion maps $C^\infty(M, R^n) \to C^r_S(M, R^n)$ for r finite; here $C^r_S(M, R^n)$ is the topological space that results when $C^r(M, R^n)$ is equipped with the Whitney C^r topology. For properties of the Whitney topology, see *Hirsch* [9.10], *Mather* [9.11], and *Michor* [9.12].

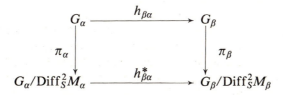

start with the function spaces $C^2(M_\alpha, R^3)$ are given some topology other than the strong topology, by following the same procedure as above we would end up giving G a different topology. We say that the set G is given the weak topology if we start by endowing each $C^2(M_\alpha, R^3)$ with the weak or the compact-open C^2 topology (see *Hirsch* [9.10] or *Michor* [9.12]), give each G_α the subspace topology (the resulting space being denoted by G_α^{W}), and let the set of Gibbs surfaces inherit its weak topology from the spaces G_α^{W}. We denote the topological space thus obtained by G^{W}. Consult *Man* [9.6] for details.

In what follows we shall regard the strong topology as the basic topology on G and study whether the naive reformulation of Gibbs's conjecture is a valid statement when G is equipped with the strong topology. As the reader might have noticed, by abuse of language we use the same symbol G to denote the space of Gibbs surfaces (with the strong topology) and its underlying set.

9.6 The Criterion (*)

For a body B immersed in a medium of fixed temperature and pressure, it is well-known that the temperature θ, the pressure p, and the Gibbs free energy density $g \equiv e - \theta\eta + p\upsilon$ will be homogeneous throughout B when the system is in equilibrium (*Man* [9.6], Theorem 4.2). The validity of the preceding proposition is based on the assumption that Gibbs surfaces be manifolds-without-boundary, which explains why we make this assumption at the outset.

Let S be a Gibbs surface which characterizes for some range of temperature and pressure the thermostatic properties of substance Σ. Suppose the constitutive functions θ, p, and g associated with S satisfy the following inequality:

$$(*) \quad \#((\theta, -p, -g)^{-1}(\underset{\sim}{y})) \leqslant 3 \,, \quad \text{for each} \quad \underset{\sim}{y} \text{ in } R^3 \,;$$

here $\#(\cdot)$ denotes the cardinal number of the set in question, and the minus signs before p and g are introduced for later convenience. It follows from the proposition above that under no circumstances within the given range of temperature and pressure could any body of substance Σ violate the first phase rule. In other words, a Gibbs surface which obeys the inequality (*) must be rule-abiding.

The subset of Gibbs surfaces singled out by the criterion (*) is clearly a proper subset of the rule-abiding surfaces. Nevertheless it turns out that this subset is already quite "large". Indeed in what follows we would characterize the "smallness" of the subset of violators by making precise how "large" the subset of surfaces which obey the inequality (*) is.

9.7 The Rule-Abiding Gibbs Surfaces. Denseness

In this section we shall outline an argument to the effect that rule-abiding surfaces are dense in G. The reader is referred to *Man* [9.6] for details.

For the present purpose it suffices to show that for each G_α those (η, υ, e) which represent rule-abiding Gibbs surfaces are dense in G_α. Let $G_\alpha^\infty \equiv G_\alpha \cap C^\infty(M_\alpha, R^3)$. Because G_α is open in $C_S^2(M_\alpha, R^3)$ (*Man* [9.6], Lemma 3.3), G_α^∞ is open in $C_S^\infty(M_\alpha, R^3)$ when it is given the subspace topology.

We define a mapping L_α: $G_\alpha^\infty \to C_S^\infty(M_\alpha, R^3)$ by

$$L_\alpha((\eta, \upsilon, e)) = (e,_\eta, e,_\upsilon, -e + \eta e,_\eta + \upsilon e,_\upsilon)$$
$$= (\theta, -p, -g)\,.$$

We call L_α the Legendre transformation on G_α^∞. We can show that L_α is a homeomorphism of G_α^∞ onto its image $L_\alpha(G_\alpha^\infty)$, which is open in $C_S^\infty(M_\alpha, R^3)$ (*Man* [9.6], Theorem 5.1 and Corollary 5.2). By appealing to the multijet transversality theorem (see *Mather* [9.13], § 3, or *Golubitsky* and *Guillemin* [9.14], Theorem 2.4.13), we can show that those $(\theta, -p, -g)$ in $L_\alpha(G_\alpha^\infty)$ which satisfy the inequality (*) are dense in $L_\alpha(G_\alpha^\infty)$. Let W_α^∞ be the preimage of such $(\theta, -p, -g)$ in G_α^∞. It follows immediately that W_α^∞ is dense in G_α^∞ (*Man* [9.6], Lemma 6.1 and 6.2). Since G_α^∞ is dense in G_α, W_α^∞ is dense in G_α. Thence we conclude that the rule-abiding surfaces are dense in G (*Man* [9.6], Theorem 6.3).

The basic trick in the above argument is to focus on Gibbs surfaces of class C^∞ so that the Legendre transformation defined above has nice properties, and to exploit the fact that Gibbs surfaces of class C^∞ are dense in G (*Man* [9.6], Theorem 3.6). Here a Gibbs surface is said to be of class C^∞ if it is a C^∞ submanifold in the $\eta - \upsilon - e$ space.

9.8 Difficulty and Refinement of the Naive Reformulation

Besides denseness, it is part of our naive reformulation of Gibbs's conjecture that rule-abiding surfaces constitute an open subset in G. The following example will convince the reader that this part of the naive reformulation is incorrect.

Example. Let $I_1 \equiv (a_1, b_1)$, $I_2 \equiv (a_2, b_2)$, $I_3 \equiv (a_3, b_3)$ be three disjoint open intervals on the real line such that $b_1 < a_2$ and $b_2 < a_3$. Let f be a C^2, locally strictly-convex function defined on the union of the three intervals. The function f has the following properties: (I) No supporting tangent line touches the graph of f at more than two points, but there is a supporting tangent line ℓ which touches the graph of f at $x_1 \in I_1$ and $x_2 \in I_2$ and is tangent to the graph of f at a_3; (II) the convex hull of f (i.e., the pointwise supremum of affine functions everywhere less than f), which is defined on (a_1, b_3), agrees with f over $(a_1, x_1]$ and (a_3, b_3), but agrees with the tangent line ℓ over $[x_1, a_3]$ (see Fig. 9.1). Within any given strong C^2 neighbourhood of f, we can easily find a C^2, locally strictly-convex function which possesses a supporting tangent line that touches its graph at three points.

In the example above, for simplicity, we consider the instance that the graph of f is a one-dimensional manifold in R^2 instead of a two-dimensional manifold in R^3. But it should be clear that counterexamples could be constructed to show that the subset of rule-abiding surfaces is not open in G.

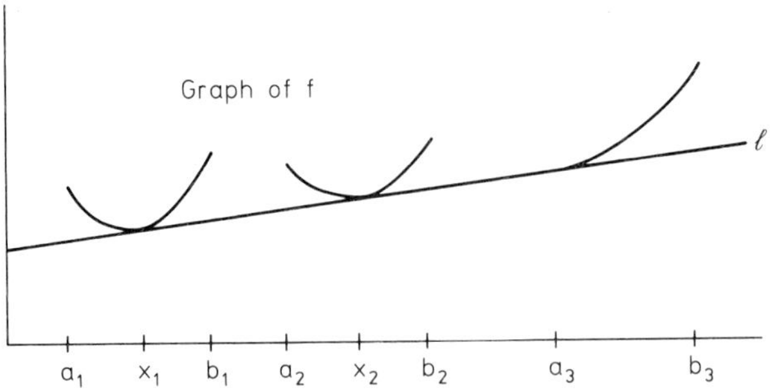

Fig. 9.1. The rule-abiding Gibbs surfaces do not constitute an open subset in G

Nevertheless we can still prove the following result (*Man* [9.6], Theorem 8.5). In what follows let $\mathscr{A}$ be the subset of rule-abiding surfaces, equipped with the subspace topology in G; $\mathscr{A}$ has a nonempty interior (see Theorem 8.1 of *Man* [9.6]).

Theorem. Let $M_\alpha^\dagger$ be a 2-dimensional *compact* C^∞ manifold-with-corners. Let $G_\alpha^\dagger$ be the subset of $C^2(M_\alpha^\dagger, R^3)$ which consists of mappings that satisfy conditions (ii) and (iii) laid down in the definition in § 4. Suppose $G_\alpha^\dagger \neq \phi$. Let M_α be an open subset of the interior of $M_\alpha^\dagger$. Let $f^\dagger \in G_\alpha^\dagger$ and let (η, υ, e): $M_\alpha \to R^3$ be defined by $\eta = pr_1 \circ (f^\dagger | M_\alpha)$, $\upsilon = pr_2 \circ (f^\dagger | M_\alpha)$ and $e = pr_3 \circ (f^\dagger | M_\alpha)$; here for $j = 1, 2, 3$, pr_j: $R^3 \to R$ is the projection to the jth coordinate.

1) Let S be the Gibbs surface represented by the 4-tuple $(M_\alpha, \eta, \upsilon, e)$. In every weak neighbourhood of S in G^W there is a rule-abiding Gibbs surface which lies in the interior of $\mathscr{A}$.
2) Suppose further that $f^\dagger$ satisfies the criterion (*). Then S itself is rule-abiding and lies in the interior of $\mathscr{A}$.

Thought-experiments in Sect. 9.3 prompted us to give a naive reformulation of Gibbs's conjecture. Although this naive reformulation is not entirely correct for the chosen topology on G (i.e., the strong topology), the theorem above does provide a rationale for our feeling about the thought-experiments: Take any compact C^2 surface $S^\dagger$ in the $\eta - \upsilon - e$ space which satisfies the adscititious constitutive inequalities, and whose boundary consists of a piecewise smooth curve. Let us denote by S the Gibbs surface that results when the boundary of $S^\dagger$ is trimmed off. Should the surface $S^\dagger$ happen to satisfy the criterion (*), the Gibbs surface S would be rule-abiding; moreover, S will remain rule-abiding after any perturbation that is sufficiently small (as defined by some strong neighbourhood of it). If the compact surface $S^\dagger$ does not satisfy the criterion (*), we can deform it slightly (to within any prescribed strong = weak neighbourhood of $S^\dagger$) to make it observe that criterion. In terms of the Gibbs surface S, the deformation required to ensure the deformed surface to be in the interior of $\mathscr{A}$ will also be as small as we please, if we use weak instead of strong neighbourhoods to define the word "small". When we form in our minds the image of a surface in R^3, we usually think of a bounded (and thence relatively compact) surface. Our intuitive feeling about perturbations certainly lacks the sophistication to distinguish the double standard of strong and weak neighbourhoods. Nor is it likely that our intuition could quickly hit upon pathological examples such as the one discussed above.

Finally let us give a condition slightly more general than the one stated in the preceding theorem but sufficient to ensure that a rule-abiding Gibbs surface be in the interior of $\mathscr{A}$. It is convenient to introduce the following

Definition. A Gibbs surface S is said to be *strongly rule-abiding* if it could be extended to become a C^2 compact manifold-with-corners $S^\dagger$ that satisfies the following conditions: (i) $S^\dagger$ admits a representation $(M^\dagger, \eta^\dagger, \upsilon^\dagger, e^\dagger)$, where $M^\dagger$ is a C^∞ compact manifold-with-corners, $((\eta^\dagger, \upsilon^\dagger), e^\dagger)$ belongs to $\mathrm{Emb}^2(M^\dagger, R^3) \cap (\mathrm{Imm}^2(M^\dagger, R^2) \times C^2(M^\dagger, R))$, and $(\eta^\dagger, \upsilon^\dagger, e^\dagger)(M^\dagger) = S^\dagger$. (ii) Let $(\theta^\dagger, -p^\dagger, -g^\dagger) \equiv (\partial e^\dagger/\partial \eta^\dagger, \partial e^\dagger/\partial \upsilon^\dagger, -e^\dagger + \eta^\dagger(\partial e^\dagger/\partial \eta^\dagger) + \upsilon^\dagger(\partial e^\dagger/\partial \upsilon^\dagger))$; $(\theta^\dagger, -p^\dagger, -g^\dagger)$ belongs to $\mathrm{Imm}^1(M^\dagger, R^3)$. (iii) $S^\dagger$ satisfies the criterion (*).

Remark. In the definition above we do not require $S^\dagger$ to satisfy the adscititious constitutive inequalities. Should $S^\dagger$ satisfy those inequalities, condition (ii) above would be superfluous. Since S is a subset of $S^\dagger$, condition (iii) implies that the Gibbs surface S is rule-abiding.

Theorem. Every strongly rule-abiding Gibbs surface lies in the interior of the subset $\mathscr{A}$ of rule-abiding surfaces (see *Man* [9.6], Theorem 8.6).

In fact there are plenty of Gibbs surfaces that are strongly rule-abiding. For further discussion on this assertion, see the paragraph following Theorem 8.6 of *Man* [9.6].

9.9 The Remaining Paradox

What follows is the central paradox about the first phase rule: While violations of the rule are conceivable in theory, it seems to be well-observed experimentally. By itself the fact that violators constitute a "very small" subset in G does not resolve the paradox. That violators are rare does not explain why we could not find museum specimens. Indeed, given that there are Gibbs surfaces which are violators, the paradox will remain as long as it is possible in principle to have one substance whose thermostatic properties are exactly those determined by a violator.

Besides the central paradox, other questions remain to be answered. First of all, it is clear that many Gibbs surfaces represent the same thermostatic properties of the same substance. For example, if a Gibbs surface S could be made to coincide with another surface S' by a translation parallel to the $\eta - e$ plane, then S and S' will represent the same properties of the same substance. "This results from the nature of the definitions of entropy and energy, which involve each an arbitrary constant", as *Gibbs* [9.15], p. 34, aptly explained. Secondly, whether a specific Gibbs surface could be used to represent the thermostatic properties of the substance Σ for a given range of temperature and pressure depends partly on the system of units chosen. These facts must be taken into account before we can interpret our theorems about Gibbs surfaces as results about substances.

For further discussion on the problems above and for a plausible resolution of the central paradox, the reader is referred to *Man* [9.6]. There he will also find further mathematical results regarding Gibbs surfaces.

References

9.1 J. W. Gibbs: "On the Equilibrium of Heterogeneous Substances", in *The Scientific Papers of J. Willard Gibbs*, Volume 1 (Dover, New York 1961) pp. 55 – 353
9.2 W. Noll: On certain convex sets of measures and on phases of reacting mixtures. Arch. Ration. Mech. Anal. **38**, 1 – 12 (1970)
9.3 A. S. Wightman: "Convexity and the Notion of Equilibrium State in Thermodynamics and Statistical Mechanics", in *Convexity in the Theory of Lattice Gases*, ed. by R. B. Israel (Princeton University Press, Princeton, NJ 1979) p. ix – lxxxv
9.4 M. Feinberg: On Gibbs' phase rule. Arch., Ration. Mech. Anal. **70**, 219 – 234 (1979)
9.5 J. E. Dunn, R. L. Fosdick: The morphology and stability of material phases. Arch. Ration. Mech. Anal. **74**, 1 – 99 (1980)
9.6 C.-S. Man: Material stability, the Gibbs conjecture and the first phase rule for substances. Arch. Ration. Mech. Anal. **91**, 1 – 53 (1985)
9.7 B. D. Coleman, W. Noll: On the thermostatics of continuous media. Arch. Ration. Mech. Anal. **4**, 97 – 128 (1959)
9.8 B. D. Coleman: On the stability of equilibrium states of general fluids. Arch. Ration. Mech. Anal. **36**, 1 – 32 (1970)
9.9 J. W. Gibbs: "A Method of Geometric Representation of the Thermodynamic Properties of Substances by Means of Surfaces", in *The Scientific Papers of J. Willard Gibbs*, Volume 1 (Dover, New York 1961) pp. 33 – 54
9.10 M. W. Hirsch: *Differential Topology* (Springer, New York Heidelberg Berlin 1976)
9.11 J. N. Mather: Stability of C^∞ mappings: II, infinitesimal stability implies stability. Ann. of Math. **89**, 254 – 291 (1969)
9.12 P. W. Michor: *Manifolds of Differentiable Mappings* (Shiva, Orpington/Kent 1980)
9.13 J. N. Mather: "Stability of C^∞ mappings: V, transversality", Adv. Math. **4**, 301 – 336 (1970)
9.14 M. Golubitsky, V. Guillemin: *Stable Mappings and their Singularities* (Springer, New York Heidelberg Berlin 1973)
9.15 J. W. Gibbs: "Graphical Methods in the Thermodynamics of Fluids", in *The Scientific Papers of J. Willard Gibbs*, Volume 1 (Dover, New York 1961) pp. 1 – 32

Part III

Special Material Systems

Chapter 10

Thermodynamics and the Constitutive Relations for Second Sound in Crystals

B. D. Coleman, M. Fabrizio, and D. R. Owen*

Summary

A derivation is given of the implications of the second law of thermodynamics for the constitutive equations of materials for which the heat flux vector $\boldsymbol{q}$ and the temperature θ obey the relation

$$\boldsymbol{T}(\theta)\dot{\boldsymbol{q}} + \boldsymbol{q} = -\boldsymbol{K}(\theta)\,\mathrm{grad}\,\theta\,, \qquad (\dagger)$$

with $\boldsymbol{T}(\theta)$ and $\boldsymbol{K}(\theta)$ non-singular second-order tensors that, as functions of θ, depend on the material under consideration. The relation (†), which is a natural generalization to anisotropic media of the relation of Cattaneo, has been used by Pao and Banerjee to describe second sound in dielectric crystals. It is here shown that when (†) holds the specific internal energy e depends not only on θ but also on $\boldsymbol{q}$; that is:

$$e = e_0(\theta) + \boldsymbol{q}\cdot\boldsymbol{A}(\theta)\boldsymbol{q}\,,$$

where e_0 is the classical or "equilibrium" internal energy, and $\boldsymbol{A}$ is determined by $\boldsymbol{K}$ and $\boldsymbol{T}$:

$$\boldsymbol{A}(\theta) = -\frac{\theta^2}{2}\frac{d}{d\theta}\left(\frac{\boldsymbol{Z}(\theta)}{\theta^2}\right), \qquad \boldsymbol{Z}(\theta) = \boldsymbol{K}(\theta)^{-1}\boldsymbol{T}(\theta)\,.$$

It is also shown that the second law implies that $\boldsymbol{Z}(\theta)$ is a symmetric tensor and that $\boldsymbol{K}(\theta)$ is positive definite. It is observed that if $\boldsymbol{Z}(\theta)$ and $\boldsymbol{Z}(\theta)^{-1}\boldsymbol{A}(\theta)$ are positive definite and $\partial e/\partial\theta$ is positive, a temperature-rate wave, i.e., a singular surface across which there is a jump in $\dot{\theta}$, will travel faster if it propagates opposite to, rather than parallel to the heat flux.

10.1 Introduction

In the classical theory of heat conduction, it is assumed that the heat flux vector $\boldsymbol{q}$ and the spatial gradient $\boldsymbol{g}$ of the temperature θ, i.e.,

$$\boldsymbol{g} = \mathrm{grad}_x\,\theta(\boldsymbol{x}, t)\,, \qquad (10.1)$$

are related by the constitutive equation,

---

* An Italian language version of this paper appeared in Volume 68 of the Rendiconti del Seminario Matematico della Università di Padova under the title: *Il secondo suono nei cristalli: Termodinamica ed equazioni costitutive.*

The research reported here was supported by the U.S. National Science Foundation and the Consiglio Nazionale delle Richerche.

$$q = -K(\theta)g \ . \tag{10.2}$$

Thermodynamical arguments (vid., e.g., [10.1]) imply that the temperature-dependent second-order tensor $K(\theta)$, called the *thermal conductivity*, is positive semi-definite, and since $K(\theta)$ is, in practice, an invertible tensor, it is positive definite.

In a now frequently cited paper published in 1948, *Cattaneo* [10.2] used a rough model yielding results having some properties in common with a result of Maxwell in the kinetic theory of gases[1] to suggest that the relation (10.2) should be replaced by one which, for the isotropic materials (namely gases) considered by Cattaneo, has the form

$$\tau\dot{q} + q = -\varkappa g \tag{10.3}$$

(with τ and $\varkappa$ positive functions of θ). Cattaneo pointed out that his constitutive relation yields field equations for θ and q that are free from the "paradox of instantaneous propagation of thermal disturbances" known to be associated with the relation (10.2). In 1963 *Chester* [10.4] observed that current theories of the physics of heat conduction in pure dielectric crystals at low temperatures suggest that there can be a range of temperatures in which a relation of the form (10.3) holds approximately with the order of magnitude of $\varkappa/\tau$ equal to $\frac{1}{3}cV^2$, with c the heat capacity at constant volume and V an average value of the phonon velocities, which depend, in general, on frequency and polarization as well as direction.[2] *Pao* and *Banerjee* [10.17] (see also, *Banerjee* and *Pao* [10.18]) have remarked that for anisotropic media the natural generalization of the relation (10.3) is

$$T(\theta)\dot{q} + q = -K(\theta)g \ , \tag{10.4}$$

with $K(\theta)$ as in (10.2) and with $T(\theta)$, like $K(\theta)$, a temperature-dependent, positive definite, second-order tensor.[3] When $q = 0$, equation (10.4) reduces to equation (10.2), and hence, in the theory of (10.4), one may call $K(\theta)$ the *steady state thermal conductivity tensor*. We call $T(\theta)$ the *tensor of relaxation times*.

---

1 For a modern presentation and extension of Maxwell's result see Chapters XIII and XVII of the treatise of *Truesdell* and *Muncaster* [10.3].

2 Many authors have proposed modifications and extensions of the relation (10.3). Of particular relevance is the derivation, for dielectric crystals, given by *Guyer* and *Krumhansl* [10.5]. Also of interest are the earlier articles on second sound by *Ward* and *Wilks* [10.6], *Dingle* [10.7], *Sussman* and *Thellung* [10.8], *Griffin* [10.9], *Prohofsky* and *Krumhansl* [10.10], and *Guyer* and *Krumhansl* [10.11], and the more recent articles of *Enz* [10.12], *Kwok* [10.13], and *Hardy* [10.14]. The problem of formulating constitutive relations that yield a finite velocity for the propagation of thermal disturbances has been discussed from another point of view by *Gurtin* and *Pipkin* [10.15] and *Morro* [10.16].

3 It is clear that the relations (10.3) and (10.4) are not invariant under time-dependent changes of frame, but this lack of invariance is not important for the problems we treat. A modification of (10.4) that *is* invariant under all changes of frame is $T(\theta)(\dot{q} - Wq) + q = -K(\theta)g$ with W either the velocity gradient or the vorticity tensor, i.e., the skew part of the velocity gradient. (As our discussion is confined to rigid bodies, in any motion of the materials we consider the velocity gradient is skew.) *Pao* and *Banerjee* [10.17, 18] considered non-rigid bodies and discussed the relations between thermal and acoustical waves in the framework of the linear theory of infinitesimal elastic deformations.

In an article [10.19] for the Archive for Rational Mechanics and Analysis, we have derived the restrictions that the second law of thermodynamics places on the constitutive relations of a class of materials that includes those considered by Cattaneo and Pao and Banerjee. We there show that the relation (10.4) with the tensors $\boldsymbol{T}(\theta)$ and $\boldsymbol{K}(\theta)$ non-singular is compatible with thermodynamics only if $\boldsymbol{K}(\theta)$ is positive definite, the tensor

$$\boldsymbol{Z}(\theta) = \boldsymbol{K}(\theta)^{-1}\boldsymbol{T}(\theta) \tag{10.5}$$

is symmetric, i.e.,

$$\boldsymbol{Z}(\theta)^T = \boldsymbol{Z}(\theta)\,, \tag{10.6}$$

and the specific internal energy e (per unit volume), the specific entropy η, and the specific Helmholtz free energy $\psi = e - \theta\eta$ are *not* given by functions of θ alone, but are instead given by functions $\tilde{e}$, $\tilde{\eta}$, and $\tilde{\psi}$ of the form[4]

$$e = \tilde{e}(\theta, \boldsymbol{q}) = e_0(\theta) + \frac{1}{\theta}\boldsymbol{q}\cdot\boldsymbol{Z}(\theta)\boldsymbol{q} - \frac{1}{2}\boldsymbol{q}\cdot\frac{d}{d\theta}\boldsymbol{Z}(\theta)\boldsymbol{q}\,, \tag{10.7a}$$

$$\eta = \tilde{\eta}(\theta, \boldsymbol{q}) = \eta_0(\theta) + \frac{1}{2\theta^2}\boldsymbol{q}\cdot\boldsymbol{Z}(\theta)\boldsymbol{q} - \frac{1}{2\theta}\boldsymbol{q}\cdot\frac{d}{d\theta}\boldsymbol{Z}(\theta)\boldsymbol{q}\,, \tag{10.7b}$$

$$\psi = \tilde{\psi}(\theta, \boldsymbol{q}) = \psi_0(\theta) + \frac{1}{2\theta}\boldsymbol{q}\cdot\boldsymbol{Z}(\theta)\boldsymbol{q}\,; \tag{10.7c}$$

these functions obey the relations

$$\partial_\theta\tilde{\psi} = -\tilde{\eta} \quad \text{and} \quad \theta\partial_\theta\tilde{\eta} = \partial_\theta\tilde{e}\,, \tag{10.8}$$

which imply the familar formulae,

$$d\psi_0/d\theta = -\eta_0 \quad \text{and} \quad d\eta_0/d\theta = c_0/\theta\,, \tag{10.9}$$

in which c_0 is the "equilibrium heat capacity", i.e.,

$$c_0(\theta) = de_0(\theta)/d\theta\,. \tag{10.10}$$

In the absence of both deformation and a supply of heat by radiation, the law of balance of energy takes the form

[4] (Footnote added to the English version, May, 1983): We have recently seen a paper by P. J. *Chen* and M. E. *Gurtin* [On second sound in materials with memory, Zeits. Angew. Math. Phys. **21**, 232 – 241 (1970)] in which they extend part of the theory of *Gurtin* and *Pipkin* [10.15] to deformable media and discuss specializations of that theory to unidimensional cases. It is clear from the example treated in their § 6 that, despite differences in language, methods, and initial assumptions, in the special case in which $\boldsymbol{T}$ and $\boldsymbol{K}$ are independent of θ and the heat flow is unidimensional, Chen and Gurtin's theory intersects ours and agrees with it in the conclusion that an equation equivalent to our equation (10.7c) can hold. Chen and Gurtin do not observe that equation (10.7c) is *implied* by equation (10.4), i.e., that (10.7c) is the *only* free energy function compatible with (10.4). Of course, the papers of *Gurtin* and *Pipkin* [10.15] and Chen and Gurtin predate not only our own work on second sound but that of *Pao* and *Banerjee* [10.17, 18] on second sound in deformable media.

$$\dot{e} + \operatorname{div} \boldsymbol{q} = 0 \ . \tag{10.11}$$

If we let

$$\boldsymbol{A}(\theta) = \frac{1}{\theta} \boldsymbol{Z}(\theta) - \frac{1}{2} \frac{d}{d\theta} \boldsymbol{Z}(\theta) = -\frac{\theta^2}{2} \frac{d}{d\theta} \left(\frac{\boldsymbol{Z}(\theta)}{\theta^2} \right), \tag{10.12}$$

then equation (10.7a) becomes

$$\tilde{e}(\theta, \boldsymbol{q}) = e_0(\theta) + \boldsymbol{q} \cdot \boldsymbol{A}(\theta) \boldsymbol{q} \ , \tag{10.13}$$

and, clearly, $\dot{e}$ in (10.11) is *not* given by the classical formula, $\dot{e} = c_0(\theta) \dot{\theta}$ (which was, of course, employed in [10.2, 4, 5, 14, 17, 18] but is instead given by the expression

$$\dot{e} = [c_0(\theta) + \boldsymbol{q} \cdot \boldsymbol{B}(\theta) \boldsymbol{q}] \dot{\theta} + 2\boldsymbol{q} \cdot \boldsymbol{A}(\theta) \dot{\boldsymbol{q}} \tag{10.14}$$

with

$$\boldsymbol{B}(\theta) = \frac{d}{d\theta} \boldsymbol{A}(\theta) \ . \tag{10.15}$$

Thus, the evolution of the heat flux and temperature fields is governed by a pair of partial differential equations,

$$\begin{gathered} \boldsymbol{T}(\theta) \dot{\boldsymbol{q}} + \boldsymbol{q} + \boldsymbol{K}(\theta) \operatorname{grad} \theta = \boldsymbol{0} \ , \\ \operatorname{div} \boldsymbol{q} + c_0(\theta) \dot{\theta} + \boldsymbol{q} \cdot \boldsymbol{B}(\theta) \boldsymbol{q} \dot{\theta} + 2\boldsymbol{q} \cdot \boldsymbol{A}(\theta) \dot{\boldsymbol{q}} = 0 \ , \end{gathered} \tag{10.16}$$

for which the tensorial coefficients $\boldsymbol{A}(\theta)$ and $\boldsymbol{B}(\theta)$ in the second equation are determined by the temperature-dependence of the coefficients $\boldsymbol{T}(\theta)$ and $\boldsymbol{K}(\theta)$ in the first. As the relations (10.12) and (10.15) do not, in general, yield $\boldsymbol{A} = \boldsymbol{B} = \boldsymbol{0}$, the second equation in (10.16) is non-linear in $\boldsymbol{q}$.

Below we give a derivation of the relations (10.6 – 8) that has certain advantages over that which we gave in [10.19]. Both derivations are set in the framework of Coleman and Owen's general theory of thermodynamical systems [10.20, 21]. Here we take a state to be a pair $(\theta, \boldsymbol{q})$, rather than $(e, \boldsymbol{q})$,[5] and we avoid the assumption that the function $\theta \mapsto \boldsymbol{e}(\theta, \boldsymbol{q})$ is invertible at fixed $\boldsymbol{q}$. In future papers Coleman and Owen will discuss circumstances under which such invertibility does not hold, and, for fixed non-zero values of $\boldsymbol{q}$, $\partial_\theta \tilde{e}(\theta, \boldsymbol{q})$ changes sign from positive to negative as θ decreases toward 0.[6]

10.2 Derivation of Thermodynamical Relations

We are concerned with materials for which the *state* at a point (or material element) can be described by giving the local temperature θ and the local heat

---

[5] In [10.19] we treat at length a general class of materials for which states are pairs (e, α), or (θ, α), whose second members, α, become $\boldsymbol{q}$ only in certain special cases such as that in which (10.4) holds.

[6] (Footnote added to the English version, May, 1983): The first of these papers has been printed: *B. D. Coleman* and *D. R. Owen*: On the nonequilibrium behavior of solids that transport heat by second sound. Comp. Math. with Appls. **9**, 529 – 546 (1983).

flux $\boldsymbol{q}$. The collection Σ of all states of an element is here a set of the form $\mathscr{I} \times \mathscr{A}$ with $\mathscr{I}$ an open interval of positive numbers and $\mathscr{A}$ an open connected subset of $\mathscr{V}^{(3)}$ containing the zero vector $\mathbf{0}$. ($\mathscr{V}^{(3)}$ is the three-dimensional Euclidean vector space in which $\boldsymbol{q}$ lies.) We write either σ or $(\theta, \boldsymbol{q})$ for the members of Σ, and, when we discuss such concepts from analysis as derivatives and line integrals, we regard Σ as a subset of the vector space $\mathbb{R} \oplus \mathscr{V}^{(3)}$ with the inner product

$$\langle(\alpha,\boldsymbol{a}),(\beta,\boldsymbol{b})\rangle = \alpha\beta+\boldsymbol{a}\cdot\boldsymbol{b} \tag{10.17}$$

("$\cdot$" is the inner product on $\mathscr{V}^{(3)}$).

With each material there are associated three continuously differentiable functions:

$$\begin{aligned} \boldsymbol{T}&: \ \mathscr{I} \to \mathrm{Lin}^*(\mathscr{V}^{(3)}) \,, \\ \boldsymbol{K}&: \ \mathscr{I} \to \mathrm{Lin}^*(\mathscr{V}^{(3)}) \,, \\ \tilde{e}&: \ \Sigma \to \mathbb{R} \,, \end{aligned} \tag{10.18}$$

with $\mathrm{Lin}^*(\mathscr{V}^{(3)})$ the set of all invertible linear transformations of $\mathscr{V}^{(3)}$ into $\mathscr{V}^{(3)}$. Each process of an element of the material is a piecewise continuous $(p-C^0)$ function P_t mapping an interval $[0, t)$, with $t>0$, into $\mathbb{R} \oplus \mathscr{V}^{(3)}$. The values of P_t are pairs $(\zeta(\xi), \boldsymbol{g}(\xi))$ in which $\zeta(\xi)$ is the time-derivative of θ and $\boldsymbol{g}(\xi)$ is the spatial gradient of θ. The set Π of all processes of a given element is defined as follows: For each $p-C^0$ function $P_t = (\zeta, \boldsymbol{g})$: $[0, t) \to \mathbb{R} \oplus \mathscr{V}^{(3)}$, let $\mathscr{D}(P_t)$ be the set of states $\sigma_0 = (\theta_0, \boldsymbol{q}_0)$ for which the equations

$$\begin{aligned} \dot{\theta} &= \zeta \,, \\ \dot{\boldsymbol{q}} &= -\boldsymbol{T}(\theta)^{-1}\boldsymbol{q} - \boldsymbol{T}(\theta)^{-1}\boldsymbol{K}(\theta)\boldsymbol{g} \,, \end{aligned} \tag{10.19}$$

with the initial conditions

$$\theta(0) = \theta_0 \,, \qquad \boldsymbol{q}(0) = \boldsymbol{q}_0 \,, \tag{10.20}$$

have a solution $\xi \mapsto (\theta(\xi), \boldsymbol{q}(\xi))$ whose values lie in Σ for all ξ in $[0, t]$; if $\mathscr{D}(P_t)$ is not empty, then P_t is a *process*, i.e.,

$$\Pi = \{P_t\colon [0,t) \to \mathbb{R} \oplus \mathscr{V}^{(3)}, p-C^0 \mid \mathscr{D}(P_t) \neq \emptyset\} \,. \tag{10.21}$$

For each pair (P_t, σ_0) with P_t in Π and $\sigma_0 = (\theta_0, \boldsymbol{q}_0)$ in $\mathscr{D}(P_t)$, the solution $\xi \mapsto (\theta(\xi), \boldsymbol{q}(\xi))$ of (10.19) obeying (10.20) is called the *parameterized trajectory in Σ corresponding to* (P_t, σ_0). The final point $(\theta(t), \boldsymbol{q}(t))$ of this trajectory is interpreted as the "state $\sigma_t = (\theta_t, \boldsymbol{q}_t)$ at the instant of completion of the process P_t". The dependence of σ_t upon the "initial state" σ_0 is indicated by writing

$$\sigma_t = \varrho_{P_t}\sigma_0 \,, \tag{10.22}$$

and the operator ϱ_{P_t} so defined is called the *state-transformation function induced by the process* P_t. Familiar theorems in the theory of differential equations tell us that, for each P_t in Π, the domain $\mathscr{D}(P_t)$ of ϱ_{P_t} is an open subset of Σ, and ϱ_{P_t} is not only single-valued but also continuous on $\mathscr{D}(P_t)$.

A pair (P_t, σ_0) for which σ_0 is in $\mathscr{D}(P_t)$ *and*

$$\varrho_{P_t}\sigma_0 = \sigma_0 \tag{10.23}$$

is called *cyclic.*

If P_t is in Π, v is in $(0, t)$, and P_v is the restriction of P_t to $[0, v)$, then P_v is also in Π, and $\mathscr{D}(P_v)$ contains $\mathscr{D}(P_t)$ as a subset. Moreover, if P_{t_1} and P_{t_2} are in Π, and if the range $\mathscr{R}(P_{t_1})$ of $\sigma_{P_{t_1}}$ intersects the domain $\mathscr{D}(P_{t_2})$ of $\varrho_{P_{t_2}}$, then the function $P_{t_1+t_2}$, defined on $[0, t_1+t_2)$ by the formula

$$P_{t_1+t_2}(\xi) = \begin{cases} P_{t_1}(\xi)\,, & \xi\in[0, t_1)\,, \\ P_{t_2}(\xi - t_1)\,, & \xi\in[t_1, t_1+t_2)\,, \end{cases} \tag{10.24}$$

is in Π, has $\mathscr{D}(P_{t_1+t_2}) = \varrho_{P_{t_1}}^{-1}(\mathscr{D}(P_{t_2}) \cap \mathscr{R}(P_{t_1}))$, and $\varrho_{P_{t_1+t_2}}\sigma = \varrho_{P_{t_2}}\varrho_{P_{t_1}}\sigma$ for each σ in $\mathscr{D}(P_{t_1+t_2})$; $P_{t_1+t_2}$ is called *the process resulting from the successive application of* (first) P_{t_1} *and* (then) P_{t_2}.

Let c be an oriented $p-C^1$ (piecewise continuously differentiable) curve that lies in Σ, and let the function $\xi \mapsto (\theta(\xi), \boldsymbol{q}(\xi))$, from $[0, t]$ into Σ, be a $p-C^1$ parameterization of c. Clearly, the function $P_t = (\zeta, \boldsymbol{g})$, defined on $[0, t)$ by the equations

$$\left.\begin{aligned} \zeta(\xi) &= \dot{\theta}(\xi)\,, \\ \boldsymbol{g}(\xi) &= -\boldsymbol{K}(\theta(\xi))^{-1}\boldsymbol{T}(\theta(\xi))\dot{\boldsymbol{q}}(\xi) - \boldsymbol{K}(\theta(\xi))^{-1}\boldsymbol{q}(\xi)\,, \end{aligned}\right\} \tag{10.25}$$

is $p-C^0$, and if this function is substituted into equation (10.19), the solution of (10.19) with initial value σ_0 will be precisely the function $\xi \mapsto (\theta(\xi), \boldsymbol{q}(\xi))$. It follows that P_t is in Π, $\xi \mapsto (\theta(\xi), \boldsymbol{q}(\xi))$ is the parameterized trajectory in Σ corresponding to the pair (P_t, σ_0), and $\varrho_{P_t}\sigma_0$ equals $(\theta(t), \boldsymbol{q}(t))$, the final point of c. In summary, we here assert, as in [10.19]:

Remark 1. If c is an oriented $p-C^1$ curve lying in Σ, then each $p-C^1$ parameterization of c is the parameterized trajectory corresponding to a unique pair (P_t, σ_0) with P_t in Π and σ_0 in $\mathscr{D}(P_t)$; σ_0 is the initial point of c; $\varrho_{P_t}\sigma_0$ is the final point of c. If c is a closed curve, then $\varrho_{P_t}\sigma_0 = \sigma_0$, and the pair (P_t, σ_0) is cyclic.

Because each pair of points in Σ can be joined by a $p-C^1$ curve, Remark 1 implies the validity of the following assertion:

Remark 2. For each pair (σ', σ'') of states in Σ, there is a process P_t in Π with σ' in $\mathscr{D}(P_t)$ and $\sigma'' = \varrho_{P_t}\sigma'$; in other words, the set $\{\sigma\in\Sigma \mid \sigma = \varrho_{P_t}\sigma_0\}$ of states "accessible" from any given state σ_0 is equal to all of Σ.

It follows from these observations that the material elements under consideration are *systems* in the sense in which the term is used in the general theory of references [10.20, 21].

We turn now to the function $\tilde{e}$. The value $e = \tilde{e}(\theta, \boldsymbol{q})$ of $\tilde{e}$ is the specific internal energy. The rate of change of e along the parameterized trajectory $\xi \mapsto (\theta(\xi), \boldsymbol{q}(\xi))$ corresponding to a pair (P_t, σ) is, for each ξ in $[0, t]$,

$$\dot{e}(\xi) = \partial_\theta \tilde{e}(\theta(\xi), \boldsymbol{q}(\xi))\,\dot{\theta}(\xi) + \partial_{\boldsymbol{q}}\tilde{e}(\theta(\xi), \boldsymbol{q}(\xi)) \cdot \dot{\boldsymbol{q}}(\xi)\,, \tag{10.26}$$

and, according to the first law of thermodynamics,

$$\dot{e}(\xi) = h(\xi) \,, \tag{10.27}$$

where h is the rate at which heat is absorbed at the material element. This rate is determined by the heat flux $\boldsymbol{q}$ and the rate r of supply of heat by radiation from sources external to the element:

$$h = r - \operatorname{div} \boldsymbol{q} \,. \tag{10.28}$$

Our assumption that e is given by a function $\tilde{e}$ of state is clearly compatible with the first law of thermodynamics. It has been customary in this subject to assume that the function $\tilde{e}$ reduces to a function of θ alone, i.e., to assume that e is independent of $\boldsymbol{q}$, but, as we shall show below, such an assumption is not compatible with thermodynamics. Using the method of *Coleman* and *Owen* [10.20],[7] we shall show that the second law implies that $\tilde{e}$ must have the form shown in equation (10.7a), with $\boldsymbol{Z}$ as in equation (10.5).

Along the parameterized trajectory $\xi \mapsto (\theta(\xi), \boldsymbol{q}(\xi))$ corresponding to a pair (P_t, σ), one may calculate the integral

$$\mathfrak{s}(P_t, \sigma) = \int_0^t \left(\frac{r}{\theta} - \operatorname{div} \frac{\boldsymbol{q}}{\theta} \right) d\xi \,; \tag{10.29}$$

for, by (10.28) and the relation

$$\operatorname{div} \frac{\boldsymbol{q}}{\theta} = \frac{1}{\theta} \operatorname{div} \boldsymbol{q} - \frac{1}{\theta^2} \boldsymbol{q} \cdot \boldsymbol{g} \,, \tag{10.30}$$

there holds

$$\mathfrak{s}(P_t, \sigma) = \int_0^t \left(\frac{h(\xi)}{\theta(\xi)} + \frac{\boldsymbol{q}(\xi) \cdot \boldsymbol{g}(\xi)}{\theta(\xi)^2} \right) d\xi \,, \tag{10.31}$$

and hence (10.26) and (10.27) yield

$$\mathfrak{s}(P_t, \sigma) = \int_0^t \left(\frac{\partial_\theta \tilde{e}(\theta(\xi), \boldsymbol{q}(\xi))}{\theta(\xi)} \dot{\theta}(\xi) + \frac{\partial_{\boldsymbol{q}} \tilde{e}(\theta(\xi), \boldsymbol{q}(\xi))}{\theta(\xi)} \cdot \dot{\boldsymbol{q}}(\xi) + \frac{\boldsymbol{q}(\xi) \cdot \boldsymbol{g}(\xi)}{\theta(\xi)^2} \right) d\xi \,. \tag{10.32}$$

We here take this last expression as the definition of $\mathfrak{s}$; it makes obvious the fact that $\mathfrak{s}$ is well defined for *every* pair (P_t, σ) with P_t in Π and σ in $\mathcal{D}(P_t)$. It is not difficult to show that, for each process P_t, the function $\sigma \mapsto \mathfrak{s}(P_t, \sigma)$ is continuous on $\mathcal{D}(P_t)$. Moreover, if $P_{t_1+t_2}$ is the result of the successive application of P_{t_1} and P_{t_2}, then for each σ in $\mathcal{D}(P_{t_1+t_2})$,

---

[7] The principal advantage of this method over that proposed by *Coleman* and *Noll* [10.1] and *Coleman* and *Mizel* [10.22, 23], and employed in the investigations of *Gurtin* and *Pipkin* [10.15], *Coleman* and *Gurtin* [10.24], and *Morro* [10.16], is that it does not require *a priori* assumptions about the regularity or even existence of entropy (or free energy) as a function of state. The method we use, i.e., that of reference [10.20], also avoids a situation encountered by *Morro* [10.16], who observed that his starting assumptions are not in accord with equipresence.

$$\mathscr{s}(P_{t_1+t_2}, \sigma) = \mathscr{s}(P_{t_1}, \sigma) + \mathscr{s}(P_{t_2}, \varrho_{P_{t_1}}\sigma) . \tag{10.33}$$

Hence, $\mathscr{s}$ is an *action* for the system (Σ, Π) in the sense of Definition 2.2 of reference [10.20]. In accord with Definition 3.1 of that paper, we say that $\mathscr{s}$ has the *Clausius property* at a state σ_0 if, for each $\varepsilon > 0$, σ_0 has an open neighborhood, $\mathscr{O}_\varepsilon(\sigma_0)$, for which

$$P_t \in \Pi , \quad \sigma_0 \in \mathscr{D}(P_t) , \quad \varrho_{P_t}\sigma_0 \in \mathscr{O}_\varepsilon(\sigma_0) \quad \text{implies}$$
$$\mathscr{s}(P_t, \sigma_0) < \varepsilon ; \tag{10.34}$$

i.e., $\mathscr{s}(P_t, \sigma_0)$ is approximately negative whenever the trajectory in Σ determined by the pair (P_t, σ_0) is approximately closed.

As in [10.19–21], we here take the second law of thermodynamics to be the assertion:

Second Law. *The action $\mathscr{s}$ has the Clausius property at least at one state in Σ.*

If follows from Remark 2 above and Remark 3.1 of reference [10.20] that *the second law here implies that $\mathscr{s}$ has the Clausius property at every state in Σ.*

If a pair (P_t, σ) is cyclic, i.e., if $\varrho_{P_t}\sigma = \sigma$, then $\varrho_{P_t}\sigma$ is in every neighborhood of σ, and, if $\mathscr{s}$ has the Clausius property at σ, $\mathscr{s}(P_t, \sigma)$ is less than every $\varepsilon > 0$, which means that $\mathscr{s}(P_t, \sigma) \leqslant 0$. Hence the second law has the following implication: *If the pair (P_t, σ) is cyclic, $\mathscr{s}(P_t, \sigma)$ is not positive:* i.e., *for each σ in Σ,*

$$P_t \in \Pi , \quad \sigma \in \mathscr{D}(P_t) , \quad \varrho_{P_t}\sigma = \sigma \quad \text{implies}$$
$$\mathscr{s}(P_t, \sigma) \leqslant 0 . \tag{10.35}$$

Let c be an oriented $p-C^1$ curve lying in Σ. Each $p-C^1$ parameterization of c is, by Remark 1, the parameterized tracjectory $\xi \mapsto (\theta(\xi), \boldsymbol{q}(\xi))$ corresponding to a pair (P_t, σ) with σ in $\mathscr{D}(P_t)$, and for this pair the equations (10.32) and (10.19) yield the formula,

$$\mathscr{s}(P_t, \sigma) = \int_0^t \left(\frac{\partial_\theta \tilde{e}(\theta, \boldsymbol{q})}{\theta} \dot{\theta} + \frac{\theta \partial_{\boldsymbol{q}} \tilde{e}(\theta, \boldsymbol{q}) - \boldsymbol{Z}(\theta)^{\mathrm{T}}\boldsymbol{q}}{\theta^2} \cdot \dot{\boldsymbol{q}} - \frac{\boldsymbol{q} \cdot \boldsymbol{K}(\theta)^{-1}\boldsymbol{q}}{\theta^2} \right) d\xi , \tag{10.36}$$

in which θ, $\boldsymbol{q}$, $\dot{\theta}$, and $\dot{\boldsymbol{q}}$ stand for $\theta(\xi)$, $\boldsymbol{q}(\xi)$, $\dot{\theta}(\xi)$ and $\dot{\boldsymbol{q}}(\xi)$. Therefore, $\mathscr{s}(P_t, \sigma)$ can be written as a sum,

$$\mathscr{s}(P_t, \sigma) = \mathscr{J}_1 + \mathscr{J}_2 , \tag{10.37}$$

of two terms, the first of which,

$$\mathscr{J}_1 = \mathscr{J}_1(c) = \int_c \left(\frac{\partial_\theta \tilde{e}(\theta, \boldsymbol{q})}{\theta} d\theta + \frac{\theta \partial_{\boldsymbol{q}} \tilde{e}(\theta, \boldsymbol{q}) - \boldsymbol{Z}(\theta)^{\mathrm{T}}\boldsymbol{q}}{\theta^2} \cdot d\boldsymbol{q} \right) , \tag{10.38}$$

is a line integral independent of the parameterization of c, and the second,

$$\mathscr{J}_2 = -\int_0^t \frac{\boldsymbol{q} \cdot \boldsymbol{K}(\theta)^{-1}\boldsymbol{q}}{\theta^2} d\xi , \tag{10.39}$$

does depend on the parameterization, but has the bound

$$|\mathscr{J}_2| \leqslant Mt, \tag{10.40}$$

where

$$M = \sup_{c} |\boldsymbol{q}\boldsymbol{K}(\theta)^{-1}\boldsymbol{q}/\theta^2| \tag{10.41}$$

is finite because $\boldsymbol{K}$ is continuous on $\mathscr{I} \subset (0, \infty)$, and c is a compact subset of $\Sigma = \mathscr{I} \times \mathscr{A}$.

Now, let c be not only an oriented $p-C^1$ curve in Σ but also a closed curve. For each $p-C^1$ parameterization of c, the corresponding pair (P_t, σ) is cyclic, and hence (10.35) and (10.37) yield

$$\mathscr{J}_1(c) + \mathscr{J}_2 \leqslant 0. \tag{10.42}$$

In view of the bound (10.40) on $\mathscr{J}_2$, the relation (10.42) can hold for *all* $p-C^1$ parameterizations of c only if

$$\mathscr{J}_1(c) \leqslant 0. \tag{10.43}$$

For the curve $-c$ that differs from c only in orientation, equation (10.38) yields $\mathscr{J}_1(-c) = -\mathscr{J}_1(c)$, but the argument that gave (10.43) yields also $\mathscr{J}_1(-c) \leqslant 0$, and, therefore,

$$\mathscr{J}_1(c) = 0. \tag{10.44}$$

As c is an arbitrary $p-C^1$ closed curve in Σ, and Σ is an open connected subset of $\mathbb{R} \oplus \mathscr{V}^{(3)}$, (10.44) and a familar theorem about the existence of potentials for vector fields yield the existence of a continuously differentiable real-valued function $\tilde{\eta}$ on Σ such that

$$\theta \partial_\theta \tilde{\eta}(\theta, \boldsymbol{q}) = \partial_\theta \tilde{e}(\theta, \boldsymbol{q}), \tag{10.45}$$

$$\theta^2 \partial_{\boldsymbol{q}} \tilde{\eta}(\theta, \boldsymbol{q}) = \theta \partial_{\boldsymbol{q}} \tilde{e}(\theta, \boldsymbol{q}) - \boldsymbol{Z}(\theta)^{\mathrm{T}} \boldsymbol{q}, \tag{10.46}$$

and, for any oriented $p-C^1$ curve c in Σ, closed or not,

$$\mathscr{J}_1(c) = \tilde{\eta}(\sigma_2) - \tilde{\eta}(\sigma_1), \tag{10.47}$$

with σ_1 the initial and σ_2 the final point of c.

Equations (10.37) and (10.47) imply that for each pair (P_t, σ) with σ in $\mathscr{D}(P_t)$,

$$s(P_t, \sigma) = \eta(\varrho_{P_t}\sigma) - \eta(\sigma) + \mathscr{J}_2, \tag{10.48}$$

where $\mathscr{J}_2$ is as in equation (10.39).

If for a given state $\sigma_0 = (\theta_0, \boldsymbol{q}_0)$ and positive number t, we let $P_t^0 = (\xi^0, \boldsymbol{g}^0)$ be the process defined on $[0, t)$ by

$$\begin{aligned} \xi^0 &\equiv 0, \\ \boldsymbol{g}^0 &\equiv -\boldsymbol{K}(\theta_0)^{-1}\boldsymbol{q}_0, \end{aligned} \tag{10.49}$$

then, the corresponding solution of (10.19) is the constant

$$\theta \equiv \theta_0 \,, \qquad q = q_0 \,, \tag{10.50}$$

then pair (P_t^0, σ_0) is cyclic, and (10.35), (10.48), and (10.39) yield

$$0 \leqslant \mathscr{s}(P_t^0, \sigma_0) = -t q_0 \cdot K(\theta_0)^{-1} q_0 / \theta_0^2 \,. \tag{10.51}$$

Thus, the second law implies that for each θ in $\mathscr{I}$,

$$q \cdot K(\theta)^{-1} q \geqslant 0 \,, \tag{10.52}$$

for all q in $\mathscr{A}$ and, as $K(\theta)^{-1}$ is an invertible tensor and $\mathscr{A}$ contains a spherical neighborhood of $\mathbf{0}$, this implies that $K(\theta)$ *is positive definite for each* θ *in* $\mathscr{I}$. Furthermore, whether or not (P_t, σ) is cyclic, (10.51) and (10.39) together imply

$$\mathscr{J}_2 \leqslant 0 \,, \tag{10.53}$$

and hence (10.48) yields

$$\mathscr{s}(P_t, \sigma) \leqslant \tilde{\eta}(\varrho_{P_t}\sigma) - \tilde{\eta}(\sigma) \,. \tag{10.54}$$

As this relation holds for all pairs (P_t, σ) with P_t in Π and σ in $\mathscr{D}(P_t)$, we may assert that $\tilde{\eta}$ is an *entropy function.*[8] Of course, the existence of this entropy function implies that $\mathscr{s}(P_t, \sigma)$ is not positive when the pair (P_t, σ) is cyclic. Employing that continuity of $\tilde{\eta}$, one may show further that (10.54) implies that $\mathscr{s}$ has the Clausius property at each state in Σ, and hence we can assert

Remark 3. The response functions T, K, and $\tilde{e}$ are compatible with the second law of thermodynamics if, and only if, there is a continuously differentiable function $\tilde{\eta}: \Sigma \to \mathbb{R}$ obeying (10.54) for all pairs (P_t, σ) with P_t in Π and σ in $\mathscr{D}(P_t)$.

Suppose now that we have, in addition to the function $\tilde{\eta}$ of equation (10.47), another function $\tilde{\tilde{\eta}}: \Sigma \to \mathbb{R}$ that is an entropy function for the same material element, i.e., that obeys the relation

$$\tilde{\tilde{\eta}}(\varrho_{P_t}\sigma) - \tilde{\tilde{\eta}}(\sigma) \geqslant \mathscr{s}(P_t, \sigma) \tag{10.55}$$

for each pair (P_t, σ) with σ in $\mathscr{D}(P_t)$. For each pair of states (σ_1, σ_2), there is an oriented $p - C^1$ curve $\mathscr{c}$ that has σ_1 as its initial point and σ_2 as its final point; for each $p - C^1$ parameterization of $\mathscr{c}$, (10.55) and (10.37) yield

$$\tilde{\tilde{\eta}}(\sigma_2) - \tilde{\tilde{\eta}}(\sigma_1) \geqslant \mathscr{J}_1(\mathscr{c}) + \mathscr{J}_2 \tag{10.56}$$

and hence, by (10.40),

$$\tilde{\tilde{\eta}}(\sigma_2) - \tilde{\tilde{\eta}}(\sigma_1) \geqslant \mathscr{J}_1(\mathscr{c}) \,. \tag{10.57}$$

If we now interchange σ_1 and σ_2, and replace $\mathscr{c}$ by the curve $-\mathscr{c}$ that differs from $\mathscr{c}$ only in orientation, the same argument gives

---

[8] It follows from (10.54) and the continuity of $\tilde{\eta}$ that $\tilde{\eta}$ is also an "upper potential" for $\mathscr{s}$ in the sense of Definition 3.2 of [10.20].

$$\tilde{\tilde{\eta}}(\sigma_1) - \tilde{\tilde{\eta}}(\sigma_2) \geqslant \mathscr{J}_1(-c) = -\mathscr{J}_1(c) , \tag{10.58}$$

which is compatible with (10.57) only if $\tilde{\tilde{\eta}}(\sigma_2) - \tilde{\tilde{\eta}}(\sigma_1) = \mathscr{J}_1(c)$, and, in view (10.47), we may conclude that

$$\tilde{\tilde{\eta}}(\sigma_2) - \tilde{\tilde{\eta}}(\sigma_1) = \tilde{\eta}(\sigma_2) - \tilde{\eta}(\sigma_1) ; \tag{10.59}$$

that is, $\tilde{\tilde{\eta}}$ *can differ from* $\tilde{\eta}$ *by only a constant.*[9]

The existence of an entropy function, $\tilde{\tilde{\eta}}$, i.e., a function from Σ to $\mathbb{R}$ obeying (10.55), implies the validity of (10.35), from which, as we have shown, there follows the existence of a continuously differential entropy function $\tilde{\eta}$ that can differ from $\tilde{\tilde{\eta}}$ by at most a constant. Thus we have

Remark 4. If a material element has an entropy function, it is continuously differentiable and unique to within a constant.

The value η of the entropy function $\tilde{\eta}$ is called, of course, the *entropy*; here $\tilde{\eta}$ is unique if we assign the value 0 to the entropy in a "standard state" σ^0. If $\mathscr{I}$ has the form $(0, \alpha)$, and the function η_0, defined on $\mathscr{I}$ by

$$\eta_0(\theta) = \tilde{\eta}(\theta, 0) , \tag{10.60}$$

has a limit as $\theta \to 0$, then a natural normalization of $\tilde{\eta}$ is obtained by putting

$$\lim_{\theta \to 0} \eta_0(\theta) = 0 . \tag{10.61}$$

The value ψ of the function $\tilde{\psi}$ defined on Σ by

$$\tilde{\psi}(\theta, \boldsymbol{q}) = \tilde{e}(\theta, \boldsymbol{q}) - \theta \tilde{\eta}(\theta, \boldsymbol{q}) \tag{10.62}$$

is the *Helmholtz free energy*. The assumed smoothness of $\tilde{e}$ and the derived smoothness of $\tilde{\eta}$ imply that $\tilde{\psi}$ is continuously differentiable, and, in view of equation (10.45), we have, throughout Σ,

$$\partial_\theta \tilde{\psi}(\theta, \boldsymbol{q}) = -\tilde{\eta}(\theta, \boldsymbol{q}) , \tag{10.63}$$

and equation (10.46) yields

$$\theta \partial_{\boldsymbol{q}} \tilde{\psi}(\theta, \boldsymbol{q}) = \boldsymbol{Z}(\theta)^{\mathrm{T}} \boldsymbol{q} . \tag{10.64}$$

This last relation tells us that ψ must have the form shown in (10.7c), and once that is known, the relation (10.63) implies that $\tilde{\eta}$ must be as shown in (10.7b). Clearly, (10.7b), (10.7c), and (10.62) imply that $\tilde{e}$ must be as in (10.7a).

Of course, (10.63) and (10.45) are the same as the relations (10.8). From (10.64) we conclude that, for each θ in $\mathscr{I}$, the function $\boldsymbol{q} \mapsto \tilde{\psi}(\theta, \boldsymbol{q})$ has a gradient of order two given by

$$\theta \partial_{\boldsymbol{q}}^2 \tilde{\psi}(\theta, \boldsymbol{q}) = \boldsymbol{Z}(\theta)^{\mathrm{T}} , \tag{10.65}$$

9 Our proofs of the existence, differentiability, and uniqueness of $\tilde{\eta}$ parallel proofs we gave in [10.19] and rest on arguments introduced in the discussion of the thermodynamics of elastic elements with heat conduction in [10.20].

which implies that $Z(\theta)$ *is symmetric for each* θ in $\mathscr{I}$, i.e., (10.6) holds.

We summarize in the following

Theorem.[10] *The second law implies that for each value of* θ:

i) *the tensor* $K(\theta)$ *is positive definite;*
ii) *the tensor* $Z(\theta) = K(\theta)^{-1}T(\theta)$ *is symmetric;*
iii) e, η, *and* ψ *are not independent of* q, *but are instead given by functions* $\tilde{e}$, $\tilde{\eta}$ *and* $\tilde{\psi}$ *that are related as shown in* (10.8) *and have the forms shown in* (10.7).

Remark 5. It is a consequence of the relations (10.45) and (10.46) and the derived smoothness of the entropy function that when $\dot{\theta}$ and $\dot{q}$ are continuous so also is $\dot{\eta}$, and

$$\begin{aligned}\dot{\eta} &= \partial_\theta \tilde{\eta}(\theta, q)\dot{\theta} + \partial_q \tilde{\eta}(\theta, q)\cdot\dot{q} \\ &= [\theta\partial_\theta \tilde{e}(\theta, q)\dot{\theta} + \theta\partial_q \tilde{e}(\theta, q)\cdot\dot{q} - q\cdot Z(\theta)\dot{q}]/\theta^2 \\ &= \dot{e}/\theta - q\cdot Z(\theta)\dot{q}/\theta^2 .\end{aligned} \tag{10.66}$$

The quantity

$$\gamma = \dot{\eta} + \operatorname{div}(q/\theta) - r/\theta \tag{10.67}$$

is called that *rate of production of entropy*; it follows from (10.66), (10.27), (10.28), and (10.4) that here

$$\gamma = q\cdot K(\theta)^{-1}q/\theta^2 . \tag{10.68}$$

The positive-definiteness of K implies that γ is not negative and vanishes only if $q = \mathbf{0}$. Thus the Clausius-Duhem inequality holds in the present theory.

Singular Surfaces

Suppose that at each point x of a region $\mathscr{R}$ of a Euclidean point space, the constitutive relations,

$$T(\theta)\dot{q} + q = -K(\theta)g , \tag{10.69}$$

$$e = \tilde{e}(\theta, q) , \tag{10.70}$$

hold with T, K, and $\tilde{e}$ the continuously differentiable functions of (10.18). Suppose further that these functions are compatible with thermodynamics and hence obey the conclusions (i), (ii), and (iii) of the theorem of the previous section, so that, in particular, $\tilde{e}$ has the form (10.7a).

We are here interested in cases in which, for some $t^* > 0$, the time-dependent fields θ, q, and r are continuous on $\mathscr{R}\times(0, t^*)$, but $\mathscr{R}\times(0, t^*)$ contains a smooth hypersurface $\mathscr{S}$ across which $\dot{\theta}$, g, q, and $\operatorname{grad}_x q$ may have jumps although they are continuous on the complement of $\mathscr{S}$. Let $(n, -U)$, with $|n| = 1$ and $U \geqslant 0$, be the normal to $\mathscr{S}$ at a point (x_0, t_0) in the interior of $\mathscr{S}$; n is the *direction of propagation* and U the *speed* of $\mathscr{S}$ at (x_0, t_0). The *jump* $[f]$ experienced by a field f (such as $\dot{\theta}$, g, etc.) as $\mathscr{S}$ "passes through the place x_0 at time t_0" is

[10] Cf. [10.19], Theorem 4.1.

$$[f] = \lim_{t\to t_0^+} f(\boldsymbol{x}_0, t) - \lim_{t\to t_0^-} f(\boldsymbol{x}_0, t) . \tag{10.71}$$

We assume that $[\dot{\theta}] \neq 0$, and we call the hypersurface $\mathscr{S}$ a *temperature-rate wave.*[11]

In [10.19] we showed that a temperature-rate wave cannot be purely transverse, i.e., cannot be such that $[\dot{\boldsymbol{q}}] \cdot \boldsymbol{n} = 0$.[12] We also showed there that U must obey a quadratic equation[13] which in the present context takes the form

$$U^2 \partial_\theta \tilde{e}(\theta, \boldsymbol{q}) + U\boldsymbol{n} \cdot \boldsymbol{Z}(\theta)^{-1} \partial_{\boldsymbol{q}} \tilde{e}(\theta, \boldsymbol{q}) - \boldsymbol{n} \cdot \boldsymbol{Z}(\theta)^{-1} \boldsymbol{n} = 0 . \tag{10.72}$$

Here $\theta = \theta(\boldsymbol{x}_0, t_0)$ and $\boldsymbol{q} = \boldsymbol{q}(\boldsymbol{x}_0, t_0)$ with $(\boldsymbol{x}_0, t_0)$ a point on $\mathscr{S}$ at which the wave speed is U and the direction of propagation is $\boldsymbol{n}$. By (10.13),

$$\partial_\theta \tilde{e}(\theta, \boldsymbol{q}) = c_0(\theta) + \boldsymbol{q} \cdot \boldsymbol{B}(\theta) \boldsymbol{q} , \tag{10.73}$$

and

$$\partial_{\boldsymbol{q}} \tilde{e}(\theta, \boldsymbol{q}) = 2\boldsymbol{A}(\theta)\boldsymbol{q} = \frac{2}{\theta} \boldsymbol{Z}(\theta)\boldsymbol{q} - \frac{d}{d\theta} \boldsymbol{Z}(\theta)\boldsymbol{q} , \tag{10.74}$$

with c_0, $\boldsymbol{B}$, $\boldsymbol{A}$, and $\boldsymbol{Z}$ as in (10.10), (10.15), (10.12), and (10.5).

Let us assume now, in accord with experience, that for each θ in $\mathscr{I}$: (I) $\boldsymbol{Z}(\theta)$ is positive definite, and (II) $c_0(\theta)$ is positive. As we have shown that $\boldsymbol{K}(\theta)$ is positive-definite, for crystals of high enough symmetry (e.g., cubic crystals) (I) is implied by the physical observation that $\boldsymbol{T}(\theta)$, the tensor of relaxation times, is positive definite. The assumption (II) that the heat capacity is positive when $\boldsymbol{q} = \boldsymbol{0}$, is obviously in accord with observation and statistical mechanical models; as $\partial_\theta \tilde{e}$ is continuous, (II) implies that, for each θ, there is a neighborhood $\mathscr{N}_\theta$ of the origin in $\mathscr{V}^{(3)}$ such that $\partial_\theta \tilde{e}(\theta, \boldsymbol{q})$ is positive for each $\boldsymbol{q}$ in $\mathscr{N}_\theta$.

From the relations (10.72–74) we read off

Remark 6. Suppose θ and $\boldsymbol{q}$ are such that $\partial_\theta \tilde{e}(\theta, \boldsymbol{q}) > 0$, and define $U_0(\theta, \boldsymbol{q}, \boldsymbol{n})$ by the relation

$$U_0(\theta, \boldsymbol{q}, \boldsymbol{n}) = \left(\frac{\boldsymbol{n} \cdot \boldsymbol{Z}(\theta)^{-1} \boldsymbol{n}}{\partial_\theta \tilde{e}(\theta, \boldsymbol{q})} \right)^{1/2} . \tag{10.75}$$

When $\boldsymbol{q}(\boldsymbol{x}_0, t_0) = \boldsymbol{0}$, i.e., when the temperature-rate wave is propagating into a region in which $\boldsymbol{q} = \boldsymbol{0}$, the speed U of the wave is

$$U(\theta, \boldsymbol{0}, \boldsymbol{n}) = U_0(\theta, \boldsymbol{0}, \boldsymbol{n}) = \sqrt{\boldsymbol{n} \cdot \boldsymbol{Z}(\theta)^{-1} \boldsymbol{n} / c_0(\theta)} . \tag{10.76}$$

---

[11] The term was introduced by *Gurtin* and *Pipkin* [10.15], and our treatment of the subject in [10.19] drew on observations made by them. See also the recent papers of *Morro* [10.16, 25] and *Cattaneo*'s now classical study [10.2] of waves of order two, i.e., surfaces across which θ and q and their first derivatives are continuous, but their second derivatives suffer jumps. Our discussion of temperature-rate waves in [10.19] was based on constitutive assumptions of greater generality than those employed here.

[12] See Remark 5.1 of [10.19]. *Cattaneo* [10.2] obtained an analogous result for waves of order two in materials that obey his theory (in which the dependence of e on q is not taken into account).

[13] Equation (5.13) of [10.19]. Analogues of this equation occur also in the papers by *Gurtin* and *Pipkin* [10.15] and *Morro* [10.16, 25].

In general, the equation (10.72) for U (with $\partial_\theta \tilde{e}(\theta, \boldsymbol{q}) > 0$) has a unique positive solution $U(\theta, \boldsymbol{q}, \boldsymbol{n})$ that can be written in the form,

$$U(\theta,\boldsymbol{q},\boldsymbol{n}) = U_0(\theta,\boldsymbol{q},\boldsymbol{n})\left[\sqrt{1+(\boldsymbol{v}\cdot\boldsymbol{n})^2} - \boldsymbol{v}\cdot\boldsymbol{n}\right] , \qquad (10.77)$$

with $\boldsymbol{v}$ in $\mathscr{V}^{(3)}$ given by

$$\begin{aligned} \boldsymbol{v} = \boldsymbol{v}(\theta,\boldsymbol{q},\boldsymbol{n}) &= \boldsymbol{Z}(\theta)^{-1}\partial_{\boldsymbol{q}}\tilde{e}(\theta,\boldsymbol{q})/2\,U_0(\theta,\boldsymbol{q},\boldsymbol{n})\,\partial_\theta\tilde{e}(\theta,\boldsymbol{q}) \\ &= \boldsymbol{Z}(\theta)^{-1}\boldsymbol{A}(\theta)\boldsymbol{q}/U_0(\theta,\boldsymbol{q},\boldsymbol{n})\,\partial_\theta\tilde{e}(\theta,\boldsymbol{q}) \\ &= \frac{2\boldsymbol{q} - \theta\boldsymbol{Z}(\theta)^{-1}(d\boldsymbol{Z}(\theta)/d\theta)\boldsymbol{q}}{2\,\theta U_0(\theta,\boldsymbol{q},\boldsymbol{n})\,\partial_\theta\tilde{e}(\theta,\boldsymbol{q})} . \end{aligned} \qquad (10.78)$$

Therefore, when, as is expected for dielectric crystals, $\boldsymbol{Z}(\theta)^{-1}\boldsymbol{A}(\theta)$ is positive definite and hence $\boldsymbol{q} \neq \boldsymbol{0}$ implies $\boldsymbol{v}\cdot\boldsymbol{q} > 0$, a temperature-rate wave propagating in the direction of the heat flux vector travels more slowly than one propagating in the opposite direction:[14]

$$\boldsymbol{q} \neq \boldsymbol{0} , \quad \boldsymbol{n} = \boldsymbol{q}/|\boldsymbol{q}| \ \text{ implies } \ U(\theta,\boldsymbol{q},\boldsymbol{n}) < U(\theta,\boldsymbol{q},-\boldsymbol{n}) . \qquad (10.79)$$

References

10.1 B. D. Coleman & W. Noll: The thermodynamics of elastic materials with heat conduction and viscosity. Arch. Rational Mech. Anal. **13**, 167 – 178 (1963).
10.2 C. Cattaneo: Sulla conduzione del calore, Atti Sem. Mat. Fis. Univ. Modena **3**, 83 – 101 (1948).
10.3 C. Truesdell & R. G. Muncaster: *Fundamentals of Maxwell's Kinetic Theory of a Simple Monatomic Gas* (Academic, New York 1980).
10.4 M. Chester: Second sound in solids. Phys. Rev. **131**, 2013 – 2015 (1963).
10.5 R. A. Guyer & J. A. Krumhansl: Solution of the linearized phonon Boltzmann equation. Phys. Rev. **148**, 766 – 778 (1966).
10.6 J. C. Ward & J. Wilks: Second sound and the thermo-mechanical effect at very low temperatures. Phil. Mag. **43**, 48 – 50 (1952).
10.7 R. B. Dingle: The velocity of second sound in various media. Proc. Roy. Soc. London **A65**, 1044 – 1050 (1952).
10.8 J. A. Sussman & A. Thellung: Thermal conductivity of perfect dielectric crystals in the absence of umklapp processes. Proc. Phys. Soc. London **81**, 1122 – 1130 (1963).
10.9 A. Griffin: On the detection of second sound in crystals by light scattering. Phys. Lett. **17**, 208 – 210 (1965).
10.10 E. W. Prohofsky & J. A. Krumhansl: Second-sound propagation in dielectric solids. Phys. Rev. **133**, A1403 – A1410 (1964).
10.11 R. A. Guyer & J. A. Krumhansl: Dispersion relation for second sound in solids. Phys. Rev. **133**, A1411 – A1417 (1964).
10.12 C. P. Enz: One particle densities, thermal propagation, and second sound in dielectric crystals. Ann. Phys. **46**, 114 – 173 (1968).
10.13 P. C. Kwok: Dispersion and damping of second sound in non-isotropic solids. Physics **3**, 221 – 229 (1967).
10.14 R. J. Hardy: Phonon Boltzmann equation and second sound in solids. Phys. Rev. B **2**, 1193 – 1207 (1970).

[14] Toward the end of their discussion of waves, *Gurtin* and *Pipkin* [10.15] make an assumption that leads them to a conclusion opposite to the present.

10.15 M. E. Gurtin & A. C. Pipkin: A general theory of heat conduction with finite wave speeds. Arch. Rational Mech. Anal. **31**, 113 – 126 (1968).

10.16 A. Morro: Wave propagation in thermo-viscous materials with hidden variables. Arch. Mech. Warszawa **32**, 145 – 161 (1980).

10.17 Y.-H. Pao & D. K. Banerjee: Thermal pulses in dielectric crystals. Lett. Appl. Eng. Sci. **1**, 35 – 41 (1973).

10.18 D. K. Banerjee & Y.-H. Pao: Thermoelastic waves in anisotropic solids. J. Acoustic. Soc. Am. **56**, 1444 – 1454 (1974).

10.19 B. D. Coleman, M. Fabrizio & D. R. Owen: On the thermodynamics of second sound in dielectric crystals. Arch. Rational Mech. Anal. **80**, 135 – 158.

10.20 B. D. Coleman & D. R. Owen: A mathematical foundation for thermodynamics. Arch. Rational Mech. Anal. **54**, 1 – 104 (1974).

10.21 B. D. Coleman & D. R. Owen: On thermodynamics and elastic-plastic materials. Arch. Rational Mech. Anal. **59**, 25 – 51 (1975); erratum, ibid., **62**, 396. On the thermodynamics of elastic-plastic materials with temperature-dependent moduli and yield stresses, ibid. **70**, 340 – 354 (1979).

10.22 B. D. Coleman & V. J. Mizel: Thermodynamics and departure from Fourier's law of heat conduction. Arch. Rational Mech. Anal. **13**, 245 – 261 (1963).

10.23 B. D. Coleman & V. J. Mizel: Existence of caloric equations of state in thermodynamics. J. Chem. Phys. **40**, 1116 – 1125 (1964).

10.24 B. D. Coleman & M. E. Gurtin: Thermodynamics with internal state variables. J. Chem. Phys. **47**, 597 – 613 (1967).

10.25 A. Morro: Acceleration waves in thermo-viscous fluids. Rend. Sem. Mat. Univ. Padova **63**, 169 – 184 (1980).

Chapter 11
Interstitial Working and a Nonclassical Continuum Thermodynamics

J. E. Dunn

11.1 Introduction

Twenty years ago *Coleman* and *Noll* [11.1] succeeded in clarifying and making rigorous a procedure by which the laws of thermodynamics could be used to deduce constitutive restrictions on a vast variety of materials. Almost from the very beginning, indeed in a paper by *Coleman* and *Mizel* [11.2] in the same year as [11.1], it began to be clear that the procedure of Coleman and Noll, when applied to the usual forms of the basic laws of thermodynamics, in many cases imposed extraordinarily severe restrictions on the long range spatial dependence allowable in constitutive quantities. Indeed, for the class of rigid heat conductors in which the energy ε, the entropy η, and the heat flux $\boldsymbol{q}$ at a particle X may depend on the current value of the temperature θ and its first n spatial gradients at X, *Coleman* and *Mizel* [11.2] showed that thermodynamics and the procedure of [11.1] allowed ε and η to depend on at most the value of θ at X – no gradients of temperature could appear in these quantities at all. Their result was soon extended by *Eringen* [11.3] to the class of deformable, elastic bodies in which the energy, entropy, heat flux, and now the Cauchy stress $\boldsymbol{T}$ at a particle X may depend on the current value of the temperature θ and its first n spatial gradients at X as well as on the current value of the deformation gradient $\boldsymbol{F}$ and its first m spatial gradients at X. Eringen showed that in the above class of materials only the current values of θ and $\boldsymbol{F}$ at X could enter into the response functions for ε, η, and $\boldsymbol{T}$ – no higher spatial gradients of θ or $\boldsymbol{F}$ were allowed. In fact, slightly before Eringen, *Gurtin* [11.4] had established the far deeper result that, even if the energy, entropy, heat flux, and stress at X were permitted to depend on the entire current temperature and deformation gradient *fields* throughout the body, nevertheless, thermodynamics and the procedure of [11.1] allowed only the restricted dependence found by Eringen.

In fact, however, thermomechanical theories in which higher gradients of the temperature and strain entered into constitutive equations were formulated, studied, and applied long ago: In 1876 *Maxwell* [11.5] used the kinetic theory of gases to derive a formula for the stress explicitly containing spatial gradients of the temperature so as to afford an explanation of the phenomena, discovered by Reynolds, of thermal transpiration, in which the presence of a temperature gradient was itself sufficient to produce motion by giving rise to a non-equilibrated system of stresses. In 1901 *Korteweg* [11.6] proposed as a way of modelling capillarity effects in fluids a constitutive equation for the stress that contained not only the usual dependence on the temperature θ and current mass density ϱ but also depended on $\operatorname{grad}\varrho$ and $\operatorname{grad}^2\varrho$, the first and second spatial gradients of the density. Specifically, Korteweg postulated a compressible fluid model in which the "elastic" or "equilibrium" part of the stress $\boldsymbol{T}$ was given by

$$\begin{aligned}\boldsymbol{T} &= \hat{\boldsymbol{T}}(\varrho, \theta, \operatorname{grad}\varrho, \operatorname{grad}^2\varrho)\ ,\\ &= (-p+\alpha\Delta\varrho+\beta|\operatorname{grad}\varrho|^2)\mathbf{1}+\delta\operatorname{grad}\varrho\otimes\operatorname{grad}\varrho+\gamma\operatorname{grad}^2\varrho\ ,\end{aligned} \tag{11.1}$$

where $\Delta\varrho \equiv \operatorname{tr}\{\operatorname{grad}^2\varrho\}$ is the Laplacian of ϱ and where p, α, β, δ, and γ are material functions of ϱ and θ.

More recently, Truesdell[1] has combined and generalized the ideas of Maxwell and Korteweg in his so-called "Maxwellian fluid". Additionally, special instances of Korteweg's form (11.1) appeared in the work of *Fixman* [11.8] and of *Felderhof* [11.9] on the critical point for gas-liquid interfaces where a very special form for the equilibrium free energy, in which $|\operatorname{grad}\varrho|$ appears quadratically, is postulated at the outset. Certain variational principles are then laid down, the momentum equations they give rise to are examined, and a form for the Cauchy stress is then inferred.[2] This approach, which essentially makes the free energy a potential for the stress, should be contrasted to that of Korteweg and Truesdell in which the stress response function is postulated a priori and is thus compatible with there being no stored energy function and corresponding variational principle at all. Indeed, there are subtle overdeterminism difficulties for, say, the general form (11.1) which the methods of [11.8, 9] escape. These have been recently studied by *Serrin* [11.10] and will be discussed in Sect. 11.5 below. Finally, Korteweg's form (11.1) has also been studied in recent years by *Blinowski* [11.11, 12], by *Aifantis* and *Serrin* [11.13, 14], by *Slemrod* [11.15 – 17], by *Hagan* and *Slemrod* [11.18], and by *Hagan* and *Serrin* (see [11.19] and the paper by Hagan in these Proceedings).

The above-mentioned works, in using only the density ϱ and its higher gradients as the measures of the deformation appearing in constitutive equations, chiefly envisioned applications to materials that are essentially fluid-like. The more general situation, when the entire deformation gradient $\boldsymbol{F}$ and certain of its higher gradients are allowed to affect the stress, heat flux, energy, etc., was considered, in ways that now seem too special (see below), by *Toupin* [11.20, 21] in his work on couple-stresses in elasticity[3] and by *Green* and *Rivlin* [11.22, 23] in their work on multipolar continuum theories.

As we have already observed, the constitutive structure of each of the above theories is incompatible with the usual laws of thermodynamics when those laws are coupled with the procedure articulated by Coleman and Noll for deducing constitutive restrictions. Following a line of thought for energetic calculations similar to one found useful by *Ericksen* [11.24] in his work on liquid crystals, and which, in fact, was suggested (but not pursued) by *Toupin* [11.20] in his work on materials with couple-stresses, *Serrin* and I [11.25] have recently presented a particularly simple and attractive resolution of the above incompatibility. Here I wish to outline and extend some of our ideas and results.

---

[1] *Truesdell*'s original work, 1948 – 1952, appears in a rather more polished form in § 125 of [11.7].

[2] See the Introduction to [11.25] for a more complete discussion.

[3] Toupin's work is also discussed in [11.25].

11.2 Classical Continuum Thermodynamics: A Limitation

We begin by recalling the conceptual underpinnings of classical, continuum thermodynamics: let E denote a three dimensional Euclidean point space, and let us identify the material elements or particles of a continuous medium, or *body*, $\mathscr{B}$ with the positions $X \in E$ they occupy in a fixed reference configuration. Denote by B the region occupied by $\mathscr{B}$ in this reference configuration and assume given a *referential mass density* $\varrho_R(\cdot): B \to (0, \infty)$ of $\mathscr{B}$ in B such that $m(P) \equiv \int_P \varrho_R dV$ is the *mass* of the subpart P of B. Along with $\mathscr{B}$ and its referential mass distribution is given the *process class* $\mathbb{P}(\mathscr{B})$ of $\mathscr{B}$ which, roughly, consists of "everything that can happen to $\mathscr{B}$" and so characterizes the material comprising $\mathscr{B}$. More precisely, the elements $\pi \in \mathbb{P}(\mathscr{B})$ are called *processes* and, classically, are certain ordered 8-tuples of functions on $B \times \mathbb{R}$,

$$\pi = \{\chi, \theta, \varepsilon, \eta, \boldsymbol{T}, \boldsymbol{q}, \boldsymbol{b}, r\},$$

where, during π, at particle X and time t

$\boldsymbol{x} = \chi(X, t) \in E$ is the *motion*,
$\theta = \theta(X, t) \in (0, \infty)$ is the *absolute temperature*,
$\varepsilon = \varepsilon(X, t)$ is the specific *internal energy* per unit mass,
$\eta = \eta(X, t)$ is the specific *entropy* per unit mass,
$\boldsymbol{T} = \boldsymbol{T}(X, t)$ is the *Cauchy stress tensor*,
$\boldsymbol{q} = \boldsymbol{q}(X, t) \in V$ is the *heat flux* vector,
$\boldsymbol{b} = \boldsymbol{b}(X, t) \in V$ is the specific *body force* per unit mass,
$r = r(X, t)$ is the *radiant heating* per unit mass,

where V is the translation space of E. Moreover, with $P_t = X(P, t)$, each process $\pi \in \mathbb{P}(\mathscr{B})$ is required to satisfy:

i) the *balance of linear momemtum*

$$\frac{d}{dt} \int_{P_t} \varrho \dot{\boldsymbol{x}}\, dv = \int_{\partial P_t} \boldsymbol{T}\boldsymbol{n}\, da + \int_{P_t} \varrho \boldsymbol{b}\, dv , \tag{11.2}$$

ii) the *balance of energy*

$$\frac{d}{dt} \int_{P_t} \varrho\{\varepsilon + \tfrac{1}{2}\dot{\boldsymbol{x}} \cdot \dot{\boldsymbol{x}}\} dv = \int_{\partial P_t} \{\dot{\boldsymbol{x}} \cdot \boldsymbol{T}\boldsymbol{n} - \boldsymbol{q} \cdot \boldsymbol{n}\} da + \int_{P_t} \varrho\{\dot{\boldsymbol{x}} \cdot \boldsymbol{b} + r\} dv , \tag{11.3}$$

iii) the *imbalance of entropy*

$$\frac{d}{dt} \int_{P_t} \varrho \eta\, dv \geqslant - \int_{\partial P_t} \frac{\boldsymbol{q} \cdot \boldsymbol{n}}{\theta} da + \int_{P_t} \varrho \frac{r}{\theta} dv , \qquad (11.4)^4$$

[4] This is usually called the Clausius-Duhem inequality and seems to have been first expressed in this general form with the crucial term involving r by *Truesdell* and *Toupin* [11.26].

for every subpart $P \subseteq B$, where $\dot{\boldsymbol{x}} = \partial \chi(X, t)/\partial t$ is the *velocity* of the particle X at time t, where $\boldsymbol{n} = \boldsymbol{n}(\boldsymbol{x}, t)$ is the outer unit normal at $\boldsymbol{x} \in \partial P_t$, and where $\varrho = \varrho(\boldsymbol{x}, t)$ is the *spatial mass density* function induced by the motion $\chi(\cdot,\cdot)$, i.e.

$$m(P) = \int_P \varrho_R \, dV = \int_{P_t} \varrho \, dv \;\; \forall \; P \subseteq B \, .$$

It, of course, thus follows that

$$\varrho = \frac{\varrho_R}{|\det \boldsymbol{F}|} \, , \tag{11.5}$$

where $\boldsymbol{F} = \boldsymbol{F}(X, t) = \nabla \chi(X, t)$[5] is the *deformation gradient* of $\chi(\cdot,\cdot)$, which we always take to be nonsingular.

When sufficient smoothness is assumed and (11.5) is taken into account, it is easily shown that (11.2–4) are equivalent to the local conditions

$$\operatorname{div} \boldsymbol{T} + \varrho \boldsymbol{b} = \varrho \ddot{\boldsymbol{x}} \, , \tag{11.6}$$

$$\varrho \dot{\varepsilon} = \boldsymbol{T} \cdot \boldsymbol{L} - \operatorname{div} \boldsymbol{q} + \varrho r \, , \tag{11.7}$$

$$\varrho \{ \dot{\psi} + \eta \dot{\theta} \} - \boldsymbol{T} \cdot \boldsymbol{L} + \frac{\boldsymbol{q} \cdot \boldsymbol{g}}{\theta} \leqslant 0 \, , \tag{11.8}$$

where $\psi \equiv \varepsilon - \theta\eta$ is the *Helmholtz free energy*, where $\boldsymbol{L} \equiv \operatorname{grad} \dot{\boldsymbol{x}}$ and $\boldsymbol{g} \equiv \operatorname{grad} \theta$ are, respectively, the spatial gradients of velocity and temperature, and where $\operatorname{div}(\cdot)$ is just the contraction of $\operatorname{grad}(\cdot)$. Finally, a superposed dot denotes the usual material time derivative of the indicated quantity. The inequality (11.8) is usually called the *dissipation inequality.*

At this point two caveats: first, while it is conventional to assume that the stress tensor $\boldsymbol{T}$ is symmetric, I will not make this assumption. Rather, I note the elementary consequence of (11.5) and (11.6) that, for any fixed point z_0 and every subpart $P \subseteq B$,

$$\int_{\partial P_t} (\boldsymbol{x} - z_0) \wedge \boldsymbol{T}\boldsymbol{n} \, da + \int_{P_t} (\boldsymbol{x} - z_0) \wedge \varrho \boldsymbol{b} \, dv = \frac{d}{dt} \int_{P_t} \varrho (\boldsymbol{x} - z_0) \wedge \dot{\boldsymbol{x}} \, dv + \int_{P_t} ax(\boldsymbol{T} - \boldsymbol{T}^{\mathrm{T}}) \, dv \, , \tag{11.9}$$

where, for any two vectors, $\boldsymbol{a}$ and $\boldsymbol{b}$, $\boldsymbol{a} \wedge \boldsymbol{b}$ denotes their usual cross product, and where, for any skew tensor $\boldsymbol{W}$, $ax(\boldsymbol{W})$ denotes the axial vector of $\boldsymbol{W}$.[6] Of course, (11.9) is just the well-known result that the resultant torque due to the surface tractions $\boldsymbol{T}\boldsymbol{n}$ and the body force $\boldsymbol{b}$ balance the rate of change of the angular momentum if and only if $ax(\boldsymbol{T} - \boldsymbol{T}^{\mathrm{T}}) \equiv 0$, i.e., if and only if $\boldsymbol{T}$ is symmetric. Here I adopt the stance taken in particle mechanics: balance of angular momentum can be proven as a theorem in that subject if one makes certain assumptions about the forces of interaction in a system of massy particles; here too balance of

[5] Throughout our work "∇" will denote differentiation with respect to the *particle* X in B while "grad" will denote differentiation with respect to the *place* $\boldsymbol{x}$ in B_t.

[6] That is, $ax(\boldsymbol{W})$ is the unique vector in V satisfying $\boldsymbol{W}\boldsymbol{a} = ax(\boldsymbol{W}) \wedge \boldsymbol{a}$ for every $\boldsymbol{a} \in V$.

angular momentum (the symmetry of $\boldsymbol{T}$) will emerge in an explicit way as reflecting now the nature of the thermomechanical interactions within $\mathscr{B}$.

Second, it should be noticed that so far the process class $\mathbb{P}(\mathscr{B})$ could be arbitrarily small, in violation of our physical prejudice that most bodies can be made to undergo a vast variety of processes. It is exactly this point of ensuring a very large process class that is the heart of the procedure of *Coleman* and *Noll* [11.1]; I now describe what might be called their Axiom of Size.

To be concrete, let us suppose that we are given a *constitutive structure* for $\mathscr{B}$; indeed, it will suffice here to consider the particular constitutive structure embodied in the assumption that the energy $\varepsilon(\boldsymbol{X}, t)$, the entropy $\eta(\boldsymbol{X}, t)$, the stress $\boldsymbol{T}(\boldsymbol{X}, t)$, and the heat flux $\boldsymbol{q}(\boldsymbol{X}, t)$ are, for every $\pi \in \mathbb{P}(\mathscr{B})$, given by smooth functions of $\boldsymbol{F}$, θ, $\nabla \boldsymbol{F}$, $\nabla^2 \boldsymbol{F}$, $\boldsymbol{g}$, and $\dot{\boldsymbol{F}}$, i.e.,

$$
\begin{aligned}
\varepsilon &= \hat{\varepsilon}(\boldsymbol{F}, \theta, \nabla \boldsymbol{F}, \nabla^2 \boldsymbol{F}, \boldsymbol{g}, \dot{\boldsymbol{F}}) , \\
\eta &= \hat{\eta}(\boldsymbol{F}, \theta, \nabla \boldsymbol{F}, \nabla^2 \boldsymbol{F}, \boldsymbol{g}, \dot{\boldsymbol{F}}) , \\
\boldsymbol{T} &= \hat{\boldsymbol{T}}(\boldsymbol{F}, \theta, \nabla \boldsymbol{F}, \nabla^2 \boldsymbol{F}, \boldsymbol{g}, \dot{\boldsymbol{F}}) , \\
\boldsymbol{q} &= \hat{\boldsymbol{q}}(\boldsymbol{F}, \theta, \nabla \boldsymbol{F}, \nabla^2 \boldsymbol{F}, \boldsymbol{g}, \dot{\boldsymbol{F}}) .
\end{aligned}
\qquad (11.10)_{1-4}
$$

Of course, once $\hat{\varepsilon}(\cdot)$ and $\hat{\eta}(\cdot)$ are given, the relation $\psi \equiv \varepsilon - \theta \eta$ determines a function $\hat{\psi}(\cdot)$ such that

$$
\psi = \hat{\psi}(\boldsymbol{F}, \theta, \nabla \boldsymbol{F}, \nabla^2 \boldsymbol{F}, \boldsymbol{g}, \dot{\boldsymbol{F}})^7 . \qquad (11.10)_5
$$

Certain of our results require the common, open domain $\mathscr{A}$ of the response functions $\hat{\varepsilon}(\cdot)$, $\hat{\eta}(\cdot)$, $\hat{\boldsymbol{T}}(\cdot)$, $\hat{\boldsymbol{q}}(\cdot)$, and $\hat{\psi}(\cdot)$ to be suitably connected; for simplicity, we shall suppose $\mathscr{A}$ to be convex. Also, we shall sometimes suppose that if $(\boldsymbol{F}, \theta, \nabla \boldsymbol{F}, \nabla^2 \boldsymbol{F}, \boldsymbol{g}, \dot{\boldsymbol{F}})$ is in $\mathscr{A}$, then so too are its associated "equilibrium" state $(\boldsymbol{F}, \theta, \nabla \boldsymbol{F}, \nabla^2 \boldsymbol{F}, 0, 0)$ and its associated "relaxed" state $(\boldsymbol{F}, \theta, 0, 0, 0, 0)$. It is important to note that $\mathscr{A}$ constitutes a tacit restriction on the process class $\mathbb{P}(\mathscr{B})$ since any process π in $\mathbb{P}(\mathscr{B})$ must be such that its associated motion and temperature field, $\boldsymbol{\chi}(\cdot,\cdot)$ and $\theta(\cdot,\cdot)$, satisfy

$$
(\nabla \boldsymbol{\chi}, \theta, \nabla^2 \boldsymbol{\chi}, \nabla^3 \boldsymbol{\chi}, \operatorname{grad} \theta, \nabla \dot{\boldsymbol{\chi}})(\boldsymbol{X}, t) \in \mathscr{A}
$$

for all $(\boldsymbol{X}, t) \in B \times \mathbb{R}$. Accordingly, two maps, $\boldsymbol{\chi}(\cdot,\cdot)$ and $\theta(\cdot,\cdot)$, will henceforth be called, respectively, a *motion* and a *temperature field* for $\mathscr{B}$ *only* if they meet this restriction.

The procedure formalized in [11.1] may now be described as follows: let *any* motion $\boldsymbol{\chi}(\cdot,\cdot)$ and *any* temperature field $\theta(\cdot,\cdot)$ be assigned on $B \times \mathbb{R}$; then (11.5) and the constitutive equations (11.10) enable one to calculate an associated density $\varrho(\cdot,\cdot)$, energy $\varepsilon(\cdot,\cdot)$, entropy $\eta(\cdot,\cdot)$, stress $\boldsymbol{T}(\cdot,\cdot)$, heat flux $\boldsymbol{q}(\cdot,\cdot)$, and

7 For simplicity we study only materials for which the *response functions* $\hat{\varepsilon}(\cdot)$, $\hat{\eta}(\cdot)$, $\hat{\boldsymbol{T}}(\cdot)$, $\hat{\boldsymbol{q}}(\cdot)$, (and then $\hat{\psi}(\cdot)$) can be taken as the same for each particule $\boldsymbol{X} \in B$. Thus, we suppose that the material comprising $\mathscr{B}$ is *homogeneous* and that B is one of its homogeneous reference configurations. We henceforth thus also suppose that the referential mass density $\varrho_R(\boldsymbol{X})$ is independent of $\boldsymbol{X}$.

free energy $\psi(\cdot,\cdot)$. Entering these fields into the local balance laws (11.6, 7), we may in turn calculate an associated body force $\boldsymbol{b}(\cdot,\cdot)$ and radiant heating $r(\cdot,\cdot)$. We thus have produced an ordered 8-tuple of functions on $B \times \mathbb{R}$, viz., and in a notation that anticipates our next axiom,

$$\pi_{(\chi,\theta)} \equiv \{\chi, \theta; \varepsilon, \eta, \boldsymbol{T}, \boldsymbol{q}, \boldsymbol{b}, r\} ;$$

we call it the ordered 8-tuple *induced* by the motion-temperature pair (χ, θ) and observe that it automatically meets balance of linear momentum and balance of energy.

The idea of *Coleman* and *Noll* [10.1] is to now make the following fundamental

Axiom of Size. *Every induced 8-tuple* $\pi_{(\chi,\theta)}$ *is a process,* i.e., *belongs to the process class* $\mathbb{P}(\mathscr{B})$ *and so must satisfy the dissipation inequality* (11.8).

This axiom ensures that $\mathbb{P}(\mathscr{B})$ is now extremely rich in processes – indeed, so rich that the only way the dissipation inequality (11.8) can be satisfied is for our choices of the response functions $\hat{\varepsilon}(\cdot)$, $\hat{\eta}(\cdot)$, $\hat{\boldsymbol{T}}(\cdot)$, $\hat{\boldsymbol{q}}(\cdot)$, and $\hat{\psi}(\cdot)$ to be correspondingly impoverished. In fact, using essentially the methods of [11.1 – 3] it is now not hard to prove

Theorem 1. *Let the classic balance laws* (11.2 – 4) *hold. If a material with the constitutive structure* (11.10) *satisfies the Axiom of Size, then it is necessary that the free energy* $\hat{\psi}(\cdot)$

i) *be independent of* $\nabla \boldsymbol{F}$, $\nabla^2 \boldsymbol{F}$, $\boldsymbol{g}$, and $\dot{\boldsymbol{F}}$, i.e.,

$$\psi = \hat{\psi}(\boldsymbol{F}, \theta) \equiv \hat{\psi}(\boldsymbol{F}, \theta, 0, 0, 0, 0) ,$$

ii) *serve as a potential for the entropy* $\hat{\eta}(\cdot)$, *the energy* $\hat{\varepsilon}(\cdot)$, *and the* ***equilibrium stress*** $\boldsymbol{T}^{\mathrm{E}}(\cdot) \equiv \hat{\boldsymbol{T}}(\cdot)\,|_{\boldsymbol{g}=\dot{\boldsymbol{F}}=0}$, *in the sense that*

$$\begin{aligned} \eta &= -\hat{\psi}_{\theta}(\boldsymbol{F}, \theta) , \\ \varepsilon &= \hat{\psi}(\boldsymbol{F}, \theta) - \theta \hat{\psi}_{\theta}(\boldsymbol{F}, \theta) , \\ \boldsymbol{T}^{\mathrm{E}} &= \varrho \hat{\psi}_{\boldsymbol{F}}(\boldsymbol{F}, \theta) \boldsymbol{F}^{\mathrm{T}} . \end{aligned}$$

Here subscripts denote partial differentiation, and, as Theorem 1 makes clear, for materials satisfying its hypotheses classical thermodynamics drastically limits the way higher order spatial interaction effects can be directly modelled within the response function for energy, entropy, and stress – indeed, the energy ε and the entropy η can never be directly influenced by $\nabla \boldsymbol{F}$, $\nabla^2 \boldsymbol{F}$, $\boldsymbol{g}$, or $\dot{\boldsymbol{F}}$, nor can the stress $\boldsymbol{T}$ be directly influenced by $\nabla \boldsymbol{F}$ and $\nabla^2 \boldsymbol{F}$ at any particle in *local equilibrium* (i.e., at which $\boldsymbol{g} = \dot{\boldsymbol{F}} = 0$). In particular, as a glance at (11.5) reveals, we have the immediate corollary result that Korteweg's model (11.1) [8] satisfies the Axiom of

[8] More precisely, Korteweg's model (11.1) for the stress $\boldsymbol{T}$ embedded within classical thermodynamics, with accompanying response functions for ε, η, and $\boldsymbol{q}$ defined on the same domain as that for $\boldsymbol{T}$.

Size only if all its non-classical coefficients α, β, δ, and γ vanish identically! A way to escape such consequences of classical thermodynamics is the subject of the next section.

11.3 A Nonclassical Continuum Thermodynamics: Interstitial Work Flux

As we have seen, if one wants a constitutive structure like, say, (11.10) in which higher gradients of $\boldsymbol{F}$ and θ genuinely enter, and if one wants something like the Axiom of Size so as to deduce constitutive restrictions[9], then the normal laws of continuum thermodynamics must be modified. The particular modification Serrin and I posited in [11.25] was one that was partially foreshadowed in theories of liquid crystals [11.24], in theories of polar media [11.22, 23] and materials with couple-stresses [11.20, 21], and in other theories meant to model mildly long range spatial interaction effects between the material elements or particles of $\mathscr{B}$ (which may be due to, here implicit, "substructures" of those elements) [11.8, 9, 11, 12]. Specifically, our desire was to preserve both the purely thermal principle of entropy imbalance (11.4) and the purely mechanical principle of linear momentum balance (11.2) along with its attendant notion that the net local contact force between the subparts of $\mathscr{B}$ is delivered in the usual way by the Cauchy stress tensor $\boldsymbol{T}$. Thus, we modified only the energy balance (11.3), and we did this by postulating, for each process π, an *interstitial work flux* $\boldsymbol{u} = \boldsymbol{u}(X, t), (X, t) \in B \times \mathbb{R}$, such that the balance of energy (11.3) for each subpart $P \subseteq B$ is replaced by

$$\frac{d}{dt} \int_{P_t} \varrho\{\varepsilon + \tfrac{1}{2}\dot{\boldsymbol{x}} \cdot \dot{\boldsymbol{x}}\} dv = \int_{\partial P_t} \{\dot{\boldsymbol{x}} \cdot \boldsymbol{T}\boldsymbol{n} + \boldsymbol{u} \cdot \boldsymbol{n} - \boldsymbol{q} \cdot \boldsymbol{n}\} da + \int_{P_t} \varrho\{\dot{\boldsymbol{x}} \cdot \boldsymbol{b} + r\} dv \,. \tag{11.3'}$$

The flux $\boldsymbol{u}$, as its name suggests, seems here to be most appropriately thought of as mechanical in nature – see (11.13). We attribute $\boldsymbol{u}$ to the longer range spatial interactions we are trying to model, and we note that it engenders *an interstitial working*, i.e., a rate of supply $\boldsymbol{u} \cdot \boldsymbol{n}$ of energy, across every material surface in $\mathscr{B}$ above and beyond the usual mechanical energy supply due to the working $\dot{\boldsymbol{x}} \cdot \boldsymbol{T}\boldsymbol{n}$ of the surface traction $\boldsymbol{T}\boldsymbol{n}$.

It is interesting to look at our introduction of $\boldsymbol{u}$ in a more abstract way. In the conventional thermodynamics which is embodied in (11.2–4), the *surface energy flux* $\boldsymbol{h}$, beyond that due to the working of the surface tractions, is just the heat flux $\boldsymbol{q}$ and is thus inextricably linked to the *surface entropy flux* $\boldsymbol{j}$ which is

---

[9] As was in essence observed long ago, the special power of the dissipation inequality (11.8) to restrict response functions relies crucially on the Axiom of Size or something very much like it. Without such an axiom, one has only the basic laws (11.6–8), all now on an equal footing, and all now serving merely as necessary conditions on the elements of $\mathbb{P}(\mathscr{B})$. With such limited structure, if $\mathbb{P}(\mathscr{B})$ is small enough, almost any response functions will be compatible with (11.6–8).

just q/θ. Our introduction of the interstitial work flux u has severed this link since now $h = q - u$ while j is still given by q/θ. Mathematically therefore our theory includes and is equivalent to either a theory in which h was kept equal to (interpreted as) the heat flux q but the entropy flux j was taken to be given by $q/\theta - k$, where k would represent a flux of entropy due to longer range spatial interactions, or a theory in which longer range spatial interactions resulted in neither h nor j having any simple dependence on (interpretation in terms of) the heat flux q (which, in fact, would then disappear from this theory[10]). Thus, although motivated differently, our theory has formal similarities to that of *Müller* [11.27].

For us then, the process class $\mathbb{P}(\mathscr{B})$ of a body will henceforth be composed of processes π which are certain ordered *9-tuples* of functions on $B \times \mathbb{R}$,

$$\pi = \{\chi, \theta, \varepsilon, \eta, T, q, u, b, r\} ,$$

with the physical interpretations and mathematical properties we have discussed above, and which satisfy the balance of linear momentum (11.2), the imbalance of entropy (11.4), and the balance of energy in the new form (11.3′). Paralleling (11.6 – 8), it is easy to show then that (11.2), (11.3′), and (11.4) are, given sufficient smoothness and (11.5), equivalent to the local conditions

$$\operatorname{div} T + \varrho b = \varrho \ddot{x} , \tag{11.6}$$

$$\varrho \dot{\varepsilon} = T \cdot L - \operatorname{div} q + \operatorname{div} u + \varrho r , \tag{11.7'}$$

$$\varrho\{\dot{\psi} + \eta \dot{\theta}\} - T \cdot L - \operatorname{div} u + \frac{q \cdot g}{\theta} \leqslant 0 . \tag{11.8'}$$

As we did with (11.8), we will henceforth refer to (11.8′) as the *dissipation inequality*.

In addition to entering u into the balance of energy, we also – and this is crucial for our theory – adjoin u to the constitutive structure of $(11.10)_{1-5}$. That is, like $\hat{\varepsilon}(\cdot)$, $\hat{\eta}(\cdot)$, $\hat{T}(\cdot)$, $\hat{q}(\cdot)$, and $\hat{\psi}(\cdot)$, there is a response function $\hat{u}(\cdot)$ giving the values of $u(X, t)$ for every process $\pi \in \mathbb{P}(\mathscr{B})$ according to

$$u = \hat{u}(F, \theta, \nabla F, \nabla^2 F, g, \dot{F}) . \tag{11.11}$$

Now the presence of $\dot{F}$ in (11.11) and our suggested interpretation of $u \cdot n$ as a rate of supply of mechanical energy might motivate one, based on the formal analogy with the working of the surface tractions $(T^{\mathrm{T}} \dot{x}) \cdot n$, to require that the interstitial working

$$u \cdot n = \hat{u}(F, \theta, \nabla F, \nabla^2 F, g, \dot{F}) \cdot n$$

[10] Note that, in terms of this last alternate, our theory can be seen as a way of reintroducing the heat flux q by *defining* $q \equiv \theta j$ and then $u = q - h$. The first alternate theory, of course, is just the reintroduction of the heat flux q by the identifications $q \equiv h$ and then $k \equiv q/\theta - j$. As these alternate but equivalent theories suggest, outside the traditional structure (11.2 – 4) the identification of the "real" heat flux is a rather subtle affair.

be linear in $\dot{F}$. We shall *not* make such an assumption, and indeed the form of $\dot{F}$'s appearance in $\hat{u}(\cdot)$ will be one of the main results of the theorems we give below.

Finally, we need an appropriately amended form of the Axiom of Size to accommodate our new theory's local laws (11.6), (11.7′), and (11.8′), as well as its constitutive structure, $(11.10)_{1-5}$ and (11.11). But this is straightforward: we repeat verbatim the first half of our earlier prescription for producing an induced 8-tuple associated with any motion $\chi(\cdot,\cdot)$ and any temperature field $\theta(\cdot,\cdot)$ and now simply observe that the constitutive equation (11.11) lets us also calculate an interstitial work flux $u(\cdot,\cdot)$ associated with $\chi(\cdot,\cdot)$ and $\theta(\cdot,\cdot)$. We now use (11.6) and (11.7′) to compute an associated body force $b(\cdot,\cdot)$ and radiant heating $r(\cdot,\cdot)$. We now have thus produced an ordered 9-tuple of functions defined on $B \times \mathbb{R}$, viz.,

$$\pi_{(\chi,\theta)} \equiv \{\chi, \theta; \varepsilon, \eta, T, q, u, b, r\},$$

which was *induced* by the motion-temperature pair (χ, θ). It now makes sense to postulate the following modified

Axiom' of Size. *Every induced 9-tuple* $\pi_{(\chi,\theta)}$ *is a process and so must satisfy the dissipation inequality* (11.8′).

We now prove a fundamental theorem delimiting the possible functional forms of $\hat{\psi}(\cdot)$ and $\hat{u}(\cdot)$. To state it, let us introduce the function $\bar{u}(\cdot)$ given by

$$\bar{u}(F, \theta, \nabla F, \nabla^2 F, g, L) \equiv \hat{u}(F, \theta, \nabla F, \nabla^2 F, g, LF).$$

Since the spatial velocity gradient L and the time rate of F are related by $L = \dot{F}F^{-1}$ during any motion, it is clear that u may be equally well computed during any process π by use of $\hat{u}(\cdot)$ or $\bar{u}(\cdot)$. Let us also recall that, for any second order tensor A on V, the adjugate of A is the unique second order tensor A^* satisfying $A^{*T}(a \wedge b) = (Aa) \wedge (Ab)$ for all a and b in V. As a consequence, one may easily show that

$$A^* = A^2 - \mathrm{I}A + \mathrm{II}\,\mathbf{1},$$

where I and II are, respectively, the first and second principal invariants of A. Finally, if $F_{i\alpha}$ and $F_{i\alpha\beta}$ denote, respectively, the Cartesian components of F and ∇F, let us define a third order tensor $\mathbb{K}$ on V by setting

$$\mathbb{K}_{ijk} = \varrho \frac{\partial \hat{\psi}}{\partial F_{j\alpha\beta}} F_{i\alpha} F_{k\beta}.$$

Note that $\mathbb{K}$ is symmetric in its first and third places.

We now have

Theorem 2. *Let the balance laws* (11.2), (11.3′), and (11.4) *hold. If a material with the constitutive structure* (11.10) *and* (11.11) *satisfies the Axiom' of Size, then it is necessary that the free energy* $\hat{\psi}(\cdot)$

i) *be independent of* $\nabla^2 F$, g, *and* $\dot{F}$, *i.e.,*

$$\psi = \hat{\psi}(F, \theta, \nabla F) \equiv \hat{\psi}(F, \theta, \nabla F, 0, 0, 0),$$

ii) *serve as a potential for the entropy* $\hat{\eta}(\cdot)$ *and the energy* $\hat{\varepsilon}(\cdot)$ *in the sense that*

$$\eta = -\hat{\psi}_\theta(\boldsymbol{F}, \theta, \nabla \boldsymbol{F}),$$
$$\varepsilon = \hat{\psi}(\boldsymbol{F}, \theta, \nabla \boldsymbol{F}) - \theta \hat{\psi}_\theta(\boldsymbol{F}, \theta, \nabla \boldsymbol{F}).$$

It is also necessary that

iii) *the interstitial work flux* $\bar{\boldsymbol{u}}(\cdot)$ *be at most of the form*

$$\begin{aligned} \boldsymbol{u} &= \bar{\boldsymbol{u}}(\boldsymbol{F}, \theta, \nabla \boldsymbol{F}, \nabla^2 \boldsymbol{F}, \boldsymbol{g}, \boldsymbol{L}), \\ &= \boldsymbol{w} + \{\mathbb{K} + \mathbb{E}\}\boldsymbol{L} + \boldsymbol{L}^* \boldsymbol{m}, \end{aligned} \qquad (11.13)^{11}$$

where the third order tensor $\mathbb{E}$ *is skew in its first and third places, and where* $\mathbb{E}$ *and the vectors* $\boldsymbol{w}$ *and* $\boldsymbol{m}$ *may depend on at most* $\boldsymbol{F}$, θ, $\nabla \boldsymbol{F}$, $\nabla^2 \boldsymbol{F}$, *and* $\boldsymbol{g}$.

Theorem 2 tells us that $\bar{\boldsymbol{u}}(\cdot)$ is at most the sum of a term $\boldsymbol{w}$ which is independent of the deformation rate, a term $\{\mathbb{K} + \mathbb{E}\}\boldsymbol{L}$ which is linear in the deformation rate and which is partially determined by the free energy, and a term $\boldsymbol{L}^*\boldsymbol{m}$ which is quadratic in the deformation rate. We will analyze each of these terms below, but for now we just note that, in a thermodynamics with no interstitial work flux $\boldsymbol{u}$, it follows from (11.13) that $\mathbb{K}$ must vanish identically (as must also $\boldsymbol{w}$, $\mathbb{E}$, and $\boldsymbol{m}$). Thus, we would then have $\hat{\psi}_{\nabla \boldsymbol{F}} \equiv 0$, and ψ would become a function of $\boldsymbol{F}$ and θ alone – certain longer range spatial interaction effects would thus drop out of (not be modelable in terms of) $\hat{\psi}(\cdot)$.

We now sketch the proof of Theorem 2.

Proof. When the chain rule is applied to the dissipation inequality (11.8′), one finds the condition

$$\begin{aligned} &\varrho[\hat{\psi}_{\boldsymbol{F}} \cdot \dot{\boldsymbol{F}} + \{\hat{\psi}_\theta + \hat{\eta}\}\dot{\theta} + \hat{\psi}_{\nabla \boldsymbol{F}} \cdot \dot{\overline{\nabla \boldsymbol{F}}} + \hat{\psi}_{\boldsymbol{g}} \cdot \dot{\boldsymbol{g}} + \hat{\psi}_{\nabla^2 \boldsymbol{F}} \cdot \dot{\overline{\nabla^2 \boldsymbol{F}}} + \hat{\psi}_{\dot{\boldsymbol{F}}} \cdot \ddot{\boldsymbol{F}}] - \hat{\boldsymbol{T}}\boldsymbol{F}^{-1^{\mathrm{T}}} \cdot \dot{\boldsymbol{F}} \\ &- \hat{\boldsymbol{u}}_\theta \cdot \boldsymbol{g} - \hat{\boldsymbol{u}}_{\boldsymbol{g}} \cdot \boldsymbol{G} - [\hat{\boldsymbol{u}}_{\boldsymbol{F}} \diamond \nabla \boldsymbol{F} + \hat{\boldsymbol{u}}_{\nabla \boldsymbol{F}} \diamond \nabla^2 \boldsymbol{F} + \hat{\boldsymbol{u}}_{\nabla^2 \boldsymbol{F}} \diamond \nabla^3 \boldsymbol{F} + \hat{\boldsymbol{u}}_{\dot{\boldsymbol{F}}} \diamond \nabla \dot{\boldsymbol{F}}] \cdot \boldsymbol{F}^{-1^{\mathrm{T}}} \\ &+ \frac{\hat{\boldsymbol{q}} \cdot \boldsymbol{g}}{\theta} \leqslant 0, \end{aligned} \qquad (11.14)$$

where all of the indicated response functions and their partial derivatives are evaluated at $(\boldsymbol{F}, \theta, \nabla \boldsymbol{F}, \nabla^2 \boldsymbol{F}, \boldsymbol{g}, \dot{\boldsymbol{F}}) \in \mathscr{I}$, where $\boldsymbol{G} = \operatorname{grad}^2 \theta$, and where for any two tensors of order n, $\boldsymbol{\Gamma}$ and $\boldsymbol{\Phi}$, we define $\boldsymbol{\Gamma} \diamond \boldsymbol{\Phi}$ to be the second order tensor such that, in Cartesian components,

$$(\boldsymbol{\Gamma} \diamond \boldsymbol{\Phi})_{ij} = \Gamma_{ipq\ldots t} \Phi_{pq\ldots tj}.$$

Now, given any point Λ in the constitutive domain $\mathscr{I}$, we can always find a motion $\chi(\cdot,\cdot)$, a temperature field $\theta(\cdot,\cdot)$, and a particle time pair $(\boldsymbol{X}_0, t_0) \in \boldsymbol{B} \times \mathbb{R}$ such that

$$(\boldsymbol{F}, \theta, \nabla \boldsymbol{F}, \nabla^2 \boldsymbol{F}, \boldsymbol{g}, \dot{\boldsymbol{F}})(\boldsymbol{X}_0, t_0) = \Lambda,$$

[11] In Cartesian components then (11.13) asserts that

$$u_i = w_i + \{\mathbb{K}_{ijk} + E_{ijk}\}L_{jk} + L^*_{ij} m_j.$$

while at (X_0, t_0) the seven quantities $\dot{\theta}, \nabla\dot{F}, \nabla^2\dot{F}, \dot{g}, \ddot{F}, G$, and $\nabla^3 F$ are independent of Λ and arbitrary (up to certain symmetries which I leave implicit). Moreover, this motion-temperature field may then be used to produce an ordered 9-tuple which, by the Axiom' of Size, will be a process and so must satisfy (11.14). But, in (11.14), $\dot{\theta}, \nabla\dot{F}, \nabla^2\dot{F}, \dot{g}, \ddot{F}, G$ and $\nabla^3 F$ appear linearly; we conclude at once that

$$\hat{\psi}_\theta + \hat{\eta} \equiv 0\,, \qquad \hat{\psi}_g \equiv 0\,,$$
$$\hat{\psi}_{\nabla^2 F} \equiv 0\,, \qquad \hat{\psi}_{\dot{F}} \equiv 0\,, \tag{11.15}$$

and

$$(\hat{\boldsymbol{u}}_{\dot{F}} \diamond \nabla\dot{F}) \cdot F^{-1^{\mathrm{T}}} \equiv \varrho\,\psi_{\nabla F} \cdot \nabla\dot{F}\,,$$
$$\hat{\boldsymbol{u}}_g \cdot G \equiv 0\,, \tag{11.16}$$
$$(\hat{\boldsymbol{u}}_{\nabla^2 F} \diamond \nabla^3 F) \cdot F^{-1^{\mathrm{T}}} \equiv 0\,,$$

where, up to their respective implicit symmetries, $\nabla\dot{F}$, G, and $\nabla^3 F$ are arbitrary.

When integrated, $(11.15)_{2,3,4}$ tell us that $\hat{\psi}(\cdot)$ is independent of g, $\nabla^2 F$, and $\dot{F}$, as we had claimed. Further, $(11.15)_1$ then tells us that $\hat{\eta}(\cdot)$ enjoys the same independence and that, in fact, $\hat{\eta} = -\hat{\psi}_\theta$. That $\varepsilon = \psi - \theta\psi_\theta$ now follows easily since $\varepsilon = \psi + \theta\eta$.

The condition $(11.16)_2$ is well known to be equivalent to the requirement that $\hat{\boldsymbol{u}}(\cdot)$ be an affine function of g with skew linear part. The conditions $(11.16)_{1,3}$ are more difficult to analyze, but we note that, in terms of the third order tensor $\mathbb{K}$ of (11.12) and the function $\bar{\boldsymbol{u}}(\cdot)$, we may write $(11.16)_1$ in the form

$$\bar{\boldsymbol{u}}_L \cdot \mathbb{G} = \mathbb{K} \cdot \mathbb{G} \tag{11.17}$$

for every third order tensor $\mathbb{G}$, symmetric in its first and third places, and where

$$(\bar{\boldsymbol{u}}_L)_{ijk} = \frac{\partial \bar{u}_i}{\partial L_{jk}}\,.$$

Observe that, by our earlier result concerning $\hat{\psi}(\cdot)$, the tensor $\mathbb{K}$ is independent of L. But (11.17) is exactly the functional restriction addressed by the representation theorem given below. We thus conclude that (11.13) holds; the proof of Theorem 2 is now complete. ■

Our proof of Theorem 2 depends critically on solving the functional restriction (11.17). We state here the relevant theorem; its proof will appear elsewhere.

A Representation Theorem. *Let V be a three dimensional inner product space, let T be the set of second order tensors on V, and let $\mathscr{D} \subseteq T$ be open and connected. If $\boldsymbol{u}(\cdot)$: $\mathscr{D} \to V$ is once continuously differentiable and satisfies*

$$\boldsymbol{u}_L(L) \cdot \mathbb{G} = \mathbb{K} \cdot \mathbb{G} \tag{A}$$

for all $L \in \mathscr{D}$, for all third order tensors $\mathbb{G}$, symmetric in their first and third places, and for some third order tensor $\mathbb{K}$, symmetric in its first and third places, then $\boldsymbol{u}(\cdot)$ is necessarily of the form

$$u = u(L) = w + \{\mathbb{K} + \mathbb{E}\}L + L^{*}m \ , \tag{B}$$

where the vectors w *and* m, and the third order tensor $\mathbb{E}$ *are independent of* L. *Additionally,* $\mathbb{E}$ *is skew in its first and third places.*

Conversely, if (B) *holds on* $\mathcal{D}$, *then* $u(\cdot)$ *is continuously differentiable and satisfies* (A).

If we use (11.13) and compute the divergence of u on any process (so $L = \operatorname{grad} v$ for some velocity field v), we find that

$$\operatorname{div} u = \operatorname{div} w + L \cdot \operatorname{div}\{\mathbb{K} + \mathbb{E}\} + L^{*\mathrm{T}} \cdot \operatorname{grad} m + \{\mathbb{K} + \mathbb{E}\} \cdot \operatorname{grad}^2 v^{\mathrm{T}} + m \cdot \operatorname{div} L^{*\mathrm{T}}$$

where the divergence of any third order tensor field $\mathbb{G} = \mathbb{G}(x)$ is defined to be the second order tensor $\operatorname{div}\mathbb{G}$ such that $A \cdot \operatorname{div}\mathbb{G} = \operatorname{div}(\mathbb{G}A)$ for all second order tensors A. Now the term $\operatorname{div} L^{*\mathrm{T}}$ is identically zero since L is a gradient; moreover, a short calculation, the definition (11.12) of $\mathbb{K}$, and the fact that $\mathbb{E}$ is skew in its first and third places shows that

$$\begin{aligned}\{\mathbb{K} + \mathbb{E}\} \cdot \operatorname{grad}^2 v^{\mathrm{T}} &\equiv \{\mathbb{K}_{ijk} + \mathbb{E}_{ijk}\} v_{j,ik} = \mathbb{K}_{ijk} v_{j,ki} \, , \\ &= \varrho \hat{\psi}_{\nabla F} \cdot \dot{\overline{\nabla F}} - \varrho(\hat{\psi}_{\nabla F} \square \nabla F) \cdot L \ ,\end{aligned}$$

where $(\hat{\psi}_{\nabla F} \square \nabla F)_{ij} \equiv (\partial \hat{\psi} / \partial F_{i\alpha\beta}) F_{j\alpha\beta}$. Consequently, we have

$$\operatorname{div} u = \operatorname{div} w + \varrho \hat{\psi}_{\nabla F} \cdot \dot{\overline{\nabla F}} + L^{*\mathrm{T}} \cdot \operatorname{grad} m + L \cdot [\operatorname{div}\{\mathbb{K} + \mathbb{E}\} - \varrho \hat{\psi}_{\nabla F} \square \nabla F] \ .$$

When the above form for $\operatorname{div} u$ is substituted into the dissipation inequality (11.8′) and use is made of (i) and (ii) of Theorem 2, one finds that

$$L \cdot [\varrho \psi_F F^{\mathrm{T}} + \varrho \psi_{\nabla F} \square \nabla F - \operatorname{div}\{\mathbb{K} + \mathbb{E}\} - T] - L^{*\mathrm{T}} \cdot \operatorname{grad} m + \frac{q \cdot g}{\theta} - \operatorname{div} w \leqslant 0 \, , \tag{11.18}$$

which we will call the *reduced dissipation inequality.* (It should be noted that the restrictions imposed by $(11.16)_{2,3}$ on w, $\mathbb{E}$, and m have been left tacit in (11.18). In Sect. 11.5, for materials of Korteweg type, we solve them completely.)

11.4 Forms and Effects of the Interstitial Work Flux

In classical thermomechanics the working of forces is always a linear form in the velocity $\dot{x}$, e.g., $(Tn) \cdot \dot{x}$ and $(\varrho b) \cdot \dot{x}$. Of the three terms w, $\{\mathbb{K} + \mathbb{E}\}L$, and $L^{*}m$ appearing in representation (11.13) for the interstitial work flux u, only the term $\{\mathbb{K} + \mathbb{E}\}L$ possesses this classic structure of linearity in the velocity field; the term w, which allows for energy fluxes even at particles in local equilibrium, and the term $L^{*}m$, which is quadratic in the velocity field, are very much nonstandard. They can, of course, always be removed by fiat: one merely postulates that $w \equiv m \equiv 0$. Moreover, the condition $w \equiv 0$ follows at once if we were to adopt the postulate that

$$u = 0 \quad \textit{if} \quad \mathscr{B} \quad \text{is } \textit{motionless} . \tag{P$_1$}$$

Similarly, both $\boldsymbol{w} \equiv 0$ and $\boldsymbol{m} \equiv 0$ would follow if we were to adopt the postulate that

$$u = 0 \quad \textit{if} \quad \mathscr{B} \quad \textit{moves rigidly} . \tag{P$_2$}$$

Indeed, since $\boldsymbol{L}$ is skew whenever $\chi(\cdot,\cdot)$ is rigid, this second postulate requires that

$$\boldsymbol{w} + \{\mathbb{K} + \mathbb{E}\}\boldsymbol{W} + \boldsymbol{W}^*\boldsymbol{m} = 0$$

for all skew tensors $\boldsymbol{W}$. As a consequence, not only must $\boldsymbol{w}$ and $\boldsymbol{m}$ both vanish, but also we must have that

$$\{\mathbb{K} + \mathbb{E}\}\boldsymbol{W} = 0$$

for every skew tensor $\boldsymbol{W}$. As we shall see in a moment, this last condition is enough to totally determine $\mathbb{E}$ in terms of $\mathbb{K}$ (see $(11.20)_2$ below), so that the postulate (P_2) is, in fact, quite strong.

Here, rather than adopt either of the above additional postulates, I prefer to take a more speculative approach and study the effects upon and interplay between the various parts of $\bar{u}(\cdot)$ under several alternative, additional, and in some cases more familiar, postulates. These postulates are, in effect, constitutive assumptions delimiting the modes of interstitial work flux supported by $\mathscr{B}$. With them the reduced dissipation inequality (11.18) still has much to tell us not only about the various parts of $\bar{u}(\cdot)$, but also about $\hat{\psi}(\cdot)$, $\hat{\boldsymbol{T}}(\cdot)$, and $\hat{\boldsymbol{q}}(\cdot)$, the response functions for, respectively, the free energy, the stress, and the heat flux.

First, recall that the velocity gradient $\boldsymbol{L}$ may be decomposed into its symmetric and skew parts. That is, $\boldsymbol{L} = \boldsymbol{D} + \boldsymbol{W}$, where $\boldsymbol{D} = \boldsymbol{D}^{\mathrm{T}}$ is the *stretching tensor*, and where $\boldsymbol{W} = -\boldsymbol{W}^{\mathrm{T}}$ is the *spin tensor*. A simple calculation now yields that

$$\boldsymbol{L}^* = \boldsymbol{D}^* + \boldsymbol{D}\boldsymbol{W} + \boldsymbol{W}\boldsymbol{D} - (\mathrm{tr}\boldsymbol{D})\boldsymbol{W} + \boldsymbol{\lambda}\otimes\boldsymbol{\lambda} , \tag{11.19}$$

where $\boldsymbol{\lambda}$, the axial vector of $\boldsymbol{W}$, is just $\frac{1}{2}$ of the *vorticity vector*, $\boldsymbol{\omega} = \mathrm{curl}\dot{\boldsymbol{x}}$. Let $\mathbb{K}_{ijk}$ and $\mathbb{E}_{ijk}$ denote, respectively, the Cartesian components of $\mathbb{K}$ and $\mathbb{E}$. We now have

Theorem 3. *The interstitial work flux $\bar{u}(\cdot)$ is independent of the spin tensor $\boldsymbol{W}$, i.e.,*

$$\bar{u}(\boldsymbol{F}, \theta, \nabla\boldsymbol{F}, \nabla^2\boldsymbol{F}, \boldsymbol{g}, \boldsymbol{L}) = \bar{u}(\boldsymbol{F}, \theta, \nabla\boldsymbol{F}, \nabla^2\boldsymbol{F}, \boldsymbol{g}, \boldsymbol{D}) , \tag{P$_3$}$$

where $\boldsymbol{D}$ is the stretching tensor, if and only if, in the representation (11.13),

$$\boldsymbol{m}(\cdot) \equiv 0 \tag{11.20$_1$}$$

and $\mathbb{E}$ is totally determined by $\mathbb{K}$ according to

$$\mathbb{E}_{ijk} = \mathbb{K}_{jki} - \mathbb{K}_{jik} . \tag{11.20$_2$}$$

Proof. As a result of (11.13) and (11.19), we see that (P_3) holds if and only if

$$\{\mathbb{K} + \mathbb{E}\}\boldsymbol{W} + \{\boldsymbol{D}\boldsymbol{W} + \boldsymbol{W}\boldsymbol{D} - (\mathrm{tr}\boldsymbol{D})\boldsymbol{W} + \boldsymbol{\lambda}\otimes\boldsymbol{\lambda}\}\boldsymbol{m} \equiv 0 .$$

If we take $\boldsymbol{D}=0$ in this and make the replacement $\boldsymbol{W} \to \alpha \boldsymbol{W}$ then, since $\boldsymbol{\lambda} \to \alpha \boldsymbol{\lambda}$, we find that

$$\alpha^2(\boldsymbol{\lambda} \otimes \boldsymbol{\lambda})\boldsymbol{m} + \alpha\{\mathbb{K} + \mathbb{E}\}\boldsymbol{W} = 0$$

for all α. Thus, (P_3) implies that

$$(\boldsymbol{\lambda} \otimes \boldsymbol{\lambda})\boldsymbol{m} \equiv 0 \quad \textit{and} \quad \{\mathbb{K} + \mathbb{E}\}\boldsymbol{W} \equiv 0 \,,$$

for all vectors $\boldsymbol{\lambda}$ and all skew tensors $\boldsymbol{W}$. That $(11.20)_1$ must hold is now clear.

In terms of the components of $\mathbb{K}$ and $\mathbb{E}$, the condition that $\{\mathbb{K} + \mathbb{E}\}\boldsymbol{W} = 0$ for all skew $\boldsymbol{W}$, takes the form

$$\mathbb{K}_{ijk} + \mathbb{E}_{ijk} = \mathbb{K}_{ikj} + \mathbb{E}_{ikj} \,.$$

When this equation is written down with the roles of k and i interchanged and then with the roles of j and i interchanged, and when the resulting three equations are summed, one finds, since $\mathbb{E}_{ijk}$ is skew in i and k, that

$$\begin{aligned} 2\,\mathbb{E}_{jik} &= \mathbb{K}_{ikj} + \mathbb{K}_{jki} + \mathbb{K}_{kij} - \mathbb{K}_{ijk} - \mathbb{K}_{jik} - \mathbb{K}_{kji} \,, \\ &= 2\,\mathbb{K}_{ikj} - 2\,\mathbb{K}_{ijk} \,, \end{aligned}$$

where the last holds because $\mathbb{K}_{ijk}$ is symmetric in i and k. We see that $(11.20)_2$ is also implied by (P_3).

Conversely, it is now clear that $(11.20)_1$ and $(11.20)_2$ together imply (P_3). ■

Theorem 3 tells us that when only the local stretching, and not the local spin, is allowed to influence the interstitial flux $\boldsymbol{u}$, then $\boldsymbol{u}$ is necessarily of the form

$$\begin{aligned} \boldsymbol{u} &= \boldsymbol{w} + \mathbb{V}\boldsymbol{D} \,, \\ &= \boldsymbol{w}(\boldsymbol{F}, \theta, \nabla\boldsymbol{F}, \nabla^2\boldsymbol{F}, \boldsymbol{g}) + \mathbb{V}(\boldsymbol{F}, \theta, \nabla\boldsymbol{F})\boldsymbol{D} \,, \end{aligned} \qquad (11.21)_1$$

where the third tensor $\mathbb{V}$ is, by (11.12), (11.13), and $(11.20)_2$, completely determined by $\hat{\psi}_{\nabla F}$ according to

$$\mathbb{V}_{ijk} = \varrho \frac{\partial \hat{\psi}}{\partial F_{j\alpha\beta}} F_{i\alpha} F_{k\beta} + \varrho \frac{\partial \hat{\psi}}{\partial F_{k\alpha\beta}} F_{i\alpha} F_{j\beta} - \varrho \frac{\partial \hat{\psi}}{\partial F_{i\alpha\beta}} F_{j\alpha} F_{k\beta} \,. \qquad (11.21)_2$$

It is not hard to show that

$$\mathbb{V}(\boldsymbol{F}, \theta, \nabla\boldsymbol{F}) = 0 \Leftrightarrow \hat{\psi}_{\nabla F}(\boldsymbol{F}, \theta, \nabla\boldsymbol{F}) = 0 \,,$$

i.e., the dynamic portion $\mathbb{V}\boldsymbol{D}$ of the interstitial work flux $\boldsymbol{u}$ vanishes identically only at points $(\boldsymbol{F}, \theta, \nabla\boldsymbol{F})$ where ψ is locally not sensitive to strain distortions. In the next section, for materials of Korteweg type, we will see that thermodynamics tightly entwines the static part $\boldsymbol{w}$ of $\boldsymbol{u}$ with the temperature gradient and the equilibrium heat flux (in fact, at any particle where either of these vanish $\operatorname{div}\boldsymbol{w}$ must also vanish[12]). Moreover, with one additional postulate (see our $(P_5)_2$), $\boldsymbol{w}$ itself turns out to necessarily vanish at any particle where $\operatorname{grad}\varrho$ vanishes. Under the

[12] See the Corollary to Theorem 11.

assumption (P_3) then, we can begin to see some basis for why the classical thermodynamics of homogeneous processes never found a need to introduce the flux $\boldsymbol{u}$.

Let ε_{ijk} be the usual permutation symbol. Dual to (P_3) is the condition of

Theorem 4. *The interstitial work flux $\bar{\boldsymbol{u}}(\cdot)$ is independent of the stretching tensor $\boldsymbol{D}$, i.e.,*

$$\bar{\boldsymbol{u}}(\boldsymbol{F},\theta,\nabla\boldsymbol{F},\nabla^2\boldsymbol{F},\boldsymbol{g},\boldsymbol{L}) = \bar{\boldsymbol{u}}(\boldsymbol{F},\theta,\nabla\boldsymbol{F},\nabla^2\boldsymbol{F},\boldsymbol{g},\boldsymbol{W}) \,, \tag{P_4}$$

where $\boldsymbol{W}$ is the spin tensor, if and only if, in the representation (11.13),

$$\boldsymbol{m}(\cdot) \equiv 0 \tag{11.22_1}$$

and the third order tensors $\mathbb{K}$ and $\mathbb{E}$ satisfy

$$\mathbb{K}_{ijk} + \mathbb{K}_{ikj} + \mathbb{K}_{jik} = 0 \,, \tag{11.22_2}$$

and

$$\mathbb{E}_{ijk} = \tfrac{1}{3}\{\mathbb{K}_{jik} - \mathbb{K}_{jki}\} + \phi\,\varepsilon_{ijk} \,, \tag{11.22_3}$$

where ϕ is a scalar-valued function of at most $\boldsymbol{F}$, θ, $\nabla\boldsymbol{F}$, $\nabla^2\boldsymbol{F}$, and $\boldsymbol{g}$.

We remark that, by (11.12), it is clear that $(11.22)_2$ constitutes a compatibility condition on the Helmholtz free energy function $\hat{\psi}(\cdot)$. Further, it is now not hard to show that $(11.22)_3$ implies that the linear part of $\bar{\boldsymbol{u}}(\cdot)$ is given by

$$\begin{aligned}\{\mathbb{K}+\mathbb{E}\}\boldsymbol{L} &= \{\mathbb{K}+\mathbb{E}\}\boldsymbol{W} \,,\\ &= \{\boldsymbol{H} - \phi\boldsymbol{1}\}\boldsymbol{\omega} \,,\end{aligned} \tag{11.23_1}$$

where $\boldsymbol{\omega}$ is the vorticity vector, and where $\boldsymbol{H}$ is the traceless tensor given by

$$H_{ps} = -\tfrac{2}{3}\mathbb{K}_{pqr}\varepsilon_{sqr} \,. \tag{11.23_2}$$

Proof. Since $\boldsymbol{W}^* = \boldsymbol{\lambda}\otimes\boldsymbol{\lambda}$, we see by (11.13) and (11.19) that (P_4) holds if and only if

$$\{\mathbb{K}+\mathbb{E}\}\boldsymbol{D} + \{\boldsymbol{D}^* + \boldsymbol{D}\boldsymbol{W} + \boldsymbol{W}\boldsymbol{D} - (\mathrm{tr}\boldsymbol{D})\boldsymbol{W}\}\boldsymbol{m} \equiv 0 \,.$$

If we take $\boldsymbol{W} = 0$ in this and make the replacement $\boldsymbol{D}\to\alpha\boldsymbol{D}$ then, since $\boldsymbol{D}^*\to\alpha^2\boldsymbol{D}^*$, we find that

$$\alpha^2\boldsymbol{D}^*\boldsymbol{m} + \alpha\{\mathbb{K}+\mathbb{E}\}\boldsymbol{D} = 0$$

for all α. It follows easily that (P_4) implies that

$$\boldsymbol{D}^*\boldsymbol{m} \equiv 0 \quad \textit{and} \quad \{\mathbb{K}+\mathbb{E}\}\boldsymbol{D} \equiv 0$$

for all symmetric tensors $\boldsymbol{D}$. If $\boldsymbol{a}$, $\boldsymbol{b}$, and $\boldsymbol{c}$ are mutually perpendicular unit vectors, and if we take $\boldsymbol{D} = \boldsymbol{a}\otimes\boldsymbol{a} + \boldsymbol{b}\otimes\boldsymbol{b}$, then, since $\boldsymbol{D}^* = \boldsymbol{c}\otimes\boldsymbol{c}$, the condition $\boldsymbol{D}^*\boldsymbol{m} \equiv 0$ gives us $(11.22)_1$ at once.

In terms of their components, the requirement on $\mathbb{K}$ and $\mathbb{E}$ that $\{\mathbb{K}+\mathbb{E}\}\boldsymbol{D} = 0$ for every symmetric tensor $\boldsymbol{D}$ means that

$$\mathbb{K}_{ijk} + \mathbb{E}_{ijk} \quad \textit{is skew in } j \textit{ and } k \,.$$

Equivalently,

$$\mathbb{E}_{ijk}+\mathbb{E}_{ikj}+\mathbb{K}_{ijk}+\mathbb{K}_{ikj}=0\,. \qquad (11.24)_1$$

If we permute indices in $(11.24)_1$, we see that

$$\mathbb{E}_{jik}+\mathbb{E}_{jki}+\mathbb{K}_{jik}+\mathbb{K}_{jki}=0\,, \qquad (11.24)_2$$

and

$$\mathbb{E}_{kji}+\mathbb{E}_{kij}+\mathbb{K}_{kji}+\mathbb{K}_{kij}=0\,. \qquad (11.24)_3$$

Then, if we add $(11.24)_{1-3}$, we reach

$$\begin{aligned}&\mathbb{E}_{ijk}+\mathbb{E}_{ikj}+\mathbb{E}_{kji}+\mathbb{E}_{kij}+\mathbb{E}_{jik}+\mathbb{E}_{jki}\\ &+\mathbb{K}_{ijk}+\mathbb{K}_{ikj}+\mathbb{K}_{kji}+\mathbb{K}_{kij}+\mathbb{K}_{jik}+\mathbb{K}_{jki}=0\,.\end{aligned}$$

Upon recalling that $\mathbb{K}_{ijk}$ is symmetric in i and k, while $\mathbb{E}_{ijk}$ is skew in i and k, we see at once that this last equation collapses to precisely $(11.22)_2$.

That (P_4) also implies $(11.22)_3$ can be shown as follows. Define

$$\mathbb{G}_{ijk}=3\,\mathbb{E}_{ijk}-\mathbb{K}_{jik}+\mathbb{K}_{jki}\,.$$

Since $\mathbb{E}_{ijk}$ is skew in i and k, it is clear that $\mathbb{G}_{ijk}$ is also skew in i and k. Further, this asymmetry of $\mathbb{E}_{ijk}$ and $(11.24)_3$ lets us write that

$$\begin{aligned}\mathbb{G}_{jik}&=3\,\mathbb{E}_{jik}-\mathbb{K}_{ijk}+\mathbb{K}_{ikj}\,,\\ &=3\{\mathbb{E}_{kji}+\mathbb{K}_{kji}+\mathbb{K}_{kij}\}-\mathbb{K}_{ijk}+\mathbb{K}_{ikj}\,,\\ &=-3\,\mathbb{E}_{ijk}+2\,\mathbb{K}_{kji}+3\,\mathbb{K}_{kij}+\mathbb{K}_{jki}\end{aligned}$$

where we have used the symmetry of $\boldsymbol{K}_{ijk}$ in i and k. But, by $(11.22)_2$,

$$\mathbb{K}_{kji}+\mathbb{K}_{kij}=-\mathbb{K}_{jki}\,,$$

and therefore

$$\begin{aligned}\mathbb{G}_{jik}&=-3\,\mathbb{E}_{ijk}+\mathbb{K}_{kij}-2\,\mathbb{K}_{jki}+\mathbb{K}_{jki}\,,\\ &=-3\,\mathbb{E}_{ijk}+\mathbb{K}_{jik}-\mathbb{K}_{jki}\,,\\ &=-\mathbb{G}_{ijk}\,.\end{aligned}$$

That is, $\mathbb{G}_{ijk}$ is also skew in i and j.

In the same way, we may show that $\mathbb{G}_{ijk}$ is skew in j and k; that is $\mathbb{G}_{ijk}$ is completely skew symmetric. As a consequence, it must be a multiple of ε_{ijk}, and this is exactly the assertion $(11.22)_3$.

Finally, it is straightforward to verify that $(11.22)_{1-3}$ and the symmetry of $\mathbb{K}_{ijk}$ in i and k suffice to ensure (P_4). ■

Theorems 3 and 4 tell us that the portion of $\boldsymbol{u}$ quadratic in the velocity necessarily vanishes whenever the interstitial work flux within $\mathscr{B}$ is totally independent of either the spin $\boldsymbol{W}$ or stretching $\boldsymbol{D}$. However, given the nonclassical nature of $\boldsymbol{u}$, such a priori assumptions on $\bar{\boldsymbol{u}}(\cdot)$ might seem premature. We look next therefore at some other and more familiar postulates which, along with the reduced dissipation inequality (11.18), could also be used to justify our dropping the quadratic part of $\boldsymbol{u}$ from the local laws (11.6), (11.7′), and (11.8′). As will become apparent, however, *the quadratic part of* $\boldsymbol{u}$ *could still play a role in the specification of boundary data for* $\mathscr{B}$.

In the reduced dissipation inequality (11.18) the spatial velocity gradient $\boldsymbol{L}=\dot{\boldsymbol{F}}\boldsymbol{F}^{-1}$ appears in a non-explicit way only in so far as it (i.e., $\dot{\boldsymbol{F}}$) enters into the response functions for the stress $\boldsymbol{T}$ and heat flux $\boldsymbol{q}$. Let us say the material comprising $\mathscr{B}$ is *elastic* whenever these two functions do not involve $\dot{\boldsymbol{F}}$. Thus, in an elastic material only $\boldsymbol{F}$, θ, $\nabla\boldsymbol{F}$, $\nabla^2\boldsymbol{F}$, and $\boldsymbol{g}$ enter into the constitutive equations for $\boldsymbol{T}$ and $\boldsymbol{q}$. By the Axiom' of Size, (11.18) must in essence hold for all values of $(\boldsymbol{F},\theta,\nabla\boldsymbol{F},\nabla^2\boldsymbol{F},\boldsymbol{g},\dot{\boldsymbol{F}})$ in the constitutive domain $\mathscr{I}$. Let T denote the set of all second order tensors on V. We now and henceforth assume that $\mathscr{I}$ is of the form $\mathscr{H}\times T$, i.e.,

$$\begin{aligned}&\textit{if}\quad (\boldsymbol{F},\theta,\nabla\boldsymbol{F},\nabla^2\boldsymbol{F},\boldsymbol{g})\in\mathscr{H}\quad\textit{and}\quad \dot{\boldsymbol{F}}\in T\,,\\ &\textit{then}\quad (\boldsymbol{F},\theta,\nabla\boldsymbol{F},\nabla^2\boldsymbol{F},\boldsymbol{g},\dot{\boldsymbol{F}})\in\mathscr{I}\end{aligned}\qquad (\mathrm{I}_1)$$

where $\mathscr{H}$ is open and convex. It is now straightforward to prove

Theorem 5. *If the material comprising* $\mathscr{B}$ *is elastic, then* (i) *in the representation* (11.13) *the vector* $\boldsymbol{m}$ *is a constant,* (ii) *the stress is determined according to*

$$\boldsymbol{T}=\varrho\,\hat{\psi}_{\boldsymbol{F}}\boldsymbol{F}^{\mathrm{T}}+\varrho\,\hat{\psi}_{\nabla\boldsymbol{F}}\square\nabla\boldsymbol{F}-\operatorname{div}\{\mathbb{K}+\mathbb{E}\}\,,$$

and (iii) *the reduced dissipation inequality is now just the requirement that*

$$\frac{\boldsymbol{q}\cdot\boldsymbol{g}}{\theta}-\operatorname{div}\boldsymbol{w}\leqslant 0\,.$$

By (i) of Theorem 5, the quadratic part, $\boldsymbol{L}^*\boldsymbol{m}=(\operatorname{grad}\boldsymbol{v})^*\boldsymbol{m}$, of $\boldsymbol{u}$ will now be divergence-free; as a consequence this term drops completely out of the local laws (11.6), (11.7′), and (11.8′). However, since we have not shown $\boldsymbol{m}=0$, the term $\boldsymbol{L}^*\boldsymbol{m}$ might be of importance for the specification of boundary data for $\mathscr{B}$. We do not pursue this matter here.

By (ii) of Theorem 5, the stress $\boldsymbol{T}$ in an elastic material is determined completely by the free energy ψ and the third order tensor $\mathbb{E}$. However, since $\mathbb{E}$ is skew in its first and third places, $\operatorname{div}\{\operatorname{div}\mathbb{E}\}\equiv 0$. As a consequence

$$\operatorname{div}\boldsymbol{T}=\operatorname{div}[\varrho\,\hat{\psi}_{\boldsymbol{F}}\boldsymbol{F}^{\mathrm{T}}+\varrho\,\hat{\psi}_{\nabla\boldsymbol{F}}\square\nabla\boldsymbol{F}-\operatorname{div}\mathbb{K}]\,,$$

and $\mathbb{E}$ drops completely out of the local form (11.6) of balance of linear momentum. Further, all three of the terms $\varrho\,\hat{\psi}_{\nabla\boldsymbol{F}}\square\nabla\boldsymbol{F}$, $\mathbb{K}$, and $\mathbb{E}$ drop out of the local form (11.7′) of the energy equation: the contribution each of these terms makes to the stress power $\boldsymbol{T}\cdot\boldsymbol{L}$ is exactly cancelled by its respective contribution to $\operatorname{div}\boldsymbol{u}$. Indeed, for elastic materials (11.7′) may be put in the form

$$\varrho\,\theta\dot{\eta}=-\operatorname{div}\boldsymbol{q}+\varrho r+\operatorname{div}\boldsymbol{w}\,,$$

i.e., except for the presence of the term due to $\boldsymbol{w}$, the energy equation for those of our materials which are elastic is of the same form as that in classic thermoelasticity.

All of the terms $\varrho\hat{\psi}_F\square\nabla F$, $\mathbb{K}$, and $\mathbb{E}$ affect crucially, however, the local stress system in $\mathscr{B}$. Indeed, by (11.9), we know that the usual form of the balance of angular momentum holds if and only if T is symmetric; here this means if and only if

$$\varrho\hat{\psi}_F F^{\mathrm{T}} + \varrho\hat{\psi}_{\nabla F}\square\nabla F - \operatorname{div}\{\mathbb{K}+\mathbb{E}\}$$

is symmetric, which necessitates a delicate interplay between the free energy and $\mathbb{E}$, and which usually need *not* hold. However, in the important special case (P_3), when u is independent of the spin tensor, the sum $\mathbb{K}+\mathbb{E}$ is just the tensor $\mathbb{V}$ of (11.21) and is obviously symmetric in its last two places. Therefore, $\operatorname{div}\{\mathbb{K}+\mathbb{E}\}$ will be symmetric and so now

$$\begin{aligned} T = T^{\mathrm{T}} &\Leftrightarrow \hat{\psi}_F F^{\mathrm{T}} + \hat{\psi}_{\nabla F}\square\nabla F \quad \textit{is symmetric}\,, \\ &\Leftrightarrow \hat{\psi}_F(F,\theta,\nabla F)\cdot WF + \hat{\psi}_{\nabla F}(F,\theta,\nabla F)\cdot W\nabla F = 0 \ \forall \text{ skew } W\,, \end{aligned}$$

i.e., if and only if

$$\hat{\psi}(QF,\theta,Q\nabla F) = \hat{\psi}(F,\theta,\nabla F)$$

for all $(F,\theta,\nabla F)$ in the domain of $\hat{\psi}(\cdot)$ and for all proper, orthogonal Q in a neighborhood of $\mathbf{1}$. This last condition, broadened to hold for arbitrary proper, orthogonal Q, is usually referred to as the *invariance of* ψ *under superimposed rigid rotations.* As we see, it ensures the symmetry of T whenever the interstitial work flux satisfies (P_3). On the other hand, the invariance of ψ under superimposed rigid rotations will generally not suffice for the symmetry of the stress in those of our materials whose interstitial work flux meets, say, (P_4).

Proof. To establish Theorem 5, note that the reduced dissipation inequality (11.18) and our assumption (I_1) yield that for all $L \in T$

$$\begin{aligned} -L^{*\mathrm{T}}\cdot[\operatorname{grad} m] + L\cdot[\varrho\hat{\psi}_F F^{\mathrm{T}} + \varrho\hat{\psi}_{\nabla F}\square\nabla F - \operatorname{div}\{\mathbb{K}+\mathbb{E}\} - T] \\ + \left[\frac{q\cdot g}{\theta} - \operatorname{div} w\right] \leqslant 0\,, \end{aligned} \tag{11.25}$$

where, since the material is elastic, each of the three bracketed terms depends on at most $(F,\theta,\nabla F,\nabla^2 F,g)\in\mathscr{H}$. If we replace L in (11.25) with αL, we arrive at a quadratic polynomial in α which must always be nonpositive. This implies that the coefficient of α^2 is itself nonpositive, i.e.,

$$L^{*\mathrm{T}}\cdot\operatorname{grad} m \geqslant 0 \ \ \forall\ L\in T\,.$$

Let e be any unit vector, let $\{e,d,f\}$ be a right handed, orthonormal basis for V, and let a and b be any vectors in V. If we set $L = a\otimes d + b\otimes f$, then a simple calculation shows that $L^* = e\otimes(a\wedge b)$; therefore $\operatorname{grad} m$ must satisfy

$$(a\wedge b)\cdot(\operatorname{grad} m)e \geqslant 0$$

for all vectors a, b, and e, $|e| = 1$, in V. Clearly, then $\operatorname{grad} m$ must vanish identically; equivalently

$$m = m(\nabla\chi(X), \theta(X), \nabla^2\chi(X), \nabla^3\chi(X), \operatorname{grad}\theta(X))$$

must be independent of X for every motion – temperature pair (χ, θ). That $m(\cdot)$: $\mathscr{H} \to V$ must be the constant map is now clear.

Next, since grad$m$ vanishes, the inequality (11.25) is linear in L. Consequently, we can violate (11.25) unless on $\mathscr{H}$

$$\varrho\hat{\psi}_F F^{\mathrm{T}} + \varrho\hat{\psi}_{\nabla F} \square \nabla F - \operatorname{div}\{\mathbb{K} + \mathbb{E}\} - T \equiv 0 ,$$

and

$$\frac{q \cdot g}{\theta} - \operatorname{div} w \leqslant 0 . \quad \blacksquare$$

Two other cases in which the term L^*m, quadratic in the velocity, drops out of the local laws (11.7′) and (11.8′) may also be formulated. To state them, let us, as we did with u, write T and q as functions, $\bar{T}(\cdot)$ and $\bar{q}(\cdot)$ respectively, of the variables $(F, \theta, \nabla F, \nabla^2 F, g) \in \mathscr{H}$ and $L \in T$. For any second order tensor M, let sym(M) and sk(M) denote, respectively, the symmetric and the skew parts of M. Lastly, note that use of (11.19) lets us write the reduced dissipation inequality in the form

$$\begin{aligned} &-[D^* - WD - DW + (\operatorname{tr} D)W + \lambda \otimes \lambda] \cdot \operatorname{grad} m \\ &+ L \cdot [\varrho\bar{\psi}_F F^{\mathrm{T}} + \varrho\hat{\psi}_{\nabla F} \square \nabla F - \operatorname{div}\{\mathbb{K} + \mathbb{E}\} - T] - \operatorname{div} w + \frac{q \cdot g}{\theta} \leqslant 0 . \end{aligned} \tag{11.26}$$

Paralleling the alternate hypotheses (P_3) and (P_4), we can now use (11.26) to prove

Theorem 6. *If the material comprising* $\mathscr{B}$ *is such that* $\bar{T}(\cdot)$ *and* $\bar{q}(\cdot)$ *are both*

A) *independent of the spin tensor* W,

or

B) *independent of the stretching tensor* D,

then in the representation (11.13) *the vector* m *is a constant. Moreover, in Case* A *the skew part of* T *is independent of* D *as well as* W *and is given by*

$$\operatorname{sk}(T) = \operatorname{sk}(\varrho\hat{\psi}_F F^{\mathrm{T}} + \varrho\hat{\psi}_{\nabla F} \square \nabla F) - \operatorname{sk}(\operatorname{div}\{\mathbb{K} + \mathbb{E}\}) ;$$

while in Case B *the symmetric part of* T *is independent of* W *as well as* D *and is given by*

$$\operatorname{sym}(T) = \operatorname{sym}(\varrho\hat{\psi}_F F^{\mathrm{T}} + \varrho\hat{\psi}_{\nabla F} \square \nabla F) - \operatorname{sym}(\operatorname{div}\{\mathbb{K} + \mathbb{E}\}) .$$

We omit the proof of Theorem 6 since it rests on arguments very much like those we have already employed. The only difficult point is in proving a lemma that $a \otimes a \cdot \operatorname{grad} m \geqslant 0$, for all vectors a in V and all motion-temperature pairs (χ, θ), necessitates $m(\cdot)$: $\mathscr{H} \to V$ being a constant map.

Theorem 6 is interesting. It asserts that in Case A, where T and q can depend on the stretching D, this dependence is not completely arbitrary: the skew part of T is essentially elastic, being independent of D and totally determined by $\hat{\psi}(\cdot)$ and $\mathbb{E}(\cdot)$. Similar remarks, of course, apply to the symmetric part of T in Case *B*. Theorem 6 also tells us that the quadratic part L^*m of u will enter into the local laws (11.7′) and (11.8′) only when $\bar{T}(\cdot)$ and/or $\bar{q}(\cdot)$ involve both the stretching D and the spin W. In this regard it is worth recalling that, within the kinetic theory of gases, both Maxwellian Iteration (see *Müller* [11.28]) and the Chapman-Enskog approximation method (see *Edelen* and *McLennan* [11.29]) lead to forms for $\bar{T}(\cdot)$ and $\bar{q}(\cdot)$ that do indeed involve both D and W. While these forms can involve other constitutive variables which are not accounted for in our present theory, it would be interesting to find the forms for $\bar{u}(\cdot)$ implied by the various models used in the kinetic theory. This problem, however, will be taken up in a later paper.

11.5 Materials of Korteweg Type

An important subclass of our general materials (11.10, 11) are the materials of *Korteweg type*. They arise when the constitutive equations of (11.10, 11) are specialized to

$$\begin{aligned} \varepsilon &= \bar{\varepsilon}(\varrho, \theta, d, S, g, L) \;, \\ \eta &= \bar{\eta}(\varrho, \theta, d, S, g, L) \;, \\ T &= \bar{T}(\varrho, \theta, d, S, g, L) \;, \\ q &= \bar{q}(\varrho, \theta, d, S, g, L) \;, \\ u &= \bar{u}(\varrho, \theta, d, S, g, L) \;, \\ \psi &= \bar{\psi}(\varrho, \theta, d, S, g, L) \;, \end{aligned} \tag{11.27}$$

where ϱ is the spatial mass density of (11.5), where $d \equiv \operatorname{grad}\varrho$, $S = S^{\mathrm{T}} \equiv \operatorname{grad}^2\varrho$, and $L = \dot{F}F^{-1} = \operatorname{grad}\dot{x}$, and where we still use $\mathscr{I}$ to denote the common domain of the response functions in (11.27). It is clear that such materials more than include the original proposal (11.1) of Korteweg. As we shall see, however, the form (11.1) is inconsistent with our general thermodynamic structure unless fairly specific relations hold among Korteweg's coefficients α, β, δ, and γ.

When Theorem 2 is applied to materials of Korteweg type, we find at once that S, g, and L drop out of $\bar{\varepsilon}(\cdot)$, $\bar{\eta}(\cdot)$, and $\bar{\psi}(\cdot)$, and indeed

$$\begin{aligned} \psi &= \bar{\psi}(\varrho, \theta, d) \;, \\ \eta &= -\bar{\psi}_\theta(\varrho, \theta, d) \;, \\ \varepsilon &= \bar{\psi}(\varrho, \theta, d) - \theta\bar{\psi}_\theta(\varrho, \theta, d) \;. \end{aligned} \tag{11.28}$$

Further, for materials of Korteweg type the representation (11.13) is just

$$\begin{aligned} u &= \bar{u}(\varrho, \theta, \cdot\;, S, g, L) \;, \\ &= w + \{\mathbb{K} + \mathbb{E}\}L + L^* m \;, \end{aligned}$$

where here the tensor $\mathbb{K}$ of (11.12) is given by

$$\mathbb{K} = -\tfrac{1}{2}\varrho^2\{\mathbf{1} \otimes \bar{\psi}_d + \bar{\psi}_d \otimes \mathbf{1}\} \;, \tag{11.29}$$

and where w, $\mathbb{E}$, and m are functions of at most ϱ, θ, d, S, and g. Finally, a straightforward calculation shows that for materials of Korteweg type

$$\hat{\psi}_F F^{\mathrm{T}} + \hat{\psi}_{\nabla F} \square \nabla F = -\{(\varrho\bar{\psi}_\varrho + \bar{\psi}_d \cdot d)\mathbf{1} + d \otimes \bar{\psi}_d\}.$$

Hence, the reduced dissipation inequality (11.18) becomes here the requirement that

$$L^{*\mathrm{T}} \cdot \operatorname{grad} m + L \cdot [\varrho(\varrho\bar{\psi}_\varrho + \bar{\psi}_d \cdot d)\mathbf{1} + \varrho d \otimes \bar{\psi}_d + \operatorname{div}\{\mathbb{K} + \mathbb{E}\} + T] + \operatorname{div} w - \frac{q \cdot g}{\theta} \geqslant 0 . \tag{11.30}$$

The expression (11.29) for $\mathbb{K}$ in materials of Korteweg type allows for some striking specializations of our general results. Theorem 3 yields

Theorem 7. *In materials of Korteweg type, the interstitial work flux $\bar{u}(\cdot)$ is independent of the spin if and only if $m(\cdot) \equiv 0$ and*

$$\mathbb{E} = \tfrac{1}{2}\varrho^2\{\mathbf{1} \otimes \bar{\psi}_d - \bar{\psi}_d \otimes \mathbf{1}\}.$$

Consequently,

$$\begin{aligned} u &= w - \varrho^2 \bar{\psi}_d (\operatorname{tr} L), \\ &= w + \varrho\dot{\varrho}\bar{\psi}_d .^{13} \end{aligned}$$

Theorem 4 is the main ingredient of

Theorem 8. *In materials of Korteweg type, the interstitial work flux $\bar{u}(\cdot)$ is independent of the stretching if and only if $\mathbb{K} \equiv m(\cdot) \equiv 0$ and $\mathbb{E}_{ijk} = \phi\varepsilon_{ijk}$. Thus, $\bar{\psi}_d(\cdot) \equiv 0$ (i.e., ψ is a function of ϱ and θ alone) and*

$$u = w - \phi \operatorname{curl} \dot{x} ,$$

where ϕ is a scalar-valued function of at most ϱ, θ, and d.

Only two things are not immediately obvious in the above theorem. First is the necessity of $\mathbb{K} \equiv \bar{\psi}_d \equiv 0$, and this follows at once from the representation (11.29) and the compatibility condition $(11.22)_2$. Second is the fact that ϕ cannot depend on S or on g, and this may be shown to follow from the general representation $(11.40)_2$ below.

Theorem 5 has the specialization

Theorem 9. *For an elastic material of Korteweg type, the vector m of (11.13) is a constant, and the stress is determined according to*

$$\begin{aligned} T &= -\varrho\{\varrho\bar{\psi}_\varrho + d \cdot \bar{\psi}_d\}\mathbf{1} - \varrho d \otimes \bar{\psi}_d - \operatorname{div}\{\mathbb{K} + \mathbb{E}\}, \\ &= -\varrho^2\{\bar{\psi}_\varrho - \tfrac{1}{2}\operatorname{div}\bar{\psi}_d\}\mathbf{1} + \tfrac{1}{2}\varrho^2 \operatorname{grad}\bar{\psi}_d^{\mathrm{T}} - \operatorname{div}\mathbb{E} . \end{aligned} \tag{11.31}$$

A simple calculation, based on the first of (11.31), shows that we can also write T in the forms

$$\begin{aligned} T &= \varrho\{\operatorname{div}(\varrho\bar{\psi}_d) - (\varrho\bar{\psi})_\varrho + \psi\}\mathbf{1} - \varrho d \otimes \bar{\psi}_d - \operatorname{div}\{\varrho^2\bar{\psi}_d \otimes \mathbf{1} + \mathbb{K} + \mathbb{E}\}, \\ &= \varrho\{\psi - \mu\}\mathbf{1} - d \otimes (\varrho\bar{\psi})_d - \operatorname{div}\{\varrho^2\bar{\psi}_d \otimes \mathbf{1} + \mathbb{K} + \mathbb{E}\}, \end{aligned} \tag{11.32}$$

---

[13] By (11.5), $\dot{\varrho} = -\varrho(\operatorname{tr} L)$.

where $\varrho^2 \bar{\psi}_d \otimes \mathbf{1} + \mathbb{K} = \frac{1}{2}\varrho^2\{\bar{\psi}_d \otimes \mathbf{1} - \mathbf{1} \otimes \bar{\psi}_d\}$ is, like $\mathbb{E}$, skew in its first and third places, and where, in terms of the Helmholtz free energy *per unit current volume* $\varrho\bar{\psi}$, we have defined the (chemical) potential

$$\mu = (\varrho\bar{\psi})_\varrho - \operatorname{div}(\varrho\bar{\psi})_d \,. \tag{11.33}$$

Before continuing our general development, it is interesting to apply (11.31) and (11.32) to the special forms for $\mathbb{E}$ found in Theorems 7 and 8. When $\mathbb{E}$ is as in Theorem 7, $(11.32)_1$ yields that

$$\boldsymbol{T} = \varrho\{\operatorname{div}(\varrho\bar{\psi}_d) - (\varrho\psi)_\varrho + \psi\}\mathbf{1} - \varrho\boldsymbol{d} \otimes \bar{\psi}_d \,.$$

From this it is clear that $\boldsymbol{T}$ will be symmetric if and only if $\bar{\psi}_d(\varrho, \theta, \boldsymbol{d})$ is parallel to $\boldsymbol{d}$, and this occurs if and only if $\psi = \psi(\varrho, \theta, M)$, $M = |\boldsymbol{d}|^2$. It is also clear that the temperature gradient, $\boldsymbol{g} = \operatorname{grad}\theta$, and the second density gradient, $\boldsymbol{S} = \operatorname{grad}^2\varrho$, will now enter into $\boldsymbol{T}$ only through the term $\operatorname{div}(\varrho\bar{\psi}_d)$; thus, they will affect only the spherical part of $\boldsymbol{T}$.

On the other hand, when $\mathbb{E}$ has the form given in Theorem 8, $(11.31)_2$ yields that

$$\begin{aligned}\boldsymbol{T} &= -\varrho^2\{\bar{\psi}_\varrho - \tfrac{1}{2}\operatorname{div}\bar{\psi}_d\}\mathbf{1} + \tfrac{1}{2}\varrho^2 \operatorname{grad}\bar{\psi}_d^{\mathrm{T}} - \operatorname{per}\{\operatorname{grad}\phi\}\,,\\ &= -\varrho^2\{\bar{\psi}_\varrho - \tfrac{1}{2}\operatorname{div}\bar{\psi}_d\}\mathbf{1} + \tfrac{1}{2}\varrho^2\{\boldsymbol{d} \otimes \bar{\psi}_{\varrho d} + \boldsymbol{g} \otimes \bar{\psi}_{\theta d} + \boldsymbol{S}\,\bar{\psi}_{dd}\}\\ &\quad - \operatorname{per}\{\phi_\varrho\boldsymbol{d} + \phi_\theta\boldsymbol{g} + \boldsymbol{S}\,\phi_d\}\,,\end{aligned}$$

where, for any vector $\boldsymbol{a}$, $\operatorname{per}\{\boldsymbol{a}\}$ is the (skew) second order tensor given by $(\operatorname{per}\{\boldsymbol{a}\})_{ij} \equiv \varepsilon_{ijk}a_k$. This form for the stress $\boldsymbol{T}$ allows, in particular, $\boldsymbol{g}$ and $\boldsymbol{S}$ to play a much richer role than was their lot in the previous special case: in the present model non-zero values for $\boldsymbol{g}$ and/or $\boldsymbol{S}$ will, in general, lead to shear stresses. Notice, however, that $\boldsymbol{T}$ will now generally fail to be symmetric. Indeed, even if ψ is independent of $\boldsymbol{d}$ altogether (as it is if we adopt all of the forms in Theorem 8), we have that

$$\boldsymbol{T} = -\varrho^2\bar{\psi}_\varrho\mathbf{1} - \operatorname{per}\{\operatorname{grad}\phi\}\,,$$

so that

$$\operatorname{sk}(\boldsymbol{T}) = -\operatorname{per}\{\operatorname{grad}\phi\}\,,$$

which vanishes if and only if $\operatorname{grad}\phi = 0$.

The above two special cases illustrates some of the vast variety of stress systems contained within the class of elastic materials of Korteweg type. Nevertheless, it is not hard to show, by $(11.32)_2$, that *every elastic material of Korteweg type meets*

$$\operatorname{div}\boldsymbol{T} = -\varrho\operatorname{grad}\mu + (\varrho\bar{\psi})_\theta\operatorname{grad}\theta\,. \tag{11.34}$$

Hence, the chemical potential μ not only determines a portion of the stress $\boldsymbol{T}$ – it also determines all of $\operatorname{div}\boldsymbol{T}$ whenever the temperature field is spatially uniform. It follows that, in isothermal problems, the equations of motion (11.6) take the form

$$-\operatorname{grad}\mu + \boldsymbol{b} = \ddot{\boldsymbol{x}}\,,$$

for every elastic material of Korteweg type. Moreover, these equations are of exactly the same structure as those governing the motion of an Eulerian perfect fluid in a barotropic flow. Consequently, whenever the body force $\boldsymbol{b}$ is derivable from a potential, $\boldsymbol{b} = -\operatorname{grad} f(\boldsymbol{x}, t)$, the acceleration field will be a gradient, and, as a result, much of classical hydrodynamics carries over at once to every elastic material of Korteweg type. More specifically still: *whenever the temperature field is uniform, the body motionless, and* $\boldsymbol{b} = -\operatorname{grad} f(\boldsymbol{x})$, *the equations of equilibrium for our materials admit of the first integral*

$$\mu + f = \text{constant} = k \,,$$

i.e.,

$$\operatorname{div}(\varrho\bar{\psi})_d - (\varrho\bar{\psi})_\varrho = f(\boldsymbol{x}) - k \,.$$

Thus, instead of the equilibrium configurations of our material being governed by 3 third order partial differential equations ($\operatorname{div}\boldsymbol{T} = \operatorname{grad} f$) for the single scalar field ϱ, which would usually result in ϱ being overdetermined[14], they are in fact governed by a single partial differential equation of second order.

The forms (11.31) – (11.34) have several other significant consequences. One of them we shall take up in the next section; here let us consider the problem of embedding Korteweg's form (11.1) into our thermodynamics of elastic materials of Korteweg type. Specifically, let the coefficients of α, β, δ, and γ of (11.1) depend on ϱ, θ, and (even) $\boldsymbol{d}$, compute the divergence of (11.1) and require it to satisfy (11.33) and (11.34) for every density field $\varrho = \varrho(\boldsymbol{x})$ and every temperature field $\theta = \theta(\boldsymbol{x})$, for some Helmholtz free energy response function $\psi = \bar{\psi}(\varrho, \theta, \boldsymbol{d})$. This is the basic idea behind

Theorem 10. *Suppose that in* (11.1) *p is a function of* ϱ *and* θ, *while* α, β, δ, *and* δ *may be functions of* ϱ, θ, *and* $\boldsymbol{d}$. *If* (11.1) *gives the stress in an elastic material of Korteweg type, then it is necessary that* (i) α, β, δ *and* γ *be functions of* ϱ *alone that meet*

$$\alpha + \gamma = \varrho c \,, \qquad \gamma' = c + \delta \,, \qquad \beta + \delta = \tfrac{1}{2}(\varrho c' - c) \tag{11.35}$$

for $c = c(\varrho)$, *and* (ii) *the associated free energy then must be of the form*

$$\psi = \bar{\psi}(\varrho, \theta, \boldsymbol{d}) = \frac{1}{2}\left(\frac{c}{\varrho}\right)|\boldsymbol{d}|^2 + \zeta(\varrho)\cdot\boldsymbol{d} + f(\varrho, \theta) \,. \tag{11.36}$$

We remark that in Theorem 10 we regard $(11.35)_1$ or $(11.35)_2$ as defining $c = c(\varrho)$; we also note that the formulae (11.35) were found earlier by *Aifantis* and *Serrin* [11.13] in their more specialized study of the variational theory of van der Waals. Finally, it should be noted that the vector-valued function $\zeta = \zeta(\varrho)$ of (11.36) may be easily shown to vanish identically if one supposes ψ to be invariant under superimposed rigid rotations.

A converse to Theorem 10 is slightly subtle to state due to the fact that, if an elastic material of Korteweg type satisfies (11.36) for some function $c = c(\varrho)$, the relations (11.35) determine the four-tuple $(\alpha, \beta, \delta, \gamma)$ only up to the addition of a

---

[14] In [11.10] *Serrin* has shown that precisely such overdeterminism occurs for Korteweg's form (11.1) unless a very special relation holds among the non-classical coefficients α, β, δ, and γ.

term of the form $(-h, -h', h', h)$, where $h = h(\varrho)$ is arbitrary. Thus, when we enter (11.36) into (11.33) and then into $(11.32)_2$, we find the formula

$$T = \{-\varrho^2 f_\varrho + d \cdot (\varrho\zeta) + (\varrho c)\Delta\varrho + \tfrac{1}{2}(\varrho c)' |d|^2\}\mathbf{1} - c d \otimes d - d \otimes (\varrho\zeta)$$
$$- \operatorname{div}\{\varrho^2 \bar{\psi}_d \otimes \mathbf{1} + \mathbb{K} + \mathbb{E}\},$$

where

$$\varrho^2 \bar{\psi}_d \otimes \mathbf{1} + \mathbb{K} = \tfrac{1}{2}(\varrho c)\{d \otimes \mathbf{1} - \mathbf{1} \otimes d\} + \tfrac{1}{2}\varrho^2\{\zeta \otimes \mathbf{1} - \mathbf{1} \otimes \zeta\}.$$

But now, the third order tensors

$$\mathbb{A} \equiv s \otimes \mathbf{1} - \mathbf{1} \otimes s,$$
$$\mathbb{B} \equiv h\{d \otimes \mathbf{1} - \mathbf{1} \otimes d\},$$

with $s = s(\varrho) \equiv \int \varrho\zeta\, d\varrho$ and $h = h(\varrho)$ arbitrary, are skew in their 1st and 3rd places and satisfy

$$\operatorname{div}\mathbb{A} = d \cdot (\varrho\zeta)\mathbf{1} - d \otimes (\varrho\zeta),$$
$$\operatorname{div}\mathbb{B} = \{h\Delta\varrho + h' |d|^2\}\mathbf{1} - h' d \otimes d - h \operatorname{grad}^2\varrho,$$

since $d = \operatorname{grad}\varrho$. Thus, our last formula for the stress T may be written as

$$T = \{-\varrho^2 f_\varrho + (\varrho c - h)\Delta\varrho + (\tfrac{1}{2}(\varrho c)' - h') |d|^2\}\mathbf{1} + (h' - c) d \otimes d$$
$$+ h \operatorname{grad}^2\varrho - \operatorname{div}\Pi,$$

which, up to the divergence of

$$\Pi \equiv \varrho^2 \bar{\psi}_d \otimes \mathbf{1} + \mathbb{K} + \mathbb{E} - \mathbb{A} - \mathbb{B},$$

is exactly Korteweg's form (11.1) with coefficients α, β, δ, and $\gamma\,(\equiv h)$ satisfying (11.35). Note, however, that if we pick $\mathbb{E}$ so as to have, say, $\Pi \equiv 0$, then the tensors $\mathbb{A}$ and $\mathbb{B}$ will contribute to the interstitial working; indeed, the linear part of the interstitial flux u will then the given by

$$\{\mathbb{K} + \mathbb{E}\}L = \{\mathbb{A} + \mathbb{B} - \varrho^2 \bar{\psi}_d \otimes \mathbf{1}\}L,$$
$$= (s + hd - \varrho^2 \bar{\psi}_d) \operatorname{tr}L - L(s + hd),$$
$$= (\operatorname{tr}D)\{\tfrac{2}{3}(s + hd - \varrho^2 \bar{\psi}_d\} - \{D - \tfrac{1}{3}(\operatorname{tr}D)\mathbf{1}\}(s + hd) - W(s + hd),$$

so that u will be sensitive to (i) the isotropic part of any stretching (through $\operatorname{tr}D$), to (ii) non-isotropic stretchings (through $D - \tfrac{1}{3}(\operatorname{tr}D)\mathbf{1}$), and to (iii) spins (through W). Note also that, since $s = s(\varrho)$, the combination $s + hd$ vanishes for all ϱ and d if and only if $s(\varrho)$ and $h(\varrho)$ both vanish identically. Thus, the presence of the term $\gamma \operatorname{grad}^2\varrho$ in Korteweg's form (11.1) requires, within the theory of elastic materials of Korteweg type, an interstitial work flux u sensitive to both spins and non-isotropic stretchings.

Of course, forms for the stress far more complex than Korteweg's (11.1) are delivered by our (11.31) and (11.32), since the third order tensor-valued function $\mathbb{E} = \mathbb{E}(\varrho, \theta, d, S, g)$ is still at our disposal. How much at our disposal, we now study. Indeed, let us return to general, not necessarily elastic, materials of

Korteweg type. As we shall now see, the reduced dissipation inequality (11.30) significantly structures each of the terms $\boldsymbol{m}$, $\mathbb{E}$, and $\boldsymbol{w}$ appearing in the interstitial work flux $\boldsymbol{u}$. Indeed, when the terms grad$\boldsymbol{m}$, div$\mathbb{E}$, and div$\boldsymbol{w}$ are expanded in (11.30) there will appear a term of the form

$$\left[\frac{\partial}{\partial \boldsymbol{g}}\{\boldsymbol{L}^*\boldsymbol{m}+\mathbb{E}\boldsymbol{L}+\boldsymbol{w}\}\right]\cdot \operatorname{grad}^2\theta\,,$$

and a term of the form

$$\left[\frac{\partial}{\partial \boldsymbol{S}}\{\boldsymbol{L}^*\boldsymbol{m}+\mathbb{E}\boldsymbol{L}+\boldsymbol{w}\}\right]\cdot \operatorname{grad}^3\varrho\,,$$

and these terms will be the only ones in (11.30) *containing* $\operatorname{grad}^2\theta$ *and* $\operatorname{grad}^3\varrho$, i.e., (11.30) will be linear in these two quantities. Invoking the Axiom' of Size, we conclude easily that

$$\left[\frac{\partial}{\partial \boldsymbol{g}}\{\boldsymbol{L}^*\boldsymbol{m}+\mathbb{E}\boldsymbol{L}+\boldsymbol{w}\}\right]\cdot \boldsymbol{G}\equiv 0\,, \tag{11.37$_1$}$$

and

$$\left[\frac{\partial}{\partial \boldsymbol{S}}\{\boldsymbol{L}^*\boldsymbol{m}+\mathbb{E}\boldsymbol{L}+\boldsymbol{w}\}\right]\cdot \mathbb{G}\equiv 0\,, \tag{11.37$_2$}$$

for all completely symmetric 2nd and 3rd order tensors, $\boldsymbol{G}$ and $\mathbb{G}$, respectively. The restrictions $(11.37)_{1,2}$ are just the form $(11.16)_{2,3}$ take when specialized to materials of Korteweg type. Here they can be analyzed in rather full detail.

Since $\{\boldsymbol{L}^*\boldsymbol{m}+\mathbb{E}\boldsymbol{L}+\boldsymbol{w}\}$ is a quadratic form in $\boldsymbol{L}$, it is clear that $(11.37)_{1,2}$ holds if and only if

$$\left[\frac{\partial}{\partial \boldsymbol{g}}\{\boldsymbol{L}^*\boldsymbol{m}\}\right]\cdot \boldsymbol{G}\equiv\left[\frac{\partial}{\partial \boldsymbol{g}}\{\mathbb{E}\boldsymbol{L}\}\right]\cdot \boldsymbol{G}\equiv\frac{\partial \boldsymbol{w}}{\partial \boldsymbol{g}}\cdot \boldsymbol{G}\equiv 0\,, \tag{11.38}$$

and

$$\left[\frac{\partial}{\partial \boldsymbol{S}}\{\boldsymbol{L}^*\boldsymbol{m}\}\cdot \mathbb{G}\equiv\left[\frac{\partial}{\partial \boldsymbol{S}}\{\mathbb{E}\boldsymbol{L}\}\right]\cdot \mathbb{G}\equiv\frac{\partial \boldsymbol{w}}{\partial \boldsymbol{S}}\cdot \mathbb{G}\equiv 0\,, \tag{11.39}$$

for all completely symmetric 2nd and 3rd order tensors, $\boldsymbol{G}$ and $\mathbb{G}$, respectively. The first of (11.38) and (11.39) are very easy to analyze: $(11.38)_1$ tells us that $\boldsymbol{L}^*\boldsymbol{m}_g$ must be skew, i.e.,

$$\boldsymbol{L}^*\boldsymbol{m}_g=-\boldsymbol{m}_g^{\mathrm{T}}\boldsymbol{L}^{*\mathrm{T}}$$

for all $\boldsymbol{L}^*$ which are adjugates. It is easy to see that this can hold only if $\boldsymbol{m}_g\equiv 0$, i.e., $\boldsymbol{m}(\varrho,\theta,\boldsymbol{d},\boldsymbol{S},\boldsymbol{g})$ is independent of $\boldsymbol{g}$. It is also independent of $\boldsymbol{S}$. To see this, let $\boldsymbol{a}$ be any vector in V and take $\mathbb{G}=\boldsymbol{a}\otimes\boldsymbol{a}\otimes\boldsymbol{a}$ in $(11.39)_1$ to reach

$$(\boldsymbol{L}^{*\mathrm{T}}\boldsymbol{a})\cdot \boldsymbol{m}_S(\boldsymbol{a}\otimes\boldsymbol{a})\equiv 0$$

for all $\boldsymbol{a}\in V$ and all $\boldsymbol{L}^*$ which are adjugates. It is not hard to show that this can hold only if $\boldsymbol{m}_S(\boldsymbol{a}\otimes\boldsymbol{a})\equiv 0$ for all $\boldsymbol{a}\in V$. Hence, by the spectral theorem, $\boldsymbol{m}_S\boldsymbol{S}'\equiv 0$ for all symmetric $\boldsymbol{S}'$, i.e., at every $\boldsymbol{S}$ the directional derivative of $\boldsymbol{m}$ with respect to S vanishes in every direction. Clearly, $m(\varrho, \theta, d, S, g)$ is independent of S as well as g:

$$m = m(\varrho, \theta, d) \; .$$

Unfortunately, the remaining four conditions of (11.38) and (11.39), restricting the dependence of $\mathbb{E}$ and w on S and g, are more complicated to analyze. However, it turns out[15] that $(11.38)_3$ and $(11.39)_3$ hold if and only if $w(\cdot)$ is of the form

$$\begin{aligned} w &= w(\varrho, \theta, d, S, g) \; , \\ &= k(\varrho, \theta, d) + r(\varrho, \theta, d)[S] + \omega_1(\varrho, \theta, d) \wedge g + \omega_2(\varrho, \theta, d)[S] \wedge g + S^* z(\varrho, \theta, d) \; , \end{aligned} \quad (11.40)_1$$

where $r(\varrho, \theta, d)[\cdot]$ and $\omega_2(\varrho, \theta, d)[\cdot]$ are linear and satisfy

$$e \cdot r(\varrho, \theta, d)[e \otimes e] \equiv 0 \equiv e \wedge \omega_2(\varrho, \theta, d)[e \otimes e] \; ,$$

respectively, for all unit vectors e. The function $w^{\dagger}(\cdot) \equiv \mathbb{E}(\cdot)L$ must, by $(11.38)_2$ and $(11.39)_2$, have the same functional form $(11.40)_1$ as $w(\cdot)$ has; however, for $w^{\dagger}(\cdot)$ the fact that L may be varied at will, along with the fact that $\mathbb{E}$ is skew in its 1st and 3rd places, allows us to show that $(11.38)_2$ and $(11.39)_2$ hold if and only if $\mathbb{E}(\cdot)L$ is of the more special form

$$\begin{aligned} \mathbb{E}L &= \mathbb{E}(\varrho, \theta, d, S, g)L \; , \\ &= \mathbb{E}^0(\varrho, \theta, d)L + \mathbb{E}^1(\varrho, \theta, d)[S]L + \{L^{\mathrm{T}} \xi(\varrho, \theta, d)\} \wedge g \; , \end{aligned} \quad (11.40)_2$$

where $\mathbb{E}^1(\varrho, \theta, d)[\cdot]$ is linear and meets

$$e \cdot \mathbb{E}^1(\varrho, \theta, d)[e \otimes e]L \equiv 0 \; ,$$

for all unit vectors e. Of course, $\mathbb{E}^0$ and $\mathbb{E}^1$ are, like $\mathbb{E}$, skew in their 1st and 3rd places.

The forms $(11.40)_{1,2}$ have several interesting consequences. First, consider our earlier expression (11.31) for the stress in an elastic material. In that expression the first two terms, $-\varrho\{\varrho \bar{\psi}_{\varrho} + d \cdot \bar{\psi}_d\}\mathbf{1}$ and $\varrho d \otimes \bar{\psi}_d$, involve only ϱ, θ, and d; it is only through the term $\operatorname{div}\{\mathbb{K} + \mathbb{E}\}$ that the temperature gradient g and the second density gradient S can enter into (can influence directly) the stress T. Moreover, the term $\operatorname{div}\mathbb{K}$ will, by (11.29), be affine in g and S, while, by $(11.40)_2$, the term $\operatorname{div}\mathbb{E}$ will contain no terms involving $\operatorname{grad}^3\varrho$ or $\operatorname{grad}^2\theta$ and will be at most quadratic in S and linear in g. Thus, for elastic materials of Korteweg type, thermodynamics regulates rather severely how g and S may influence the stress, although it does clearly allow for a much richer influence than Korteweg's (11.1).

Perhaps even more interesting are the consequences of the representation $(11.40)_1$ for w, the static part of the interstitial flux u. Indeed, let us define

$$w^{\mathrm{E}} = w^{\mathrm{E}}(\varrho, \theta, d, S) \equiv w(\varrho, \theta, d, S, 0) \; ,$$

so that w^{E} is the "equilibrium part" of w. By $(11.40)_1$, we see that

[15] For the details, see Appendices B and C of [11.25].

$$w^{\mathrm{E}} = k(\varrho, \theta, d) + r(\varrho, \theta, d)[S] + S^{*}z(\varrho, \theta, d)\ ,$$

and that

$$w = w^{\mathrm{E}} + (\omega_1 + \omega_2) \wedge g\ .$$

Further, by $(11.38)_3$ and $(11.39)_3$ directly, we see that, for any density field $\varrho = \varrho(x, t)$ and for any temperature field $\theta = \theta(x, t)$,

$$\begin{aligned} \operatorname{div} w &= w_\varrho \cdot d + w_\theta \cdot g + w_d \cdot S\ , \\ &= w_\varrho^{\mathrm{E}} \cdot d + w_d^{\mathrm{E}} \cdot S + w_\theta^{\mathrm{E}} \cdot g + (\omega_1 + \omega_2)_\varrho \wedge g \cdot d + \{(\omega_1 + \omega_2) \wedge g\}_d \cdot S\ , \end{aligned} \tag{11.41}$$

where we have used $(11.40)_1$ and the definition of w^{E}.

Now enter (11.41) into the reduced dissipation inequality (11.30) and select $g = L = 0$. We thus find the restriction

$$w_\varrho^{\mathrm{E}} \cdot d + w_d^{\mathrm{E}} \cdot S \geqslant 0 \tag{11.42}$$

for all (ϱ, θ, d, S) in the domain of $w^{\mathrm{E}}(\cdot)$. Since $w^{\mathrm{E}}(\cdot)$ is quadratic in S, the inequality (11.42) is cubic in S. Moreover, if

we assume that the domain of $w^{\mathrm{E}}(\cdot)$ *is of the form* $\quad \mathscr{R} \times \mathscr{S} \times T_{\mathrm{s}}\ , \quad (\mathrm{I}_2)$

where $\mathscr{R} \subseteq (0, \infty) \times (0, \infty)$ is open and connected, where $\mathscr{S} \subseteq V$ is star-shaped with respect to the zero vector, and where T_{s} is the set of symmetric, second order tensors on V, then it turns out that (11.42) holds if and only if

$$w_\varrho^{\mathrm{E}} \cdot d + w_d^{\mathrm{E}} \cdot S \equiv 0$$

for all $(\varrho, \theta, d, S) \in \mathscr{R} \times \mathscr{S} \times T_{\mathrm{s}}$. (More detailed restrictions on the parts, k, r, and z, of w^{E} are, of course, entailed by (11.42), but we defer discussion of these for the moment.)

For any density field $\varrho(x, t)$ and for any temperature field $\theta(x, t)$, (11.41) now gives that

$$\begin{aligned} \operatorname{div} w &= w_\theta^{\mathrm{E}} \cdot g + (\omega_1 + \omega_2)_\varrho \wedge g \cdot d + \{(\omega_1 + \omega_2) \wedge g\}_d \cdot S\ , \\ &= g \cdot \{w_\theta^{\mathrm{E}} + d \wedge (\omega_1 + \omega_2)_\varrho - \operatorname{per}\{(\omega_1 + \omega_2)_d S\}\}\ , \end{aligned} \tag{11.43}$$

where, for any second order tensor A, $(\operatorname{per}\{A\})_i \equiv \varepsilon_{ijk} A_{jk}$. Thus, *thermodynamics alone forces* $\operatorname{div} w$ *to be a linear function of the temperature gradient* g. We thus begin to get a little clearer picture of the possible effects of the static interstitial flux w: $\operatorname{div} w$ vanishes completely whenever the temperature field is spatially uniform.

Moreover, insert (11.43) into the reduced dissipation inequality (11.30), replace L and g in (11.30) with, respectively, αL and αg where $\alpha > 0$, and then divide the resulting inequality by α and let $\alpha \downarrow 0$. This yields that, for all L and g,

$$\begin{aligned} &L \cdot [\varrho(\varrho \bar{\psi}_\varrho + d \cdot \bar{\psi}_d) 1 + \varrho d \otimes \bar{\psi}_d + \operatorname{div}^{\mathrm{E}}\{\mathbb{K} + \mathbb{E}\} + T^{\mathrm{E}}] \\ &+ g \cdot \left[w_\theta^{\mathrm{E}} + d \wedge (\omega_1 + \omega_2)_\varrho - \operatorname{per}\{(\omega_1 + \omega_2)_d S\} - \frac{q^{\mathrm{E}}}{\theta} \right] \geqslant 0\ , \end{aligned} \tag{11.44}$$

where

$$\operatorname{div}^{E}\{\mathbb{K}+\mathbb{E}\} \equiv [\operatorname{div}\{\mathbb{K}+\mathbb{E}\}]\,|_{g=0}\,,$$

and where the *equilibrium stress* $\boldsymbol{T}^{E}$ and the *equilibrium heat flux* $\boldsymbol{q}^{E}$ are defined by

$$\boldsymbol{T}^{E} = \bar{\boldsymbol{T}}^{E}(\varrho,\theta,\boldsymbol{d},\boldsymbol{S}) \equiv \bar{\boldsymbol{T}}(\varrho,\theta,\boldsymbol{d},\boldsymbol{S},0,0)$$

and

$$\boldsymbol{q}^{E} = \bar{\boldsymbol{q}}^{E}(\varrho,\theta,\boldsymbol{d},\boldsymbol{S}) \equiv \bar{\boldsymbol{q}}(\varrho,\theta,\boldsymbol{d},\boldsymbol{S},0,0)\,,$$

respectively.

Since $\boldsymbol{L}$ and $\boldsymbol{g}$ both appear linearly in (11.44), it is clear that we have now proven

Theorem 11. *For any material of Korteweg type, the equilibrium stress and the equilibrium heat flux are given by*

$$\boldsymbol{T}^{E} = -\varrho(\varrho\bar{\psi}_{\varrho}+\boldsymbol{d}\cdot\bar{\psi}_{\boldsymbol{d}})\boldsymbol{1} - \varrho\boldsymbol{d}\otimes\bar{\psi}_{\boldsymbol{d}} - \operatorname{div}^{E}\{\mathbb{K}+\mathbb{E}\}\,,$$

and (11.45)

$$\boldsymbol{q}^{E} = \theta\{\boldsymbol{w}^{E}_{\theta} + \boldsymbol{d}\wedge(\omega_1+\omega_2)_{\varrho} - \operatorname{per}\{(\omega_1+\omega_2)_{\boldsymbol{d}}\boldsymbol{S}\}\}\,,$$

respectively.

Corollary. *For any material of Korteweg type, the static interstitial flux* $\boldsymbol{w}$ *satisfies*

$$\operatorname{div}\boldsymbol{w} = \frac{\boldsymbol{q}^{E}\cdot\boldsymbol{g}}{\theta}$$

in every thermodynamic process.

Proof. Enter $(11.45)_2$ into (11.43). ■

Thus, whether $\boldsymbol{g}$ is zero or not, the divergence of $\boldsymbol{w}$ vanishes identically in any material of Korteweg type in which the equilibrium heat flux $\boldsymbol{q}^{E}$ vanishes identically. In such materials $\operatorname{div}\boldsymbol{w}$ drops completely out of the local balance of energy (11.7′).

The form $(11.45)_1$ for the equilibrium stress $\boldsymbol{T}^{E}$ is also interesting: except for the term $\operatorname{div}^{E}\{\mathbb{K}+\mathbb{E}\}$ replacing $\operatorname{div}\{\mathbb{K}+\mathbb{E}\}$, $(11.45)_1$ is exactly the same as the expression we found in Theorem 9 for the stress in any elastic material of Korteweg type. Thus, if we put $\boldsymbol{g}=0$ in them, all our remarks about general elastic stress systems carry over to general equilibrium stress systems in arbitrary materials of Korteweg type.

We could now enter (11.43) and $(11.45)_{1,2}$ into the reduced dissipation inequality (11.30) and then derived some further thermodynamic restrictions on $\bar{\boldsymbol{T}}(\cdot)$, $\bar{\boldsymbol{q}}(\cdot)$, $\mathbb{E}(\cdot)$, and $\boldsymbol{m}(\cdot)$ along the lines Serrin and I presented in Sect. 4 of [11.25]. However, I will not pursue this now.

Throughout our analysis so far, we have avoided any postulate to the effect that the interstitial working $\boldsymbol{u}\cdot\boldsymbol{n}$ is invariant under superimposed rigid motions. Indeed, since it is well-known that such a postulate would imply that the interstitial work flux $\bar{\boldsymbol{u}}(\cdot)$ is independent of the spin tensor $\boldsymbol{W}$, an appeal to Theorem 3 would then tell us that $\boldsymbol{m}(\cdot)\equiv 0$ and that $\mathbb{E}$ is totally determined by $\mathbb{K}$ according to $(11.20)_2$. Such a postulate would then be too strong: while Theorem 3 delimits an interesting special case of the form of $\bar{\boldsymbol{u}}(\cdot)$, we find little reason to believe that the interstitial fluxes allowed by nature don't outstrip Theorem 3's compass.

A much weaker, and we think more reasonable, postulate is that $\boldsymbol{u}\cdot\boldsymbol{n}$ *be invariant under superimposed static rigid rotations*, or, equivalently, that $\bar{\boldsymbol{u}}(\cdot)$ satisfy

$$\bar{\boldsymbol{u}}(\boldsymbol{QF},\theta,\boldsymbol{Q}\nabla\boldsymbol{F},\boldsymbol{Q}\nabla^2\boldsymbol{F},\boldsymbol{Qg},\boldsymbol{QLQ}^{\mathrm{T}})=\boldsymbol{Q}\bar{\boldsymbol{u}}(\boldsymbol{F},\theta,\nabla\boldsymbol{F},\nabla^2\boldsymbol{F},\boldsymbol{g},\boldsymbol{L})\,, \qquad (\mathrm{P}_5)_1$$

for all proper orthogonal tensors $\boldsymbol{Q}$, and for all $(\boldsymbol{F},\theta,\nabla\boldsymbol{F},\nabla^2\boldsymbol{F},\boldsymbol{g},\boldsymbol{L})\in\mathscr{I}$. For materials of Korteweg type, $\boldsymbol{u}\cdot\boldsymbol{n}$ is invariant under superimposed static rigid rotations if and only if

$$\bar{\boldsymbol{u}}(\varrho,\theta,\boldsymbol{Qd},\boldsymbol{QSQ}^{\mathrm{T}},\boldsymbol{Qg},\boldsymbol{QLQ}^{\mathrm{T}})=\boldsymbol{Q}\bar{\boldsymbol{u}}(\varrho,\theta,\boldsymbol{d},\boldsymbol{S},\boldsymbol{g},\boldsymbol{L})\,, \qquad (\mathrm{P}_5)_2$$

for all proper orthogonal tensors $\boldsymbol{Q}$, and for all $(\varrho,\theta,\boldsymbol{d},\boldsymbol{S},\boldsymbol{g},\boldsymbol{L})\in\mathscr{I}$. Note that to even state $(\mathrm{P}_5)_1$ or $(\mathrm{P}_5)_2$ constitutes a tacit structure assumption on the corresponding constitutive domain $\mathscr{I}$. I close the present section by considering some of the implications of $(\mathrm{P}_5)_2$ for our earlier results on materials of Korteweg type.

Due to the way in which $\boldsymbol{L}$ enters into the form $\boldsymbol{u}=\boldsymbol{w}+\{\mathbb{K}+\mathbb{E}\}\boldsymbol{L}+\boldsymbol{L}^*\boldsymbol{m}$, it is easy to see that $(\mathrm{P}_5)_2$ holds if and only if

$$\begin{aligned}\boldsymbol{w}(\varrho,\theta,\boldsymbol{Qd},\boldsymbol{QSQ}^{\mathrm{T}},\boldsymbol{Qg})&=\boldsymbol{Qw}(\varrho,\theta,\boldsymbol{d},\boldsymbol{S},\boldsymbol{g})\,,\\ \{\mathbb{K}(\varrho,\theta,\boldsymbol{Qd})+\mathbb{E}(\varrho,\theta,\boldsymbol{Qd},\boldsymbol{QSQ}^{\mathrm{T}},\boldsymbol{Qg})\}\boldsymbol{QLQ}^{\mathrm{T}}&=\boldsymbol{Q}\{\mathbb{K}(\varrho,\theta,\boldsymbol{d})+\mathbb{E}(\varrho,\theta,\boldsymbol{d},\boldsymbol{S},\boldsymbol{g})\}\boldsymbol{L}\,,\\ \boldsymbol{m}(\varrho,\theta,\boldsymbol{Qd})&=\boldsymbol{Qm}(\varrho,\theta,\boldsymbol{d})\,,\end{aligned} \qquad (11.46)$$

for all proper orthogonal $\boldsymbol{Q}$, for all $(\varrho,\theta,\boldsymbol{d},\boldsymbol{S},\boldsymbol{g})$ in the domain of $\boldsymbol{w}(\cdot)$ and $\mathbb{E}(\cdot)$, for all $(\varrho,\theta,\boldsymbol{d})$ in the domain of $\mathbb{K}(\cdot)$ and $\boldsymbol{m}(\cdot)$, and for all tensors $\boldsymbol{L}$. Note that in writing $(11.46)_3$ we have used our thermodynamic analysis which revealed $\boldsymbol{m}$ to be independent of $\boldsymbol{S}$ and $\boldsymbol{g}$, as is $\mathbb{K}$ in $(11.46)_2$ by $(11.28)_1$ and (11.29).

The consequence of $(11.46)_3$ is well-known: $\boldsymbol{m}(\cdot)$ must be of the form

$$\boldsymbol{m}=m(\varrho,\theta,M)\boldsymbol{d}$$

for some scalar-valued function $m(\cdot)$ of ϱ, θ and $M\equiv|\boldsymbol{d}|^2$. The conditions $(11.46)_1$ and $(11.46)_2$ are considerably more difficult. However, it may be shown[16] that $\boldsymbol{w}(\cdot)$ meets the representation $(11.40)_1$, the thermodynamic restriction (11.42), the domain assumption (I_2), and $(11.46)_1$ if and only if

$$\boldsymbol{w}=\boldsymbol{w}(\varrho,\theta,\boldsymbol{d},\boldsymbol{S},\boldsymbol{g})=r\boldsymbol{d}\wedge(\boldsymbol{Sd})+\omega_1\boldsymbol{d}\wedge\boldsymbol{g}+\omega_2(\boldsymbol{Sd})\wedge\boldsymbol{g}\,, \qquad (11.47)$$

where $r(\cdot)$, $\omega_1(\cdot)$, and $\omega_2(\cdot)$ are scalar-valued functions of ϱ, θ, and $M=|\boldsymbol{d}|^2$. (Thus, in our earlier representation $(11.40)_1$, the functions $\boldsymbol{k}(\cdot)$ and $\boldsymbol{z}(\cdot)$ now vanish altogether, while the functions $\boldsymbol{r}(\cdot)$, $\omega_1(\cdot)$, and $\omega_2(\cdot)$ now have delightfully simple forms.)

We read off at once from (11.47) that, in every material of Korteweg type, *the static part* $\boldsymbol{w}$ *of the interstitial flux* (i) *vanishes at any particle where* $\boldsymbol{d}=\operatorname{grad}\varrho$ *vanishes,* (ii) *is, in rough terms, quadratic in the departure of the fields* $\varrho(\boldsymbol{x},t)$ *and* $\theta(\boldsymbol{x},t)$ *from uniformity and* (iii) *vanishes identically for any material possessing a center of symmetry for* $\boldsymbol{w}(\cdot)$, *in the sense that*

---

[16] See Appendix B of [11.25] for the details.

$$w(\varrho, \theta, -d, S, -g) = -w(\varrho, \theta, d, S, g) .$$

We thus have found strong reasons why the classical thermodynamics of homogeneous processes never found a need for the introduction of w. Finally, if we enter (11.47) into $(11.45)_2$, we find the simple relation

$$q^{\mathrm{E}} = \theta(r_\theta + \omega_{2\varrho} - 2\,\omega_{1M})\, d \wedge (Sd) .$$

Turning now to $(11.46)_2$ and using the representation $(11.40)_2$ for $\mathbb{E}(\cdot)$, it is not hard to show that $(11.46)_2$ holds if and only if

$$\begin{aligned} \mathbb{K}(\varrho, \theta, Qd)QLQ^{\mathrm{T}} &= Q\,\mathbb{K}(\varrho, \theta, d)L , \\ \mathbb{E}^0(\varrho, \theta, Qd)QLQ^{\mathrm{T}} &= Q\,\mathbb{E}^0(\varrho, \theta, d)L , \\ \mathbb{E}^1(\varrho, \theta, Qd)[QSQ^{\mathrm{T}}]QLQ^{\mathrm{T}} &= Q\,\mathbb{E}^1(\varrho, \theta, d)[S]L , \\ \xi(\varrho, \theta, Qd) &= Q\,\xi(\varrho, \theta, d) , \end{aligned} \tag{11.48}$$

where we have also used that $\mathbb{K}$ is symmetric and $\mathbb{E}^0$ is skew in their 1st and 3rd places. From $(11.48)_4$ we see that

$$\xi = \xi(\varrho, \theta, M)d \qquad (11.49)_1$$

for a scalar-valued function $\xi(\cdot)$. Next, a short calculation based on (11.29) shows that $(11.48)_1$ holds if and only if

$$\bar{\psi}_d(\varrho, \theta, Qd) = Q\,\bar{\psi}_d(\varrho, \theta, d) ,$$

and this is easily integrated to yield

$$\bar{\psi}(\varrho, \theta, Qd) = \bar{\psi}(\varrho, \theta, d)$$

for all proper, orthogonal Q and for all (ϱ, θ, d) in the domain of $\bar{\psi}(\cdot)$. Thus, $u \cdot n$ invariant under superimposed static rigid rotations requires that ψ also be so invariant. As a consequence

$$\psi = \psi(\varrho, \theta, M) ,$$

where, as before, $M = |d|^2$.

The conditions $(11.48)_{2,3}$ may also be analyzed. One finds that a third order tensor-valued function $\mathbb{E}^0(\cdot)$ of (ϱ, θ, d), with values which are skew in their 1st and 3rd places, satisfies $(11.48)_2$ if and only if, for every second order tensor L,

$$\begin{aligned} \mathbb{E}^0 L &= \mathbb{E}^0(\varrho, \theta, d)L , \\ &= A\,\mathrm{per}\{L\} + B(L^{\mathrm{T}}d) \wedge d + C\{1 \otimes d \quad d \otimes 1\}L , \end{aligned} \qquad (11.49)_2$$

where $A(\cdot)$, $B(\cdot)$, and $C(\cdot)$ are scalar-valued functions of ϱ, θ, and $M = |d|^2$. Of course, when $L = \mathrm{grad}\,\dot{x}$, the term $A\,\mathrm{per}\{L\}$ is just $-A\,\omega$, where ω is the vorticity vector.

Similarly, it can be shown that a third order tensor-valued function $\mathbb{E}^1(\cdot)$ of (ϱ, θ, d, S), with values which are skew in their 1st and 3rd places, which is linear in S and meets

$$e \cdot \mathbb{E}^1(\varrho, \theta, d)[e \otimes e]L \equiv 0$$

for all unit vectors $\boldsymbol{e}$ and all second order tensors $\boldsymbol{L}$, satisfies $(11.48)_3$ if and only if

$$\begin{aligned}\mathbb{E}^1\boldsymbol{L} &= \mathbb{E}^1(\varrho,\theta,\boldsymbol{d})[\boldsymbol{S}]\boldsymbol{L}\ ,\\ &= F\,\mathrm{per}\{\boldsymbol{SL}\} + G(\boldsymbol{L}^{\mathrm{T}}\boldsymbol{d})\wedge(\boldsymbol{Sd}) + H[(\mathrm{tr}\boldsymbol{S})\{\boldsymbol{1}\otimes\boldsymbol{d}-\boldsymbol{d}\otimes\boldsymbol{1}\} - \boldsymbol{S}\{\boldsymbol{1}\otimes\boldsymbol{d}-\boldsymbol{d}\otimes\boldsymbol{1}\}\\ &\quad - \{\boldsymbol{1}\otimes(\boldsymbol{Sd}) - \boldsymbol{d}\otimes\boldsymbol{S}\}]\boldsymbol{L}\ , \end{aligned} \tag{11.49$_3$}$$

where $F(\cdot)$, $G(\cdot)$, and $H(\cdot)$ are scalar-valued functions of ϱ, θ, and $M = |\boldsymbol{d}|^2$. Note that, when $\boldsymbol{S} = \boldsymbol{1}$ in $(11.49)_3$, $\mathbb{E}^1\boldsymbol{L}$ reduces to exactly the same form as $\mathbb{E}^0\boldsymbol{L}$ in $(11.49)_2$.

We have thus found the most general form of the interstitial work flux $\boldsymbol{u} = \boldsymbol{w} + \{\mathbb{K} + \mathbb{E}\}\boldsymbol{L} + \boldsymbol{L}^*\boldsymbol{m}$ in all materials of Korteweg type for which the interstitial working $\boldsymbol{u}\cdot\boldsymbol{n}$ is invariant under superimposed static rigid rotations. Further, in the case that the material also possesses a center of symmetry for $\boldsymbol{w}(\cdot)$ and $\mathbb{E}(\cdot)$, so that both

$$\boldsymbol{w}(\varrho,\theta,-\boldsymbol{d},\boldsymbol{S},-\boldsymbol{g}) = -\boldsymbol{w}(\varrho,\theta,\boldsymbol{d},\boldsymbol{S},\boldsymbol{g})$$

and

$$\mathbb{E}(\varrho,\theta,-\boldsymbol{d},\boldsymbol{S},-\boldsymbol{g})\boldsymbol{L} = -\mathbb{E}(\varrho,\theta,\boldsymbol{d},\boldsymbol{S},\boldsymbol{g})\boldsymbol{L}\ ,$$

then, alongside

$$\boldsymbol{w}(\cdot) \equiv 0\ ,$$

we have

$$\begin{aligned}\mathbb{E} &= \mathbb{E}(\varrho,\theta,\boldsymbol{d},\boldsymbol{S},\boldsymbol{g})\ ,\\ &= C\{\boldsymbol{1}\otimes\boldsymbol{d}-\boldsymbol{d}\otimes\boldsymbol{1}\} + H[(\mathrm{tr}\boldsymbol{S})\{\boldsymbol{1}\otimes\boldsymbol{d}-\boldsymbol{d}\otimes\boldsymbol{1}\} - \boldsymbol{S}\{\boldsymbol{1}\otimes\boldsymbol{d}-\boldsymbol{d}\otimes\boldsymbol{1}\} - \{\boldsymbol{1}\otimes\boldsymbol{Sd}-\boldsymbol{d}\otimes\boldsymbol{S}\}]\ ,\end{aligned}$$

i.e., the scalar-valued functions $\xi(\cdot)$, $A(\cdot)$, $B(\cdot)$, $F(\cdot)$, and $G(\cdot)$ of $(11.49)_{1,2,3}$ vanish identically.

11.6 An Application: Rules Like Maxwell's

Let $v = \varrho^{-1}$ denote specific volume, and suppose that at some uniform, constant temperature θ a body $\mathscr{B}$ consists of a liquid in equilibrium with its vapor. Suppose also that, however they may be distributed throughout $\mathscr{B}$, both liquid and vapor have uniform specific volumes v_1 and v_2, respectively. Finally, let $p = p^{\mathrm{H}}(v,\theta)$ denote the pressure function governing the homogeneous, equilibrium, fluid states of $\mathscr{B}$. If $p^{\mathrm{H}}(\cdot,\theta)$ is defined on $[v_1, v_2]$, *Maxwell's rule* is the assertion that the saturation pressure (i.e., the real pressure) p^{R} of the mixture is uniform and is related to (is determined by) $p^{\mathrm{H}}(\cdot,\theta)$ according to

$$\int_{v_1}^{v_2}\{p^{H}(v,\theta) - p^{\mathrm{R}}\}dv = 0 \quad \textit{and} \quad p^{\mathrm{H}}(v_1,\theta) = p^{\mathrm{R}} = p^{\mathrm{H}}(v_2,\theta)\ . \tag{11.50}$$

Of course, $(11.50)_1$ just asserts that, between v_1 and v_2, p^{R} is such that the area under $p^{\mathrm{H}}(\cdot,\theta) - p^{\mathrm{R}}$ equals the area above $p^{\mathrm{H}}(\cdot,\theta) - p^{\mathrm{R}}$; $(11.50)_{2,3}$ just assert that $p^{\mathrm{H}}(\cdot,\theta) - p^{\mathrm{R}}$ vanishes at the endpoints v_1 and v_2. For sufficiently nice functions $p^{\mathrm{H}}(\cdot,\theta)$, the system (11.50) of 3 equations for the three unknowns v_1, v_2, and p^{R} has a unique, non-trivial (i.e., $v_1 \neq v_2$) solution.

In the special context of one dimensional equilibrium problems for Korteweg's form (11.1), *Aifantis* and *Serrin* [11.13] noted that if the coefficients α, β, δ, and γ were such that

$$\beta+\delta=\frac{1}{2}\varrho^2\left\{\frac{\alpha+\gamma}{\varrho^2}\right\}',$$

(with $(11.35)_1$, this is precisely our $(11.35)_3$) then the value $\bar{p}$ of the stress in the axial direction x (constant since $\operatorname{div}\boldsymbol{T}=0$) and the pressure function $p=p(v,\theta)$ of (11.1) satisfied

$$\int_{x_1}^{x_2}\{p(v(x),\theta)-\bar{p}\}\frac{dv(x)}{dx}\,dx=0\,, \tag{11.51}$$

where $v=v(x)\equiv\varrho(x)^{-1}$ is the solution of the equilibrium equations, and where x_1 and x_2 are any two places where dv/dx vanishes. Equivalently,

$$\int_{v_1}^{v_2}\{p(v,\theta)-\bar{p}\}dv=0\,, \tag{11.52}$$

where $v_i=v(x_i)$, $i=1,2$. With the identifications $p(\cdot,\cdot)=p^{\mathrm{H}}(\cdot,\cdot)$ and $\bar{p}=p^{\mathrm{R}}$, we see that (11.52) is of the same form as $(11.50)_1$ in Maxwell's rule. Moreover, for those equilibrium solutions that are called "transitions" in [11.13], one can show that $dv(\bar{x})/dx=0$ implies that $p(v(\bar{x}),\theta)=\bar{p}$; thus, for transitions the integrand in (11.52) will now vanish at each of the endpoints, v_1 and v_2, making the analogy between (11.51, 52) and Maxwell's rule (11.50) complete.

For every elastic material of Korteweg type, I now give a broad generalization of the above result. The key to our *vector* Maxwell-like rules is an interesting integral identity which holds for every elastic material of Korteweg type experiencing any density field and any temperature field whatsoever.

When we turn to generalize the result of [11.13], we see that, while (11.52) is not particularly suggestive, (11.51) invites us to study the expression

$$\int_R\{\boldsymbol{T}^{\mathrm{H}}(\varrho,\theta)-\bar{\boldsymbol{T}}(\varrho,\theta,\boldsymbol{d},\boldsymbol{S},\boldsymbol{g})\}\operatorname{grad}v\,dv\,,$$

where $\varrho=\varrho(\boldsymbol{x})$ and $\theta=\theta(\boldsymbol{x})$ are, respectively, arbitrary density and temperature fields whose domain of definition includes the region R, and where $\boldsymbol{T}^{\mathrm{H}}(\varrho,\theta)$ is the stress response function for homogeneous, equilibrium states of our material, i.e.,

$$\begin{aligned}\boldsymbol{T}^{\mathrm{H}}(\varrho,\theta)&\equiv\bar{\boldsymbol{T}}(\varrho,\theta,0,0,0)\,,\\&=-\varrho^2\bar{\psi}_\varrho(\varrho,\theta,0)\,\boldsymbol{1}\,,\end{aligned} \tag{11.53}$$

where we have used $(11.31)_1$. Now, by the divergence theorem,

$$\int_R\boldsymbol{T}(\boldsymbol{x})\operatorname{grad}f(\boldsymbol{x})dv=\int_{\partial R}f\boldsymbol{T}\boldsymbol{n}\,da-\int_R f\operatorname{div}\boldsymbol{T}\,dv\,,$$

for any smooth tensor field $\boldsymbol{T}(\boldsymbol{x})$ and any smooth scalar field $f(\boldsymbol{x})$. Hence, with

$$T(x) \equiv \bar{T}(\varrho(x), \theta(x), \operatorname{grad} \varrho(x), \operatorname{grad}^2 \varrho(x), \operatorname{grad} \theta(x)) ,$$

and with $f(x) \equiv v(x) \equiv \varrho(x)^{-1}$, we find that

$$\int_R \bar{T}(\varrho, \theta, d, S, g) \operatorname{grad} v \, dv = \int_{\partial R} \frac{1}{\varrho} T n \, da - \int_R \frac{1}{\varrho} \operatorname{div} \bar{T} \, dv .$$

This, with (11.32 – 34), gives that

$$\begin{aligned} \int_R \bar{T} \operatorname{grad} v \, dv &= \int_{\partial R} \left[\{\bar{\psi} - \mu\} \mathbf{1} - d \otimes \bar{\psi}_d - \frac{1}{\varrho} \operatorname{div} \mathbb{D} \right] n \, da + \int_R \{\operatorname{grad} \mu - \bar{\psi}_\theta \operatorname{grad} \theta\} dv , \\ &= \int_{\partial R} \left\{ \bar{\psi} \mathbf{1} - d \otimes \bar{\psi}_d - \frac{1}{\varrho} \operatorname{div} \mathbb{D} \right\} n \, da - \int_R \bar{\psi}_\theta \operatorname{grad} \theta \, dv , \end{aligned}$$

where $\mathbb{D} \equiv \varrho^2 \bar{\psi}_d \otimes \mathbf{1} + \mathbb{K} + \mathbb{E}$, and where we have once again used the divergence theorem. Since $\eta = \bar{\eta}(\varrho, \theta, d) = -\bar{\psi}_\theta(\varrho, \theta, d)$, we have how established the key identity

$$\int_R \{\bar{T} \operatorname{grad} v - \bar{\eta} \operatorname{grad} \theta\} dv = \int_{\partial R} \left\{ \bar{\psi} \mathbf{1} - d \otimes \bar{\psi}_d - \frac{1}{\varrho} \operatorname{div} \mathbb{D} \right\} n \, da . \quad (11.54)$$

We emphasize that (11.54) holds for every elastic material of Korteweg type, for every density field $\varrho = \varrho(x)$ and every temperature field $\theta = \theta(x)$, and for every region R in the domain of definition of $\varrho(\cdot)$ and $\theta(\cdot)$.

Next, by (11.53), we see that

$$\begin{aligned} T^{\mathrm{H}}(\varrho(x), \theta(x)) \operatorname{grad} v(x) &= -\varrho^2(x) \bar{\psi}_\varrho(\varrho(x), \theta(x), 0) \operatorname{grad} v(x) , \\ &= \bar{\psi}_\varrho(\varrho(x), \theta(x), 0) \operatorname{grad} \varrho(x) , \\ &= \operatorname{grad} \bar{\psi}(\varrho(x), \theta(x), 0) - \bar{\psi}_\theta(\varrho(x), \theta(x), 0) \operatorname{grad} \theta(x) . \end{aligned}$$

Therefore, for any $\varrho(\cdot)$ and $\theta(\cdot)$,

$$\int_R \{T^{\mathrm{H}} \operatorname{grad} v - \eta^{\mathrm{H}} \operatorname{grad} \theta\} dv = \int_{\partial R} \bar{\psi}(\varrho, \theta, 0) n \, da , \quad (11.55)$$

where $T^{\mathrm{H}} = T^{\mathrm{H}}(\varrho(x), \theta(x))$, and where $\eta^{\mathrm{H}} = \eta^{\mathrm{H}}(\varrho(x), \theta(x))$, with $\eta^{\mathrm{H}}(\cdot, \cdot)$ being the entropy function for homogeneous, equilibrium states of $\mathscr{B}$, i.e., $\eta^{\mathrm{H}}(\varrho, \theta) \equiv \bar{\eta}(\varrho, \theta, 0) = -\bar{\psi}_\theta(\varrho, \theta, 0)$.

We have only to subtract (11.54) from (11.55) to complete the proof of the first half of

Theorem 12. *For any elastic material of Korteweg type, for every density field $\varrho(\cdot)$, and for every temperature field $\theta(\cdot)$:*

$$\begin{aligned} &\int_R \{(T^{\mathrm{H}} - \bar{T}) \operatorname{grad} v - (\eta^{\mathrm{H}} - \bar{\eta}) \operatorname{grad} \theta\} dv \\ &= \int_{\partial R} \left\{ (\bar{\psi}(\varrho, \theta, 0) - \bar{\psi}(\varrho, \theta, d)) \mathbf{1} + d \otimes \bar{\psi}_d + \frac{1}{\varrho} \operatorname{div} \mathbb{D} \right\} n \, da , \quad (11.56)_1 \end{aligned}$$

where R is any region in the domain of $\varrho(\cdot)$ and $\theta(\cdot)$. Moreover, if on the boundary of R $\operatorname{grad}\varrho(\cdot)$ vanishes and $\operatorname{grad}\theta(\cdot)$ is parallel to the outer unit normal $\boldsymbol{n}(\cdot)$, then

$$\int_R \{(\boldsymbol{T}^{\mathrm{H}}-\bar{\boldsymbol{T}})\operatorname{grad} v-(\eta^{\mathrm{H}}-\eta)\operatorname{grad}\theta\}dv=0\,. \qquad (11.56)_2$$

Additionally, with $g\equiv\boldsymbol{n}\cdot\operatorname{grad}\theta$ and $s\equiv\boldsymbol{n}\cdot(\operatorname{grad}^2\varrho)\boldsymbol{n}$, the tractions, $\boldsymbol{T}^{\mathrm{H}}\boldsymbol{n}$ and $\bar{\boldsymbol{T}}\boldsymbol{n}$, on ∂R now satisfy

$$\bar{\boldsymbol{T}}\boldsymbol{n}-\boldsymbol{T}^{\mathrm{H}}\boldsymbol{n}=\varrho^2\{g(\boldsymbol{n}\cdot\bar{\psi}^0_{\theta d})+s(\boldsymbol{n}\cdot\bar{\psi}^0_{dd}\boldsymbol{n})\}\boldsymbol{n}\,, \qquad (11.56)_3$$

where $\bar{\psi}^0_{\theta d}\equiv\bar{\psi}_{\theta d}(\varrho,\theta,0)$ and $\bar{\psi}^0_{dd}\equiv\bar{\psi}_{dd}(\varrho,\theta,0)$.

We remark that, if ψ is assumed to be invariant under superimposed rigid rotations, then it follows that $\bar{\psi}_{\theta d}(\varrho,\theta,0)$ must vanish and $\bar{\psi}_{dd}(\varrho,\theta,0)$ must be spherical, i.e., $\bar{\psi}_{dd}(\varrho,\theta,0)=\sigma(\varrho,\theta)\mathbf{1}$. Thus, in this case $(11.56)_3$ is just

$$\bar{\boldsymbol{T}}\boldsymbol{n}-\boldsymbol{T}^{\mathrm{H}}\boldsymbol{n}=s\sigma\varrho^2\boldsymbol{n}\,. \qquad (11.56)_4$$

The identities $(11.56)_{1-4}$ are our desired vector analogs of Maxwell's rule (11.50); they include (11.51) and (11.52) as a special case.

Proof: We have already proven $(11.56)_1$. It is clear that $(11.56)_2$ will now follow if we can show that $\{\operatorname{div}\mathbb{D}\}\boldsymbol{n}=0$ on ∂R. Suppose therefore that on ∂R

$$\operatorname{grad}\varrho(\boldsymbol{x})=0 \quad \textit{and} \quad \operatorname{grad}\theta(\boldsymbol{x})=g(\boldsymbol{x})\boldsymbol{n}(\boldsymbol{x})\,,$$

where $g(\cdot)$ is some scalar field. As is well-known, it follows from the first of these that on ∂R

$$\operatorname{grad}^2\varrho(\boldsymbol{x})=s(\boldsymbol{x})\boldsymbol{n}(\boldsymbol{x})\otimes\boldsymbol{n}(\boldsymbol{x})\,,$$

for some scalar field $s(\cdot)$.

Now, since $\mathbb{D}\equiv\varrho^2\bar{\psi}_d\otimes\mathbf{1}+\mathbb{K}+\mathbb{E}$, we see from (11.29) that we may write

$$\begin{aligned}\operatorname{div}\mathbb{D}&=\operatorname{div}\{\tfrac{1}{2}\varrho^2(\bar{\psi}_d\otimes\mathbf{1}-\mathbf{1}\otimes\bar{\psi}_d)\}+\operatorname{div}\mathbb{E}\,,\\&=\varrho\{(\boldsymbol{d}\cdot\bar{\psi}_d)\mathbf{1}-\boldsymbol{d}\otimes\bar{\psi}_d\}+\tfrac{1}{2}\varrho^2\{(\operatorname{div}\bar{\psi}_d)\mathbf{1}-\operatorname{grad}\bar{\psi}_d^{\mathrm{T}}\}+\operatorname{div}\mathbb{E}\,,\end{aligned}$$

where $\operatorname{grad}\bar{\psi}_d=\bar{\psi}_{\varrho d}\otimes\boldsymbol{d}+\bar{\psi}_{\theta d}\otimes\boldsymbol{g}+\bar{\psi}_{dd}\boldsymbol{S}$. To compute $\operatorname{div}\mathbb{E}$, use $(11.38)_2$ and $(11.39)_2$. Thus, for any constant tensor $\boldsymbol{A}$,

$$\begin{aligned}\boldsymbol{A}\cdot\operatorname{div}\mathbb{E}&=\operatorname{div}\{\mathbb{E}\boldsymbol{A}\}\,,\\&=\boldsymbol{d}\cdot\{\mathbb{E}\boldsymbol{A}\}_\varrho+\boldsymbol{g}\cdot\{\mathbb{E}\boldsymbol{A}\}_\theta+\boldsymbol{S}\cdot\{\mathbb{E}\boldsymbol{A}\}_d\,.\end{aligned}$$

Therefore, for any vector $\boldsymbol{a}$, for any vector $\boldsymbol{n}$, and at any point where $\boldsymbol{d}=\operatorname{grad}\varrho$ vanishes,

$$\begin{aligned}\boldsymbol{a}\cdot\{\operatorname{div}\mathbb{D}\}\boldsymbol{n}=\boldsymbol{a}\cdot\tfrac{1}{2}\varrho^2\{(\boldsymbol{g}\cdot\bar{\psi}_{\theta d}+\boldsymbol{S}\cdot\bar{\psi}_{dd})\boldsymbol{n}-\boldsymbol{g}(\boldsymbol{n}\cdot\bar{\psi}_{\theta d})-\boldsymbol{S}\bar{\psi}_{dd}\boldsymbol{n}\}\\+\boldsymbol{g}\cdot\{\mathbb{E}\boldsymbol{a}\otimes\boldsymbol{n}\}_\theta+\boldsymbol{S}\cdot\{\mathbb{E}\boldsymbol{a}\otimes\boldsymbol{n}\}_d\,.\end{aligned}$$

But, on ∂R, with $\boldsymbol{n}$ the outer unit normal, we have $\boldsymbol{g}=g\boldsymbol{n}$ and $\boldsymbol{S}=s\boldsymbol{n}\otimes\boldsymbol{n}$. Hence, for any vector $\boldsymbol{a}$, we have that on ∂R

$$\begin{aligned} a \cdot \{\operatorname{div} \mathbb{D}\}n &= a \cdot \tfrac{1}{2}\varrho^2\{\{g(n \cdot \bar{\psi}_{\theta d}) + s(n \cdot \psi_{dd} n)\}n - gn(n \cdot \bar{\psi}_{\theta d}) - sn(n \cdot \bar{\psi}_{dd} n)\} \\ &\quad + gn \cdot \{\mathbb{E}a \otimes n\}_\theta + sn \cdot \{\mathbb{E}a \otimes n\}_d n \,, \\ &= 0 \,, \end{aligned}$$

since $\mathbb{E}(\cdot)$ is skew in its 1st and 3rd places. Thus, for fields $\varrho(\cdot)$ and $\theta(\cdot)$ as described, we have proven that $\{\operatorname{div} \mathbb{D}\}n$ vanishes everywhere on ∂R. This, with $(11.56)_1$, proves $(11.56)_2$.

To prove $(11.56)_3$, note that $\{\operatorname{div} \mathbb{D}\}n = 0$ on ∂R is equivalent to

$$\begin{aligned} \{\operatorname{div}\{\mathbb{K} + \mathbb{E}\}\}n &= -\{\operatorname{div}\{\varrho^2 \bar{\psi}_d \otimes \mathbf{1}\}\}n \,, \\ &= -\{2\varrho d \cdot \bar{\psi}_d + \varrho^2 \operatorname{div} \bar{\psi}_d\}n \,. \end{aligned}$$

Thus, on ∂R

$$\{\operatorname{div}\{\mathbb{K} + \mathbb{E}\}\}n = -\varrho^2\{g(n \cdot \bar{\psi}_{\theta d}) + s(n \cdot \bar{\psi}_{dd} n)\}n \,,$$

where $\bar{\psi}_{\theta d}$ and $\bar{\psi}_{dd}$ are, of course, evaluated at $(\varrho, \theta, d) = (\varrho(x), \theta(x), 0)$. Upon entering this last into $(11.31)_1$ and appealing to the definition (11.53) of T^{H}, we see that we have established $(11.56)_3$. ■

References

11.1 B. D. Coleman, W. Noll: The thermodynamics of elastic materials with heat conduction and viscosity. Arch. Ration. Mech. Anal. **13**, 167 – 178 (1963)

11.2 B. D. Coleman, V. J. Mizel: Thermodynamics and departures from Fourier's law of heat conduction. Arch. Ration. Mech. Anal. **13**, 245 – 261 (1963)

11.3 A. C. Eringen: A unified theory of thermomechanical materials. Int. J. Eng. Sci. **4**, 179 – 202 (1966)

11.4 M. Gurtin: Thermodynamics and the possibility of spatial interaction in elastic materials. Arch. Ration. Mech. Anal. **19**, 339 – 352 (1965)

11.5 J. C. Maxwell: On stresses in rarified gases arising from inequalities of temperature. Philos. Trans. Roy. Soc. London **170**, 231 – 256 (1879)

11.6 D. J. Korteweg: Sur la forme que prennent les équations du mouvement des fluides si l'on tient compte des forces capillaires causées par des variations de densité considérables mais continues et sur la théorie de la capillarité dans l'hypothèse d'une variation continue de la densité. Arch. Neerl. Sci. Exactes Nat. **6**, (2) 1 – 24 (1901)

11.7 C. Truesdell, W. Noll: *The Non-Linear Field Theories of Mechanics.* Handbuch der Physik, Vol. III/3, ed. by S. Flügge (Springer, Berlin Heidelberg New York 1965)

11.8 M. Fixman: Transport coefficients in the gas critical region. J. Chem. Phys. **47**, 2808 – 2818 (1967)

11.9 B. U. Felderhof: Dynamics of the diffuse gas-liquid interface near the critical point. Physica **48**, 541 – 560 (1970)

11.10 J. Serrin: The form of interfacial surfaces in Korteweg's theory of phase equilibria. Q. Appl. Math. **41**, 357 – 364 (1983)

11.11 A. Blinowski: On the surface behavior of gradient-sensitive liquids. Arch. Mech. **25**, 259 – 268 (1973)

11.12 A. Blinowski: On the order of magnitude of the gradient-of-density dependent part of an elastic potential in liquids. Arch. Mech. **25**, 833 – 849 (1973)

11.13 E. C. Aifantis, J. Serrin: The mechanical theory of fluid interfaces and Maxwell's rule. J. Colloid. Interface Sci. **96**, 517 – 529 (1983)

11.14 E. C. Aifantis, J. Serrin: Equilibrium solutions in the mechanical theory of fluid microstructures. J. Colloid. Interface Sci. **96**, 530 – 547 (1983)

11.15 M. Slemrod: Admissibility criteria for propagating phase boundaries in a van der Waals fluid. Arch. Ration. Mech. Anal. **81**, 301 – 315 (1983)
11.16 M. Slemrod: Dynamic phase transitions in a van der Waals fluid. To appear, J. Diff. Eq.
11.17 M. Slemrod: An Admissibility Criterion for Fluids Exhibiting Phase Transitions, in *Nonlinear Partial Differential Equations*, ed. by J. Ball. NATO Advanced Study Institute (Plenum, New York 1982) pp. 423 – 432
11.18 R. Hagan, M. Slemrod: The viscosity-capillarity admissibility criterion for shocks and phase transitions. Arch. Ration. Mech. Anal. **83**, 333 – 361 (1983)
11.19 R. Hagan, J. Serrin: Dynamic phase transitions in Korteweg type fluids. In preparation.
11.20 R. A. Toupin: Elastic materials with couple-stresses. Arch. Ration. Mech. Anal. **11**, 385 – 414 (1962)
11.21 R. A. Toupin: Theories of elasticity with couple-stress. Arch. Ration. Mech. Anal. **17**, 85 – 112 (1964)
11.22 A. E. Green, R. S. Rivlin: Simple force and stress multipoles. Arch. Ration. Mech. Anal. **16**, 325 – 353 (1964)
11.23 A. E. Green, R. S. Rivlin: Multipolar continuum mechanics. Arch. Ration. Mech. Anal. **17**, 113 – 147 (1964)
11.24 J. L. Ericksen: Conservation laws for liquid crystals. Trans. Soc. Rheol. **5**, 23 – 34 (1961)
11.25 J. E. Dunn, J. Serrin: On the thermomechanics of interstitial working. Arch. Ration. Mech. Anal. **88**, 95 – 133 (1985)
11.26 C. Truesdell, R. A. Toupin: "The Classical Field Theories", in *Principles of Classical Mechanics and Field Theory,* Handbuch der Physik, Vol. III/1, ed. by S. Flügge (Springer, Berlin Heidelberg New York 1960) p. 226
11.27 I. Müller: On the entropy inequality. Arch. Ration. Mech. Anal. **26**, 118 – 141 (1967)
11.28 I. Müller: On the frame dependence of stress and heat flux. Arch. Ration. Mech. Anal. **45**, 241 – 250 (1972)
11.29 D. G. B. Edelen, J. A. McLennan: Material indifference: a principle or a convenience. Int. J. Eng. Sci. **11**, 813 – 817 (1973)

Chapter 12
Phase Transformations and Non-Elliptic Free Energy Functions

R. D. James

12.1 Introduction

The occurrence of a transformation from one solid form to another is often made evident by a spontaneous change of shape of parts of the crystal. In an unloaded body, this happens at a certain transformation temperature θ_0.

Let us describe these changes of shape in the following way. Suppose that the transformation occurs when the body is cooled to θ_0. At a temperature just above θ_0, we label particles in the body by points $x \in \mathscr{R} \subset \mathbb{R}^3$. The deformation which produces the change of shape is conveniently described by a function y_0 mapping $\mathscr{R}$ into $\mathbb{R}^3$ which gives the position $y_0(x)$ of the particle x. Typically, part of the crystal, say $\mathscr{P}_\mathrm{p}$, does not transform: $y_0(x) = x$ for $x \in \mathscr{P}_\mathrm{p}$. This part[1] is associated with the parent phase. The remaining part $\mathscr{P}_\mathrm{t} = \mathscr{R} - \mathscr{P}_\mathrm{p}$ is associated with the transformed phase. In twinning transformations the parent and transformed phases have the same crystal structure, but this is not usually the case. A reasonable idealization of observation is that ∇y_0 suffers a discontinuity on $\partial \mathscr{P}_\mathrm{t}$. In unloaded bodies at θ_0, $\mathscr{P}_\mathrm{t}$ is often found to consist of several polygonal subregions on each of which y_0 is reasonably regarded as a piecewise linear function. See the photographs of *Saburi* and *Wayman* [12.1], for example.

If diffusion occurs at the transformation temperature, the concepts introduced above are not sufficiently general to describe the transformation. It would be natural in this case to turn to a mixture theory with motions $y^1(x,t), \ldots, y^\nu(x,t)$ associated with each constituent. The diffusionless transformations described above are termed martensitic, polymorphic, or displacive. They tend to occur in metals at high pressure or low temperature. In minerals they occur in profusion, the common minerals – quartz, calcite, the feldspars – each having several diffusionless transformations at various pressures and temperatures. Some steels undergo a martensitic transformation during rapid cooling which, however, competes with a diffusional transformation favored by lower cooling rates. Constitutive relations for these steels would therefore have to be more complicated than those studied here.

After some kinematic preliminaries (Sect. 12.2), we study constitutive equations in which the free energy function depends upon the deformation gradient and temperature. Forms of this function appropriate to solids which change

---

[1] There are cases – the $\alpha-\beta$ transformation in quartz for example – in which the phases do not co-exist at the transformation temperature in an unloaded body. The reason for this is a certain failure of compatibility explained in Sect. 12.2.

phase are described in Sect. 12.3. These forms give rise to systems of nonlinear PDE's whose type may change from point to point.

The stress-free transformation temperature is not generally the transformation temperature for a loaded body (or a body subject to electromagnetic or gravitational fields). In fact, even the arrangement of phases turns out to be highly sensitive to the loading device in solids. Thus, the behavior of a loaded body provides an ideal test case for the constitutive assumptions. In Sect. 12.5 and 6, we study the stability of some simple piecewise linear deformations in a dead loaded body. A particularly delicate question is whether the parent phase can be recovered in a transformed crystal by applying some system of dead loads. This question is explored in detail for the case where the parent phase has cubic symmetry; here a certain technique of averaging over the symmetry group turns out to be useful.

12.2 Kinematics of Co-Existence

One way in which transformations in fluids differ from those in solids is in the restrictions imposed by conditions of continuity. Referring to the description of the change of shape in a phase transformation given in the Introduction, we shall say that a deformation y is *coherent* if y is a continuous function on $\mathscr{R}$. In this paper a *phase boundary* will be a surface in $\mathscr{R}$ of non-zero discontinuity of ∇y. Suppose we have a coherent deformation y and a smooth phase boundary $\mathscr{S}$ which divides $\mathscr{R}$ into two open regions $\mathscr{R}^+$ and $\mathscr{R}^-$. Assume that ∇y is continuous separately on $\mathscr{R}^+$ and on $\mathscr{R}^-$ and has limiting valves on $\mathscr{S}$ from either side. Let F^+ and F^- be limiting values of $F = \nabla y$ at $x \in \mathscr{S}$ where a unit normal to $\mathscr{S}$ is n. F^+ and F^- are subject to well-known jump conditions; there is a vector a, the *amplitude*, such that

$$F^+ - F^- = a \otimes n \, . \tag{12.1}$$

The relation (12.1) does not play a significant role in the study of fluid phases because for (pure) fluids, the constitutive relations are only sensitive to the density ϱ, or, equivalently, to $\det F$, which (12.1) does not restrict. To see this, we take the determinant of (12.1) and get

$$\det F^+ = (\det F^-)(1 + \hat{a} \cdot n) \, , \tag{12.2}$$

where $\hat{a} = (F^-)^{-1} a$. Given $\varrho^+ = \varrho_0 (\det F^+)^{-1}$ and $\varrho^- = \varrho_0 (\det F^-)^{-1}$, the vector a can always be chosen to satisfy (12.2). Moreover, given ϱ^+, ϱ^- and n, we can choose F^- consistent with ϱ^-, choose a to solve (12.2), define F^+ by (12.1), and conclude that $\varrho^+ = \varrho_0 (\det F^+)^{-1}$. Still more generally, the special deformation $y(x) = x + \eta(x) v$, η being a continuous scalar function on a sufficiently small neighborhood of $\mathscr{S}$, and v being a constant vector, can be chosen to match any two smooth density fields in $\mathscr{R}^+$ and $\mathscr{R}^-$ near $\mathscr{S}$.

In solids, the constitutive equations are sensitive to F (at least) so that (12.1) plays a role, especially at places where several phase boundaries meet. To be specific, consider a point $c \in \mathscr{R}$ and a sufficiently small sphere $\mathscr{P}$ with center c. If several planar phase boundaries pass through c, they will intersect $\partial\mathscr{P}$ in arcs of great circles (edges). In general, consider e edges on $\partial\mathscr{P}$ joined at v vertices, so as to partition $\partial\mathscr{P}$ into f faces. Assume that the endpoints of each edge are distinct vertices, that each edge is connected, that the edges intersect only a their endpoints and that the system of edges and vertices is a connected set of points. Formally, we have given a multigraph (in the terminology of *Harary* [12.2]) consisting of arcs of great circles on $\partial\mathscr{P}$. Euler's relation $f-e+v=2$ holds. Now assume that each vertex is joined to c by a line segment. A sector of a plane is bounded by an edge and the line segments which join its terminal vertices to c. These e sectors, termed *interfaces*, serve to divide $\mathscr{P}$ into f open regions, $\mathscr{P}_1, \ldots, \mathscr{P}_f$. We will call $\mathscr{P}_1, \ldots, \mathscr{P}_f$ a *partition* of $\mathscr{P}$, and we will denote partitions by hollow letters, e.g. $\mathbb{P}$. See Fig. 12.1 for some examples of partitions.

I shall investigate the following question: given a partition $\mathbb{P}$, does there exist a coherent deformation y defined on $\mathscr{P}$ for which ∇y is continuous on $\bar{\mathscr{P}}_i$ separately, $i=1,\ldots,f$, and for which $\partial\bar{\mathscr{P}}_i$, $i=1,\ldots,f$, are phase boundaries? Also, I shall assume that each interface supports a non-zero discontinuity of the deformation gradient, even in limit as c as approached along an interface. This assumption is not unreasonable in view of the crystallographic considerations of Sect. 12.3. We will say that a partition is *coherent* if there exists a coherent deformation satisfying the restrictions given above.

We have confined attention to partitions with plane interfaces only to make the definition of a partition easy to state. In fact, a partition is coherent if and only if any partition diffeomorphic to it is coherent. Thus, the results shall apply to many arrangements with curved interfaces, but not to those with cusps.

Let $\mathbb{P}$ be a coherent partition and let y be the appropriate coherent deformation. Let the e interfaces be denoted by $\mathscr{I}_1, \ldots, \mathscr{I}_e$ and let n_j be a unit normal to $\mathscr{I}_j$. Let F_i denote the limiting value of ∇y as c is approached from within $\mathscr{P}_i$. Then, we necessarily have jump conditions of the form (12.1) for each interface. These jump conditions are easily organized by the use of an incidence matrix η_{ij} defined by

$$\eta_{ij} = \begin{cases} 1 & \text{if} \quad \mathscr{I}_i \subset \partial\mathscr{P}_j\,, \\ 0 & \text{if} \quad \mathscr{I}_i \not\subset \partial\mathscr{P}_j\,. \end{cases} \tag{12.3}$$

The matrix η_{ij} has e rows and f columns and two ones in each row. Let $\bar{\eta}_{ij}$ be obtained from η_{ij} by changing the second 1 in each row to -1. The jump conditions referred to above are the following: there exist non-zero amplitudes $a_1 \ldots, a_e$ such that

$$\sum_{j=1}^{f} \bar{\eta}_{ij} F_j = a_i \otimes n_i\,, \qquad i=1,\ldots,e\,. \tag{12.4}$$

Conversely, given a partition $\mathbb{P}$ with associated normals $n_1, \ldots, n_e$ and a modified incidence matrix $\bar{\eta}_{ij}$, suppose there are tensors $F_1, \ldots, F_f$ and vectors $a_1, \ldots, a_e$ which satisfy (12.4). Then, $\mathbb{P}$ is coherent. In fact, the deformation $y(x) = F_i(x-c)$, $x \in \mathscr{P}_i$, $i=1,\ldots,f$, is continuous.

It should be noted that the equations (12.4) are not sufficient for the existence of a continuous deformation of more complicated arrangements than partitions of a sphere (e.g. see [12.3, Fig. 3]). Also, the only physically meaningful deformations are invertible ones. We shall show presently that every coherent partition admits an invertible, continuous deformation. Whether or not there is an invertible deformation which satisfies the additional restrictions arising from crystallography and stability like those discussed in Sect. 12.3 is a delicate matter.

The Fredholm alternative implies (12.4) has solutions $(F_1, \ldots, F_f;\ a_1, \ldots, a_e)$ if and only if a certain homogeneous system of linear equations involving only $(a_1, \ldots, a_e)$ is satisfied. The easiest way to derive this system is by row reduction of the equations (12.4) to triangular form. Since $\bar{\eta}_{ij}$ has one 1 and one -1 and all the rest zeros in every row, it can be row reduced in a way which preserves its form at very stage, except that rows of zeros appear. The row reduction to triangular form can be accomplished by a subgroup of the usual elementary operations, that subgroup consisting of the operations

i) multiplication of a row by -1,
ii) exchange of two rows,
iii) addition of two rows.

Let η^*_{ij} by the upper triangular form of $\bar{\eta}_{ij}$. If $\bar{\eta}_{ij}$ has rank r, then the last $(e-r)$ rows of η^*_{ij} will be composed of zeros. If the same elementary operations used to triangularize $\bar{\eta}_{ij}$ are applied to the whole system (12.4), then the last $(e-r)$ rows yield the homogeneous linear equations mentioned above:

$$\sum_{j=1}^{e} \xi_{ij} a_j \otimes n_j = 0\ , \qquad i = 1, \ldots, (e-r)\ . \tag{12.5}$$

Because of the way (12.4) was reduced, the entries of ξ_{ij} are either 0, 1 or -1.

The equations (12.5) have solutions $(a_1, \ldots, a_e)$ if and only if the system (12.4) has solutions $(F_1, \ldots, F_f;\ a_1, \ldots, a_e)$, Suppose (12.5) has solutions $(a_1, \ldots, a_e)$. To get $(F_1, \ldots, F_f)$ satisfying (12.4), we return to the triangularized system. We may choose the last deformation gradient in the triangularized system arbitrarily ($f-r$ always equals 1). Let us choose it to have a positive determinant. The remaining r deformation gradients are then determined by the first r rows of the triangularized system; since each of these r equations will be of the form

$$\begin{aligned} F_k - F_l = \ldots &\text{homogeneous in } (a_1, \ldots, a_e)\ , \\ &F_l \text{ known}\ , \\ &\det F_l > 0\ , \end{aligned} \tag{12.6}$$

then the remaining deformation gradients can also be chosen to have positive determinants. That is, the system (12.5) is homogeneous, so if $(a_1, \ldots, a_e)$ satisfies it, then so does $(\lambda a_1, \ldots, \lambda a_e)$ for any scalar λ. Thus, the right hand side of $(12.6)_1$ can be made arbitrarily small by choosing $|\lambda|$ arbitrarily small. It follows that each F_k can be chosen with a positive determinant.

The whole procedure is easily carried out by hand even for very complicated local partitions since $\bar{\eta}_{ij}$ retains its general form during the row reduction.

The analysis of the equations (12.4) leads to two interesting properties of coherent partitions. The first is that many, but not all, partitions have parallel amplitudes: $a_i = \lambda_i a$, $i = 1, \ldots, e$. The second is their scarcity, suggesting that they can be classified in some way beginning with the simplest ones.

To make this idea precise, we need a definition of "simplest". A natural definition is supplied by Euler's formula which holds for any partition. Excluding the trivial cases $f \leqslant 1$ and $v \leqslant 1$, we must have

$$\begin{aligned} e &\geqslant f\,, \\ e &\geqslant v\,. \end{aligned} \tag{12.7}$$

Thus, the number e of interfaces present is a natural measure of simplicity.

Given a certain e, we can choose f and v consistent with Euler's relation and ask if there are any coherent partitions with e edges, f faces, and v vertices. I have done this for $e = 2, 3, \ldots, 7$. Beyond $e = 7$ the analysis becomes extremely tedious, even though the analysis of any given partition with e much larger than 7 is quite easy. The results are summarized by Table 12.1 and Fig. 12.1.

The notation × × × × in Table 12.1 means that there are no coherent partitions with the given values of (e, f, v) except possibly ones with removable vertices.

Table 12.1. Kinematics of partitions. The notation × × × means that there are no coherent local partitions with the given (e, f, v) except possibily those with removable vertices

e	f	v	Morphology	Restrictions
2	2	2	Fig. 12.1a	n_1 parallel to n_2
3	2	3	× × × ×	
3	3	2	Fig. 12.1b	No two of n_1, n_2, n_3
4	2	4	× × × ×	parallel
4	3	3	× × × ×	
4	4	2	Fig. 12.1c	None
5	2	5	× × × ×	
5	3	4	× × × ×	
5	4	3	× × × ×	
5	5	2	Fig. 12.1d	None
6	2	6	× × × ×	
6	3	5	× × × ×	
6	4	4	Fig. 12.1e and f	See footnote[a]
6	5	3	× × × ×	
6	6	2	Fig. 12.1g	
7	2	7	× × × ×	
7	3	6	× × × ×	
7	4	5	× × × ×	
7	5	4	Figs. 12.1h–k	See footnote[a]
7	6	3	× × × ×	
7	7	2	Fig. 12.1l	
⋮	⋮	⋮	⋮	⋮

[a] The restrictions of $e = f = 3$, $v = 2$ are satisfied at all vertices where three edges meet.

A *removable vertex* is a vertex v_k with the following two properties:

i) exactly two edges meet at v_k;
ii) those edges do not share another vertex in addition to v_k.

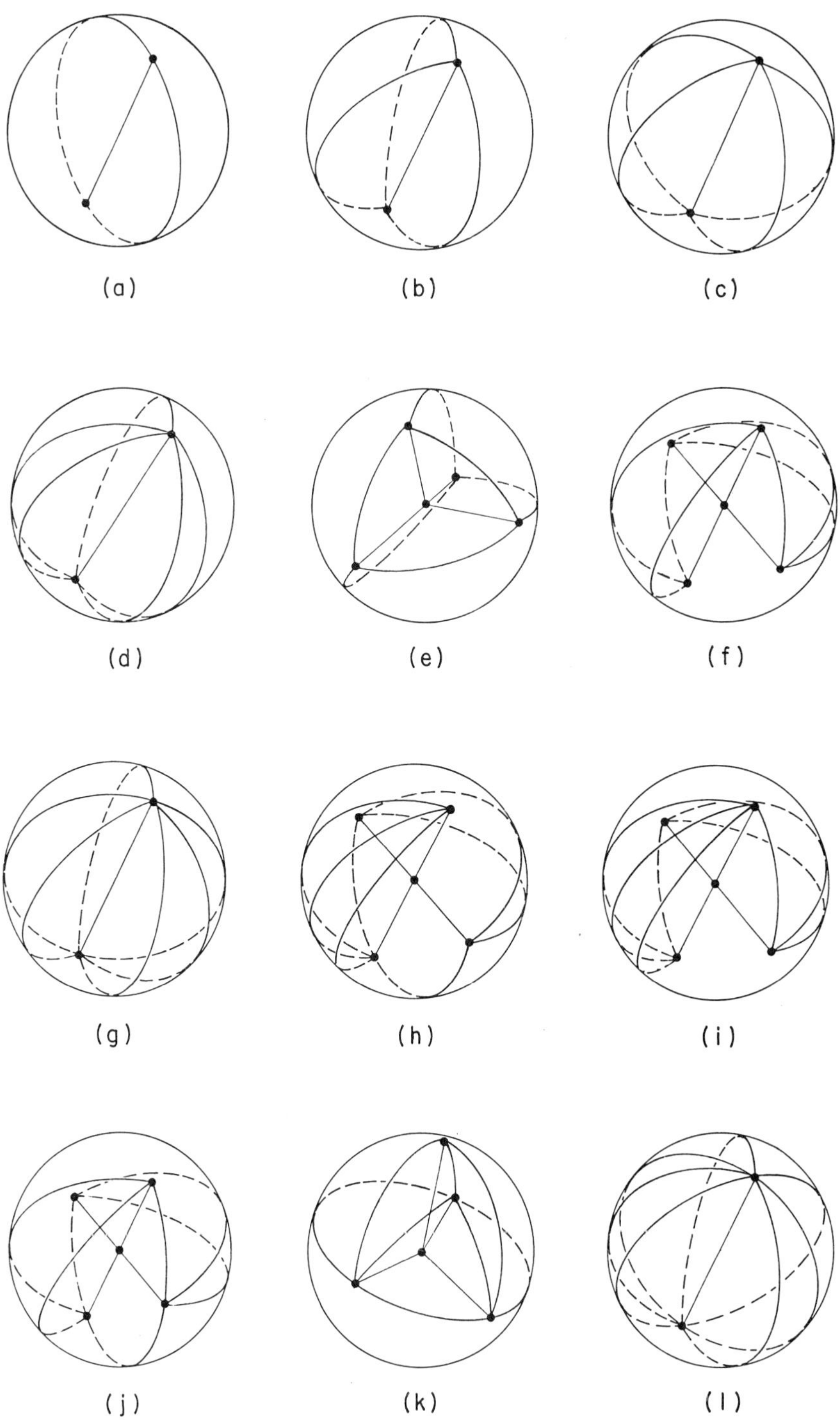

Fig. 12.1a – l. Coherent partitions with up to seven surfaces of discontinuity

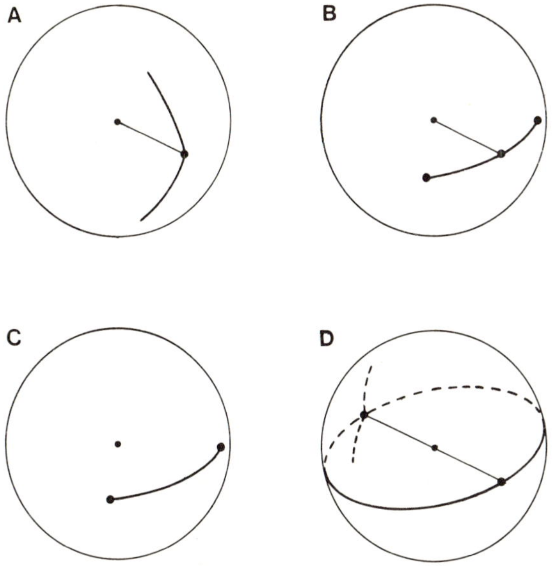

Fig. 12.2 A – D. Removable vertices. **A** is excluded by the restriction on $e = f = v = 2$ listed in Table 12.1. **B** can be changed to **C** without affecting coherence. The vertex in the foreground of **D** is not removable

That the notation $\times\times\times\times$ is reasonable follows from the result for $e = f = v = 2$. That result is the following: At any line segment where exactly two sectors of planes meet in a coherent partition, the two planes must be parallel. However, if in a coherent partition two parallel planes meet at a line segment, we may simply remove that line segment, thereby decreasing e and v by one, and not change the equations (12.4) in any essential[2] way. However, if by removing the vertex we leave an edge which does not connected distinct vertices, then we have excluded the partition. Since there is no reason to exclude such partitions (Fig. 12.1 a is one) I have imposed (ii) above. We illustrate this in Fig. 12.2.

Incidentally, the result for $e = 2$ gives us some confidence that the assumptions like limiting values exist etc. are reasonable. If a transformation is coherent, we should never see only two surfaces of discontinuity meeting at a corner. After perusing a wide variety of microstructures, I find that this property is generally true. Those few that do show pairs of surfaces incident at a corner are either very close to cusps (the corner angle is very small), or arise from transformations whose coherence is in doubt.

Figure 12.1 gives the simplest (up to $e = 7$) coherent partitions. They fall naturally into twelve classes. *Each class is defined explicitly by naming all vertices, edges and faces on $\partial \mathcal{P}$, by listing the vertices on the boundary of each edge, by listing the edges on the boundary of each face and by accounting for the restrictions listed in Table 12.1.* They found at first by the application of the rules governing partitions, by the use of the properties of arcs of great circles,

[2] A redundant equation in that system is omitted.

and finally by the solution of the equations (12.4). Later, I made use of graph theory. It turns out that the definition of a partition corresponds in part to the definition of a multigraph. *Harary* [12.2] gives a list of all possible graphs with given (small) numbers of e and v. I generalized this list to multigraphs by systematically adding edges. I transferred members of the new list to the surface of the sphere one by one, and then I used properties of great circles to exclude some. The remaining ones where then analyzed using the equations (12.4). This was done in a systematic way so that no partitions were omitted.

It is evident that coherence places strong restrictions on the possible arrangements. For example, Fig. 12.1 d (with no restrictions on the included angles) represents all coherent partitions with 5 interfaces, whereas many other partitions with 5 interfaces could be drawn.

Finally we describe another interesting class of coherent arrangements which are defined on a compact set in the interior of $\mathscr{R}$, outside of which there can be prescribed an arbitrary homogeneous deformation. As such, they may be useful for calculations on nucleation. Consider a polyhedron $\mathscr{P} \subset \mathscr{R}$ with f faces. Suppose $\mathscr{P}$ is star-shaped with respect to some point c. Divide $\mathscr{P}$ up into f open regions $\mathscr{P}_1, \ldots, \mathscr{P}_f$ by drawing lines from the vertices of $\mathscr{P}$ to c, and by filling in the triangles. Assume that adjacent faces of $\mathscr{P}$ are not parallel. Let d_i be the (shortest) distance from c to the ith face of $\mathscr{P}$ and let n_i be outward unit normal the ith face of $\mathscr{P}$.

Then, there is a coherent deformation of $\mathscr{R}$ which is an arbitrarily prescribed homogeneous deformation on $\mathscr{R} - \mathscr{P}$. In fact, the deformation

$$
\begin{aligned}
y(x) &= Fx + b\,, && x \in \mathscr{R} - \mathscr{P} \\
&= Fx + b + a\left(1 - \frac{1}{d_i}(x \cdot n_i)\right), && x \in \mathscr{P}_i
\end{aligned}
\tag{12.10}
$$

is continuous, and every continuous piecewise linear deformation of $\mathscr{R}$, with constant gradients on the obvious regions, is of the form (12.10). Note that (12.10) yields deformations with parallel amplitudes.

12.3 Non-Elliptic Free Energy Functions for Materials Which Change Phase

The transformations considered here are recognized by a change of shape. It is natural, as a first guess, to try a constitutive relation in which the specific free energy is a function of the deformation gradient and temperature:

$$
\phi = \tilde{\phi}(F, \theta)\,. \tag{12.11}
$$

Given a suitable deformation y defined on a reference configuration $\mathscr{R}$ and a constant temperature θ, the free energy of y is defined by

$$
\int_{\mathscr{R}} \tilde{\phi}(\nabla y(x), \theta)\, dV\,. \tag{12.12}
$$

We shall be interested in the effect of various loads on the stability and arrangement of the phases. Let a functional $\mathscr{L}[y]$ be the potential energy of the loading device on a suitable class of deformations. Familiar examples of these potentials are:

i) A hard loading device – the potential vanishes,

$$\mathscr{L} \equiv 0\ , \tag{12.13}$$

and all deformations satisfy given boundary conditions of place.

ii) A dead loading device – there is a vector-valued function t assigned on $\partial \mathscr{R}$ and

$$\mathscr{L}[y] = \int_{\partial \mathscr{R}} t(x) \cdot y(x)\, dA\ . \tag{12.14}$$

iii) Loading by an hydrostatic pressure – there is an assigned constant p and

$$\mathscr{L}[y] = \int_{\mathscr{R}} -p \det \nabla y(x)\, dV$$

$$(= -p\,\mathrm{Vol}(y(\mathscr{R})))\ . \tag{12.15}$$

Various combinations of these loading devices are used in mixed problems. Following *Gibbs* [12.4], we shall say that $\tilde{y}$ is *stable* if the total free energy

$$\Phi[y] = \int_{\mathscr{R}} \tilde{\phi}(\nabla y(x), \theta)\, dV - \mathscr{L}[y] \tag{12.16}$$

is minimized at $\tilde{y}$, i.e.

$$\Phi[\tilde{y}] \leqslant \Phi[y]\ , \tag{12.17}$$

for all deformations y. A discussion of metastability is given in [12.5]; in this formal treatment, I concentrate on absolute minimizers with the reservation that some deformations observed in experiment may possibly correspond to relative minima.

A stable deformation $\tilde{y}$ will satisfy equilibrium equations

$$\mathrm{Div}\, \frac{\partial \tilde{\phi}}{\partial F}(\nabla \tilde{y}, \theta) = 0 \tag{12.18}$$

in each phase, and jump conditions

$$(\tilde{\phi}_F(F^+, \theta) - \tilde{\phi}_F(F^-, \theta))\, n = 0\ , \tag{12.19}$$

$$\tilde{\phi}(F^+, \theta) - \tilde{\phi}(F^-, \theta) - (F^+ - F^-) \cdot \tilde{\phi}_F(F^-, \theta) = 0\ , \tag{12.20}$$

on each phase boundary. These conditions are discussed by *Gurtin* [12.6]. We note that (12.19) implies that $\tilde{\phi}$ cannot satisfy the conditions of strong ellipticity (e.g. *Hayes* [12.7]) everywhere in its domain if $F^+ = F^- + a \otimes n$, according to an argument of *Knowles* and *Sternberg* [12.8]. Thus, typically (12.18) will not be one type.

The free energy function shall be objective,

$$\tilde{\phi}(RF, \theta) = \tilde{\phi}(F, \theta) \quad \forall F,\ \theta,\ \mathscr{R} \in \mathscr{O}\ , \tag{12.21}$$

$\mathcal{O}$ being the group of proper orthogonal tensors. Since we shall be concerned with crystals, the form of $\tilde{\phi}$ will reflect the symmetry of $\mathcal{R}$. This is formalized by assigning a group g^ν and by assuming that

$$\tilde{\phi}(FQ, \theta) = \tilde{\phi}(F, \theta) \quad \forall F, \forall \theta, \forall Q \in g^\nu . \tag{12.22}$$

We will confine attention to finite groups g^ν of order ν which are subgroups of the proper orthogonal group (so-called point groups), although it should be noted that molecular studies like those of *Ericksen* [12.9], *Parry* [12.10] and *Pitteri* [12.11] give rise to larger groups in a natural way.

The behavior of a body which undergoes a diffusionless phase transformation as it is cooled to θ_0 is roughly described as follows. Above θ_0, the unloaded body deforms homogeneously with changes of temperature. At θ_0, part of the crystal changes shape. Commonly, this is indicated by platelets and needles of the transformed phase which grow into the body from its boundary.

Let us associate $\mathcal{R}$ with the parent phase just above θ_0. Then, g^ν is associated with the symmetry of the parent phase. Assume that the deformation of the parent phase above θ_0 under zero loads is given up to an inessential rigid motion by

$$y_p(x) = F_p(\theta)x , \quad x \in \mathcal{R} , \tag{12.23}$$

with $F_p(\theta_0) = 1$. Since the parent phase appears stable for $\theta > \theta_0$, it is natural to assume that y_p minimizes the total free energy with $\mathcal{L} \equiv 0$. A short calculation using (12.23) and (12.17) then shows that $F_p(\theta)$ necessarily minimizes $\tilde{\phi}(\cdot, \theta)$ itself. That is, for each fixed $\theta > \theta_0$,

$$\tilde{\phi}(F, \theta) \geqslant \tilde{\phi}(F_p(\theta), \theta) \quad \forall F , \tag{12.24}$$

and (12.24) is also sufficient that y_p minimize the total free energy. By assumption $F_p(\theta_0) = 1$.

Often, the deformation gradients in the unloaded transformed phase at θ_0 are found to be of the form RF_tQ, where $\mathcal{R} \in \mathcal{O}$, $Q \in g^\nu$ and F_t is a fixed tensor. We also expect an exchange of stability at θ_0. Thus it is natural to assume the existence of a function $F_t(\theta)$ which minimizes ϕ for each fixed $\theta < \theta_0$:

$$\tilde{\phi}(F, \theta) \geqslant \tilde{\phi}(F_t(\theta), \theta) \quad \forall F . \tag{12.25}$$

According to (12.24) and (12.25), we have

$$\tilde{\phi}(F_t(\theta_0), \theta_0) = \tilde{\phi}(1, \theta_0) . \tag{12.26}$$

For transformations considered of "first order", $F_t(\theta_0) \neq 1$. "Coherent phase transformations" are characterized by the assumption $F_t(\theta_0) = 1 + a \otimes n$.

In summary, we shall consider free energy functions $\tilde{\phi}(F, \theta)$ which satisfy (12.21, 22), (12.24) for $\theta \geqslant \theta_0$, and (12.25) for $\theta \leqslant \theta_0$.

12.4 Significance of Points of Convexity of the Free Energy

In this section we consider the behavior of a loaded body having a free energy function with the properties described in the preceding section. First, we study the stability of some piecewise linear deformations in a dead loading device.

Let $\mathscr{R}$ be divided into f polygonal regions $\mathscr{R}_1, \ldots, \mathscr{R}_f$, and consider a continuous piecewise linear deformation $\tilde{y}$ of $\mathscr{R}$:

$$\tilde{y}(x) = F_i x\,, \qquad x \in \mathscr{R}_i\,. \tag{12.27}$$

For example, $\mathscr{R}$ could be one of the partitioned spheres of Fig. 12.1 (or any subset of a partitioned sphere) in which case (12.4) is satisfied by $F_1, \ldots, F_f$. Let the traction $\tilde{t}(x)$, $x \in \partial\mathscr{R}$, be defined by

$$\tilde{t}(x) = \frac{\partial\phi}{\partial F}(\nabla\tilde{y}(x), \theta)\,n(x)\,, \qquad x \in \partial\mathscr{R}\,, \tag{12.28}$$

$n(x)$ being the unit outward normal to $\partial\mathscr{R}$ at x. That is, let $\tilde{t}(x)$ be exactly the Piola-Kirchhoff traction given by $\tilde{y}$. According to Sect. 12.3, $\tilde{y}$ is *stable in a dead loading device* with assigned traction $\tilde{t}$ if $\tilde{y}$ minimizes

$$\Phi[y] = \int_{\mathscr{R}} \phi(\nabla y(x), \theta)\,dV - \int_{\partial\mathscr{R}} \tilde{t}(x)\cdot y(x)\,dA \tag{12.29}$$

among all continuous piecewise differentiable deformations.

Schematically, we assert that $\tilde{y}$ minimizes Φ and look for necessary conditions. After using the divergence theorem on integrals over $\partial\mathscr{R}$ and (12.28), it is found that the calculations are greatly simplified by a rather severe restriction on the arrangement of $\mathscr{R}_1, \ldots, \mathscr{R}_f$. To state this restriction, we let the phase "A borders on B" mean that $\partial A \cap \partial B$ has non-zero two dimensional area. The restriction mentioned above is embodied in

Definition 1. *The continuous piecewise linear deformation given in* (12.27) *is simple if $\mathscr{R}_i$ borders on $\mathscr{R}_j$ for every i and j in* $\{1, \ldots, f\}$.

For example, continuous piecewise linear deformations on the partitions of Fig. 12.1 a, b or e are simple.

Theorem 1. *Let a continuous, simple, piecewise linear deformation $\tilde{y}$ of the form* (12.27) *be given. Let the Piola-Kirchhoff traction be given by* (12.28). *If $\tilde{y}$ is stable in a dead loading device with assigned traction $\tilde{t}$, then*

i) *The Piola-Kirchhoff stress is constant on all of $\mathscr{R}$,*

$$\frac{\partial\phi}{\partial F}(F_i, \theta) = T = \text{const}\,, \qquad i = 1, \ldots, f\,, \qquad \text{and} \tag{12.30}$$

ii) *each F_i is a point of convexity of $\phi(\cdot, \theta)$:*

$$\phi(G, \theta) - \phi(F_i, \theta) - (G - F_i)\cdot\phi_F(F_i, \theta) \geqslant 0 \tag{12.31}$$

for all G and $i = 1, \ldots, f$.

Conversely, if (12.30) *and* (12.31) *are satisfied by the gradients of a (not necessarily simple) continuous, piecewise linear* $\tilde{y}$ *of the form* (12.27) *and* $\tilde{t}$ *is defined by* (12.28), *then* $\tilde{y}$ *is stable in a dead loading device.*

Theorem 1 has a beautiful geometric interpretation due to *Gibbs* [12.12]. A hyperplane of slope T is pushed up against the "energy surface", the graph of $\phi(F, \theta)$ vs. F, until it just touches this surface. If it touches at certain tensors $F_1, \ldots, F_f$ and these are gradients of a continuous deformation $\tilde{y}$, then $\tilde{y}$ is stable in the dead loading device.

We remark that while the Piola-Kirchhoff stress is the same on each region in deformations governed by Theorem 1, the Cauchy stress is not. However, each of the f Cauchy stresses is symmetric. A short calculation based on these symmetries gives

Lemma 1. *Suppose the continuous, piecewise linear deformation* $\tilde{y}$ *of the form* (12.27) *yields a constant Piola-Kirchhoff stress:*

$$\frac{\partial \phi}{\partial F}(F_i, \theta) = T = \text{const}, \qquad i = 1, \ldots, f .$$

Let $i \neq j$ *belong to* $\{1, \ldots, f\}$, *and let* R_i *border on* R_j *across a plane interface with normal* n *and amplitude* a. *Then*

$$Tn \parallel a ;$$

the Piola-Kirchhoff traction on the interface is parallel to the amplitude.

Needless to say, we have tried to relax Definition 1. Our first attempt was to simply assign deformation gradients which were points of convexity (which indeed would have implied that any continuous deformation constructed from them would be stable in a dead loading device) but which corresponded to different Piola-Kirchhoff stresses. This attempt failed because of

Lemma 2. *If* F_1 *and* F_2 *are points of convexity of* $\phi(\cdot, \theta)$, *if*

$$\phi_F(F_1, \theta)\, n = \phi_F(F_2, \theta)\, n ,$$

and if

$$F_2 - F_1 = a \otimes n ,$$

then

$$\phi_F(F_1, \theta) = \phi_F(F_2, \theta) .$$

Finally, we note that if the conditions (12.35) and (12.36) are satisfied by a (not necessarily simple) continuous piecewise linear deformation, then it is stable in a wide variety of loading devices including the hard device and various mixed loading devices.

12.5 Geometry of the Domain of the Free Energy

To find points of convexity $F_1, \ldots, F_f, \ldots$ of the free energy corresponding to a fixed Piola-Kirchhoff stress T, we simply minimize (over F) the *excess function*,

$$\phi(F, \theta) - F \cdot T . \tag{12.32}$$

The existence of minima is guaranteed, for example, by mild growth conditions on ϕ which do not contradict any of the assumptions of Sect. 12.3. In this section we look for regions in the set of tensors with positive determinant (F-space, for brevity) which consist of points of convexity of $\phi(\cdot, \theta)$.

We begin with the case $\theta = \theta_0$ and $T = 0$. Then, the points of convexity are simply minima of $\phi(\cdot, \theta_0)$. By assumptions (12.24) and (12.25), any rotation R and any tensor of the form RF_tQ, $R \in \mathcal{O}$, $Q \in g^\nu$, are such minima. In general this gives rise to $n+1$ compact manifolds in F-space. Generally, these manifolds may intersect at various values of F; in particular cases this can often be easily decided by calculating the appropriate symmetric strain tensors, i.e. F^TF and FF^T. Also transversal intersections can be located without difficulty by calculating tangent planes.

We shall work out the details of this geometry for coherent transformations. We focus on this case because in coherent transformations, it is often observed that only ν deformation gradients are found on regions which border the parent phase, while the assumptions given above suggest the possibility of infinitely many. To this end, assume that $F_t = 1 + a \otimes n$ for some fixed vectors a and n. According to Sect. 12.2, the gradient of a continuous deformation is defined on a region which borders the parent phase if it is of the form $1 + b \otimes m$, for some vectors b and m. To be the gradient of a simple, piecewise linear, continuous deformation which minimizes the free energy under zero loads, it also must be of the form $R(1 + a \otimes n)Q$ for some $R \in \mathcal{O}$ and $Q \in g^\nu$. This leads to the restriction

$$R(1 + a \otimes n)Q = 1 + b \otimes m . \tag{12.33}$$

If we premultiply (12.33) by Q, postmultiply by Q^T and let $\bar{R} = QR$, we get

$$\bar{R}(1 + a \otimes n) = 1 + Qb \otimes Qm . \tag{12.34}$$

Equation (12.34) shows that Q enters (12.33) in an essentially trivial way; given vectors $(\boldsymbol{b}, \boldsymbol{m})$ which satisfy (12.33) with $Q = 1$, all others can be obtained by applying the finite group g^ν according to the rule

$$\begin{aligned} b &= Q\boldsymbol{b} \\ m &= Q\boldsymbol{m} , \quad Q \in g^\nu . \end{aligned} \tag{12.35}$$

Thus, we will have all solutions of (12.33) for the given a and n if we find all $(\bar{R}, \boldsymbol{b}, \boldsymbol{m})$, $\bar{R} \in \mathcal{O}$, which solve

$$\bar{R}(1 + a \otimes n) = 1 + \boldsymbol{b} \otimes \boldsymbol{m} . \tag{12.36}$$

This has been solved in [12.13]. In fact, there are exactly two nontrivial ($\bar{R} \neq 1$) solutions of the form $(\bar{R}, \pm\boldsymbol{b}, \pm\boldsymbol{m})$ if a is not parallel to n. If a is parallel to n there are no nontrivial solutions. Formulae for the solutions are somewhat lengthy and are given in [12.13]. Many martensitic transformations occur on so-called irrational planes; this means that Qn, $Q \in g^\nu$ produces ν distinct normals.

In this case the result given above suggests that we should observe more than ν distinct deformation gradients (in general, 2ν) bordering on the parent phase. The reason we do not do this is that the experimentally determined a and n often have just the relation which makes $\boldsymbol{b} = Q\boldsymbol{a}$ and $\boldsymbol{m} = Q\boldsymbol{n}$ for some $Q \in g^{\nu}$.

Now we look for points of convexity corresponding to stressed states at $\theta = \theta_0$. The interpretation of Gibbs given just after Theorem 1 suggests the following line of attack. Let orbit $(\bar{R}, RF_t Q)$ denote all tensors which are either rotations or are of the form $\bar{R}F_t Q$ where $\bar{R}$ is a rotation and $Q \in g^{\nu}$. Let

$$\mathscr{H} = \text{convex hull } (\text{orbit } (\bar{R}, RF_t Q)) \ . \tag{12.37}$$

The theory of convex sets shows that each $F \in \mathscr{H}$ can be written as a *finite* convex combination of members of orbit $(R, \bar{R}F_t Q)$, viz.,

$$F \in \mathscr{H} \Rightarrow F = \sum_{k=1}^{m} \lambda_k G_k \tag{12.38}$$

where $\lambda_k \geqslant 0$, $k = 1, \ldots, m$,

$$\sum_{k=1}^{m} \lambda_k = 1 \ , \tag{12.39}$$

and $G_k \in \text{orbit } (R, \bar{R}F_t Q)$, $k = 1, \ldots, m$. The significance of $\mathscr{H}$ stems from

Lemma 3. *If* $G \in \mathscr{H}$ *and* G *is point of convexity of* $\phi(\cdot, \theta_0)$, *then the stress vanishes at* G:

$$\frac{\partial \phi}{\partial F}(G, \theta_0) = 0 \ . \tag{12.40}$$

Thus, all deformation gradients belonging to simple, stable, continuous, loaded, piecewise linear deformations will lie outside of $\mathscr{H}$.

By the definition of $\mathscr{H}$, $1 \in \mathscr{H}$. Suppose that 1 does not lie on $\partial \mathscr{H}$. Then, there are no tensors sufficiently close to 1 which lie outside of $\mathscr{H}$. By Lemma 3, no deformation gradient near 1 is a point of convexity of $\phi(\cdot, \theta_0)$. In this sense, the (homogeneously deformed) parent phase cannot be recovered by applying any system of dead loads at θ_0.

Thus, it is of interest to decide whether $1 \in \partial \mathscr{H}$ or not. In the next section, we find necessary and sufficient conditions under which $1 \in \partial \mathscr{H}$ in the case where g^{ν} is the cubic group.

12.6 Special Analysis for the Case of a Cubic Parent Phase

We consider the question of whether $1 \in \partial \mathscr{H}$ for the case where g^{ν} is the cubic group ($\nu = 24$). g^{24} consists of all rotations which map a cube into itself. It might seem at first that with so many rotations available, we would never have $1 \in \partial \mathscr{H}$, but this turns out not to be the case.

An arbitrary member G of $\mathscr{H}$ can be written in the form

$$G = \sum_{k} \lambda_k R_k + \sum_{l,m} \lambda_{lm} \bar{R}_l F_t Q_m \tag{12.41}$$

where

$$\lambda_k \geqslant 0\,, \qquad \lambda_{lm} \geqslant 0 \tag{12.42}$$

and

$$\sum_k \lambda_k + \sum_{l,m} \lambda_{lm} = 1\,. \tag{12.43}$$

We take the trace of (12.41) and get

$$\operatorname{tr} G = \sum_k \lambda_k \operatorname{tr} R_k + \sum_{l,m} \lambda_{lm} \operatorname{tr} \bar{R}_l F_t Q_m \tag{12.44}$$

which is a convex combination of scalars. Note that

$$\operatorname{tr} R_k = 1 + 2\cos\theta_k \leqslant 3 \tag{12.45}$$

and

$$\operatorname{tr} \bar{R}_l F_t Q_m \leqslant \operatorname{tr} U_t\,, \tag{12.46}$$

where F_t has the polar decomposition $R_t U_t$. Equation (12.46) follows from the fact that if $\det F > 0$ and F has the polar decomposition RU, then

$$\max_{R \in \mathcal{O}} \{\operatorname{tr} RF\} = \operatorname{tr} U\,, \tag{12.47}$$

and a short calculation.

We conclude that if $\operatorname{tr} U_t \leqslant 3$, then $\operatorname{tr} G \leqslant 3$ for every $G \in \mathscr{H}$. But clearly there are tensors near 1 with trace greater than 3. Thus, we have

Lemma 4. *If* $\operatorname{tr} U_t \leqslant 3$, *then* $1 \in \partial\mathscr{H}$.

In fact, the converse of this lemma is also true. To prove this, we build up any tensor near 1 using special convex combinations. Assume that $\operatorname{tr} U_t > 3$. First note that since 1 and the 180° rotation $-1 + 2e \otimes e$ belong[3] to $\mathscr{H}$, then

$$\tfrac{1}{2} \cdot 1 + \tfrac{1}{2}(-1 + 2e \otimes e) = e \otimes e \in \mathscr{H}\,. \tag{12.48}$$

Thus, for any orthonormal set $\{e_i\}$, and for scalars $\lambda_i \geqslant 0$, $\Sigma\lambda_i = 1$, we have

$$\Sigma \lambda_i e_i \otimes e_i \in \mathscr{H}\,. \tag{12.49}$$

The conclusion (12.49) shows that any positive symmetric tensor whose eigenvalues add up to 1 is in $\mathscr{H}$.

At this point we need a tensor in $\mathscr{H}$ whose trace is greater than 3. Consider the following average[4] over the cubic symmetry group:

$$A = \frac{1}{24} \sum_{i=1}^{24} Q_i (R F_t) Q_i^{\mathrm{T}}\,. \tag{12.50}$$

Here $Q_1, \ldots, Q_{24}$ is a list of the elements of g^{24} and R is some rotation. Notice that $Q_k A Q_k^{\mathrm{T}} = A$ for each $k = 1, \ldots, 24$. It is known [12.14] that the only such invariant tensors under the cubic group are dilatations. Thus,

---

3 This conclusion restricts the proof to odd dimensions, since $-1 + 2e \otimes e$ is not a rotation in even dimensions.

4 These averages are discussed by *Weyl* [12.15].

$$A = \alpha 1 \ , \tag{12.51}$$

for some scalar α. To evaluate α, we take the trace of (12.50) and get

$$3\,\alpha = \operatorname{tr} R F_{\mathrm{t}} \, . \tag{12.52}$$

R is any rotation, so far. Choose it so that it just cancels the rotation in the polar decomposition of F_{t}. Then we get

$$\alpha = \tfrac{1}{3} \operatorname{tr} U_{\mathrm{t}} \, . \tag{12.53}$$

Since $\operatorname{tr} U_{\mathrm{t}} > 3$, then

$$\alpha 1 \in \mathscr{H} \quad \text{for some} \quad \alpha > 1 \, . \tag{12.54}$$

Now form the convex combination

$$\lambda_1 \alpha 1 + \sum_{i=2}^{4} \lambda_i e_i \otimes e_i \, . \tag{12.55}$$

A routine calculation shows that any positive symmetric tensor near 1 is delivered by (12.55). Since by definition $R\mathscr{H} = \mathscr{H}$ for any $R \in \mathscr{O}$, then by use of the polar decomposition theorem, every tensor near 1 belongs to $\mathscr{H}$. Thus, we have

Lemma 5. *If* $\operatorname{tr} U_{\mathrm{t}} > 3$, *then* $1 \notin \partial \mathscr{H}$.

Many martensitic transformations occur in cubic crystals. Also, it is common to have $F_{\mathrm{t}} = 1 + a \otimes n$ with $a \cdot n \ll |a|^2$. See the data of *Saburi* and *Wayman* [12.1], for example. In this case Lemma 5 applies, so that one should not be able to recover a homogeneous deformation $y = Fx$, with F near 1. Indeed, Wayman finds that a specimen under simple tension at θ_0 always transforms completely to the transformed phase. This conclusion can be reached by a simpler argument for simple tension. Also, it must be admitted that Wayman's loading device is not really a dead loading device. Nevertheless, the arguments given above seem to give essentially the right behavior under loads. It would be nice understand the effect of various loading devices; certainly some homogeneous deformations $y = Fx$, $x \in \mathscr{R}$, having non-zero stress with $F \in \mathscr{H}$ will be stable in a hard loading device.

Similar arguments as those presented here can be applied to other groups g^v and to the case $\theta \neq \theta_0$.

Acknowledgement. This research was partly supported by the National Science Foundation under the grant MEA-8209303 and by the Materials Research Laboratory at Brown University.

References

12.1 T. Saburi, C. M. Wayman: The shape memory mechanism and related phenomena in Ag-45 at. % Cd. Acta Metall **28**, 1 – 14 (1980)
12.2 F. Harary: *Graph Theory* (Addison-Wesley, London 1969)
12.3 R. D. James: Stress-free joints and polycrystals. Arch. Ration. Mech. Anal. **86**, 13 – 37 (1984)

12.4 J. W. Gibbs: On the equilibrium of heterogeneous substances. Trans. Conn. Acad. III, 108 – 248 (1876); 343 – 524 (1878)

12.5 R. D. James: Finite deformation by mechanical twinning. Arch. Ration. Mech. Anal. **77**, 143 – 176 (1981)

12.6 M. E. Gurtin: Two phase deformations of eleastic solids. Arch. Ration. Mech. Anal. **84**, 1 – 29 (1983)

12.7 M. Hayes: Static implications of the strong-ellipticity condition. Arch. Ration. Mech. Anal. **33**, 181 – 191 (1969)

12.8 J. K. Knowles, Eli Sternberg: On the failure of ellipticity and the emergence of discontinuous deformation gradients in plane finite elastostatics. J. Elasticity **8**, 329 (1978)

12.9 J. L. Ericksen: Continuous martensitic transitions in thermoelastic solids. J. Therm. Stresses **4**, 107 – 119 (1981)

12.10 G. P. Parry: On phase transitions involving internal strain. Int. J. Solids Struct. **17**, 361 (1981)

12.11 M. Pitteri: Reconciliation of local and global symmetries of crystals. To appear in J. Elasticity

12.12 J. W. Gibbs: Graphical methods in the thermodynamics of fluids. Trans. Conn. Acad. II, 309 – 342 (1873)

12.13 R. D. James: Mechanics of coherent phase transformations in solids. Brown Univ. Tech. Rep. (1982)

12.14 B. D. Coleman, W. Noll: Material symmetry and thermostatic inequalities in finite elastic deformations. Arch. Ration. Mech. Anal. **15**, 87 – 111 (1964)

12.15 H. Weyl: *The Classical Groups* (Princeton Univ. Press, Princeton 1946)

Chapter 13
Dynamic Changes of Phase in a van der Waals Fluid

R. Hagan and J. Serrin

13.1 Introduction

Korteweg's theory of capillarity [13.1, 2] has recently been used to find conditions for equilibrium between liquid and vapor phases of a van der Waals fluid (see [13.3]). Subsequently *Slemrod* [13.4] extended this approach to study dynamic changes of phase in a van der Waals fluid, under the assumption of isothermal motion. This study was further extended by *Hagan* and *Slemrod* in [13.5]. The next logical step was to drop the assumption of isothermal motion. This was done for a van der Waals fluid in a paper by *Slemrod* [13.6]. He showed the existence of a shock layer that converts vapor to liquid and the existence of a shock layer that converts liquid to vapor, under assumptions that render the motion nearly isothermal. These assumptions are that the specific heat capacity at constant volume is large, the coefficients of heat conduction and viscosity are of the same small order μ, and the coefficients in the capillarity terms of the stress are of order μ^2.

One of the problems that complicates the study of dynamic changes of phase is the incompatibility of the classical Korteweg stress with the Clausius-Duhem inequality [13.7]. Recently, however, a modified Korteweg theory has been developed by *Dunn* and *Serrin* [13.8] that is compatible with the Clausius-Duhem inequality. In this theory they posit the existence of a rate of supply of mechanical energy, the interstitial working, which takes into account the working of longer range interactions. With this additional term in the energy balance it is possible to derive a constitutive relation of stress that depends on spatial gradients of the density and still satisfies the Clausius-Duhem inequality. These spatial gradient terms are used to model the effects of interfacial capillarity and at the same time allow the existence of static phase transitions [13.1, 9]. That is, if we were to use the classical form of Navier and Stokes for the stress then we would find that some (shock layer type) dynamic phase transitions exist, but no static ones; see also [13.10, 11].

In this paper we use a special form of the modified Korteweg theory contained in [13.8]. The Clausius-Duhem inequality then gives a direct proof of the increase of entropy across a shock layer and it also provides a Liapunov function, in the sense of LaSalle [13.12], for the shock layer equations.

In Sect. 13.2 we derive the shock layer equations and the increase of entropy theorem. In Sect. 13.3 we examine some of the properties of the Hugoniot curves for a van der Waals fluid. In Sect. 13.4 we state sufficient conditions to guarantee the existence of a compressive shock layer in a van der Waals fluid.

At the close of Sect. 13.4 we show by example the theoretical possibility of dynamic changes of phase for van der Waals fluids, as well as the more typical gas-gas shock layer transitions. In [13.10] we have also observed that similar methods can be need to treat fluids with general equations of states; the restriction to van der Waals fluids in this article is thus by no means essential.

13.2 Basic Equations

We consider one-dimensional steady flow of a van der Waals fluid (see Fig. 13.1). The flow may be thought of as taking place in a cylinder of uniform cross-section parallel to the x-axis. We will always assume that the fluid velocity u is positive so that the fluid flows from left to right. The *absolute temperature* will be denoted by θ, and the *specific volume* by v.

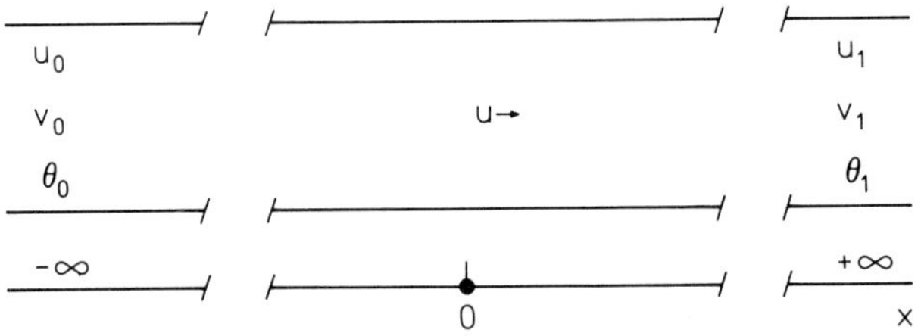

Fig. 13.1. The one-dimensional flow regime

We will seek a smooth solution $(u(x), v(x), \theta(x))$, $x \in \mathbb{R}$, of the equations of motion satisfying the following boundary conditions

$$(u(x), v(x), \theta(x)) \to \begin{cases} (u_0, v_0, \theta_0) & \text{as} \quad x \to -\infty \\ (u_1, v_1, \theta_1) & \text{as} \quad x \to +\infty \,, \end{cases} \tag{13.1}$$
$$(u(x)', v(x)', \theta(x)') \to (0,0,0) \qquad \text{as} \quad x \to \pm\infty \,.$$

Such a solution is called a *shock layer.* Of particular interest to us in this paper is the question of the existence of a shock layer when (v_0, θ_0) and (v_1, θ_1) belong to different phases of the fluid. Shock layers of this type will be called *dynamic phase transitions.*

The balance laws for mass, momentum, and energy for a one-dimensional steady flow in the absence of external body forces and radiant heating are

$$(v^{-1}u)' = 0 \,, \tag{13.2}$$

$$v^{-1}uu' = T'_{xx} \,, \tag{13.3}$$

$$v^{-1}u\varepsilon' = T_{xx}u' + k' + q' \,, \tag{13.4}$$

where $(\,)' = d(\,)/dx$. In addition to the balance laws we also have the Clausius-Duhem inequality,

$$v^{-1}u\eta' \geqslant \left(\frac{q}{\theta}\right)' \,. \tag{13.5}$$

Here T_{xx} is the x component of stress in the x direction, ε is the specific internal energy, η is the specific entropy, k is the interstitial working (see [13.8]) and q is the heat flux.

If f is a function of v, θ, it will be convenient to set $f_1 = f(v_1, \theta_1), f_0 = f(v_0, \theta_0)$ and $[f] = f_1 - f_0$.

Theorem 1.1. (The increase of entropy across a shock layer.) Suppose $q = 0$ whenever $\theta' = 0$. Then $\eta_1 \geqslant \eta_0$ for any shock layer.

Proof. Integration of (13.2) gives

$$u(x) = mv(x)\,, \tag{13.6}$$

for some constant m. Furthermore $m > 0$ since $v > 0$, and $u > 0$ by assumption. We may now integrate (13.5) from $x = -\infty$ to $x = +\infty$ to obtain

$$m(\eta_1 - \eta_0) \geqslant \frac{q}{\theta}\Bigg]_{-\infty}^{\infty}\,. \tag{13.7}$$

But $\theta' = 0$ at $x = \pm\infty$, whence

$$\eta_1 \geqslant \eta_0\,. \tag{13.8}$$

In addition to the balance laws and the Clausius-Duhem inequality we need to specify the constitutive structure of the fluid. For this purpose we shall use a special Korteweg stress theory developed by *Dunn* and *Serrin* [13.8]. The main features of this theory which make it useful for phase transition theory are the preservation of the Clausius-Duhem inequality and the occurrence of higher spatial derivatives of the density in the constitutive relation for stress.

The constitutive relations for the stress are found in equation (1.27) of [13.8], to which one must add the standard Navier-Stokes viscous stress. This yields (when specialized to one-dimensional flow, with $v = 1/\varrho$ as the principal variable instead of ϱ)

$$T_{xx} = -p + (\lambda + 2\mu)u' - \frac{\partial\sigma}{\partial v}\frac{(v')^2}{2} - \frac{\partial\sigma}{\partial\theta}v'\theta' - \sigma v''\,, \tag{13.9}$$

where $\sigma = \varrho^3 c$ and c is the surface tension coefficient, a differentiable function of ϱ, θ. For the heat flux we adopt the Fourier Law

$$q = \varkappa\theta' \tag{13.10}$$

and for the interstitial working term k we use (1.24) of [13.8]. Here the appropriate form of the free energy (already used in obtaining (13.9) above) is

$$\psi = \bar{\psi}(\varrho, \theta, d) = \tilde{\psi}(\varrho, \theta) + \frac{1}{2}\frac{c}{\varrho}|d|^2 \tag{13.11}$$

where $d = \operatorname{grad}\varrho$. Thus by (1.24) we have $k = c\dot{\varrho}d + w$. When these formulas are specialized to one-dimensional flow we get $w = 0$ and

$$\psi = \tilde{\psi}(\varrho, \theta) + \tfrac{1}{2}\sigma(v')^2, \qquad k = m\sigma(v)^2\,. \tag{13.12}$$

The associated entropy η and energy ε are obtained from (1.23) of [13.8]. In accord with the previous choice of ψ, these become

$$\eta = \tilde{\eta} - \frac{1}{2} \frac{\partial \sigma}{\partial \theta} (v')^2 , \tag{13.13}$$

$$\varepsilon = \tilde{\varepsilon} + \frac{1}{2} \left(\sigma - \theta \frac{\partial \sigma}{\partial \theta} \right) (v')^2 , \tag{13.14}$$

where $\tilde{\eta}$ and $\tilde{\varepsilon}$ are the equilibrium entropy and internal energy. They are continuously differentiable functions of v and θ only, as is the pressure p. Furthermore, these quantities satisfy the classical Maxwell relationships of thermodynamics (see equations (13.29–31) below). In general σ, λ, μ, $\varkappa$ are positive continuous functions of v and θ, with σ in particular being of class C^1.

In this paper we shall restrict ourselves to the case where p and $\tilde{\varepsilon}$ satisfy the hypotheses

$$p = p(v, \theta) = \frac{R\theta}{v-b} - \frac{a}{v^2} \quad \text{for all} \quad (v, \theta) \in \Omega , \tag{H1}$$

$$c_v = \frac{\partial \tilde{\varepsilon}}{\partial \theta} > 0 \quad \text{for all} \quad (v, \theta) \in \Omega , \tag{H2}$$

where Ω is defined as $\{(v, \theta)\colon b < v < \infty,\ 0 < \theta < \infty\}$. Here R, b, and a are positive constants and c_v is the specific heat at constant volume. (H1) is of course the van der Waals equation of state.

Substituting (13.7) into (13.3) and (13.4) yields

$$m u' = T'_{xx} , \tag{13.15}$$

$$m \varepsilon' = (T_{xx} u)' - T'_{xx} u + k' + q' . \tag{13.16}$$

We can now put (13.15) in (13.16) to obtain

$$m(\varepsilon + \tfrac{1}{2} u^2)' = (T_{xx} u)' + k' + q' . \tag{13.17}$$

Next we integrate (13.15) and (13.17) from $-\infty$ to x and apply the boundary condition (13.1). This gives (with (13.18) being used to simplify (13.19))

$$m(u - u_0) = T_{xx} - T_{xx_0} , \tag{13.18}$$

$$m(\varepsilon - \tilde{\varepsilon}_0 - \tfrac{1}{2}(u - u_0)^2) = -p_0 (u - u_0) + k + q . \tag{13.19}$$

Finally insert (13.6), (13.9), (13.10), (13.12) and (13.14) into (13.18) and (13.19) to obtain

$$\sigma v'' + \frac{\partial \sigma}{\partial \theta} v' \theta' + \frac{\partial \sigma}{\partial v} \frac{(v')^2}{2} - m(\lambda + 2\mu) v' + p - p_0 + m^2 (v - v_0) = 0 , \tag{13.20}$$

$$\varkappa \theta' = m \left\{ -\frac{1}{2} \frac{\partial}{\partial \theta} (\theta \sigma)(v')^2 + \tilde{\varepsilon} - \tilde{\varepsilon}_0 + p_0 (v - v_0) - \frac{1}{2} m^2 (v - v_0)^2 \right\} . \tag{13.21}$$

It is convenient to define $w = v'$,

$$L(v, \theta) = p(v, \theta) - p_0 + m^2(v - v_0) , \qquad (13.22)$$

and

$$M(v, \theta) = \tilde{\varepsilon}(v, \theta) - \tilde{\varepsilon}_0 + p_0(v - v_0) - \tfrac{1}{2}m^2(v - v_0)^2 \qquad (13.23)$$

and to write (13.20) and (13.21) as a system of three first order ordinary differential equations, namely

$$v' = w , \qquad (13.24a)$$

$$\sigma w' = m(\lambda + 2\mu) w - \frac{1}{2}\frac{\partial \sigma}{\partial v} w^2 - \frac{\partial \sigma}{\partial \theta} w\theta' - L(v, \theta) , \qquad (13.24b)$$

$$\varkappa \theta' = -\frac{m}{2}\frac{\partial}{\partial \theta}(\theta\sigma) w^2 + mM(v, \theta) . \qquad (13.24c)$$

We shall refer to this system as the shock layer equations.

Lemma 1.2. Given (u_0, v_0, θ_0) and (u_1, v_1, θ_1) with (v_0, θ_0), $(v_1, \theta_1) \in \Omega$, a shock layer exists satisfying (13.1) if and only if there exists a solution of (13.24) satisfying

$$(v(x), w(x), \theta(x)) \to \begin{cases} (v_0, 0, \theta_0) & \text{as} \quad x \to -\infty \\ (v_1, 0, \theta_1) & \text{as} \quad x \to +\infty \end{cases} \qquad (13.25)$$

and (13.6) holds. Naturally also $(v'(x), w'(x), \theta'(x)) \to 0$ as $x \to \pm\infty$, so that $(v_0, 0, \theta_0)$ and $(v_1, 0, \theta_1)$ are critical points of the system (13.24).

Lemma 1.3. (The Rankine-Hugoniot jump conditions). A necessary condition for a shock layer to exist satisfying (13.1) is that the Rankine-Hugoniot jump conditions are satisfied:

$$[u] = m[v] , \qquad (13.26a)$$

$$[p] + m^2[v] = 0 , \qquad (13.26b)$$

$$[\tilde{\varepsilon}] + \tfrac{1}{2}(p_1 + p_0)[v] = 0 . \qquad (13.26c)$$

Proof. If a shock layer exists satisfying (13.1) then there exists a solution of (13.24) satisfying (13.25) and (13.6). Let $x \to \infty$. Then from (13.24b) and (13.24c)

$$M_1 = L_1 = 0 ; \qquad (13.27)$$

hence $[p] + m^2[v] = 0$ and $[\tilde{\varepsilon}] + p_0[v] - \frac{1}{2}m^2[v]^2 = 0$. Therefore $[\tilde{e}] + \frac{1}{2}(p_1 + p_0) \cdot [v] = 0$. Now let $x \to \pm\infty$ in (13.6); then $u_1 = mv_1$ and $u_0 = mv_0$. Therefore $[u] = m[v]$.

We can solve the van der Waals equation of state for θ in terms of v and p, algebraically. Hence we can define $\hat{\varepsilon}(v, p) = \tilde{\varepsilon}(v, \theta)$. We may then put

$$H(v, p; v_0, p_0) = \hat{\varepsilon}(v, p) - \hat{\varepsilon}(v_0, p_0) + \tfrac{1}{2}(p + p_0)(v - v_0) . \qquad (13.28)$$

The curve in the $v-p$ plane consisting of all (v, p) satisfying $H(v, p; v_0, p_0)=0$ is called the *Hugoniot curve generated by* (v_0, p_0). Note that any state (v_1, p_1) lying on the intersection of the Hugoniot curve with the straight line given by $p=p_0-m^2(v-v_0)$ will satisfy (13.26b, c) or equivalently (13.27). The corresponding jump $[u]$ is then given directly by (13.26a). Thus all conditions of (13.26) are satisfied.

It is convenient at this point to group together several thermodynamic identities which will be useful in the following sections. First, we have the standard Gibbs identity

$$\theta d\tilde{\eta}=d\tilde{\varepsilon}+p dv \tag{13.29}$$

and the Maxwell relations

$$\frac{\partial \tilde{\varepsilon}}{\partial v}=\theta \frac{\partial p}{\partial \theta}-p \tag{13.30}$$

$$\frac{\partial \tilde{\eta}}{\partial v}=\frac{\partial p}{\partial \theta}, \quad \frac{\partial \tilde{\eta}}{\partial \theta}=\frac{c_v}{\theta}. \tag{13.31}$$

In addition, we shall need the formula

$$\left.\frac{\partial p}{\partial v}\right|_{\tilde{\eta}}=\frac{\partial p}{\partial v}-\frac{\theta}{c_v}\left(\frac{\partial p}{\partial \theta}\right)^2, \tag{13.32}$$

which follows from the chain rule

$$\frac{\partial p}{\partial v}=\left.\frac{\partial p}{\partial v}\right|_{\tilde{\eta}}+\left.\frac{\partial p}{\partial \tilde{\eta}}\right|_v \frac{\partial \tilde{\eta}}{\partial v}$$

together with (13.31) and the relation

$$\left.\frac{\partial p}{\partial \tilde{\eta}}\right|_v=\frac{\partial p}{\partial \theta} \Big/ \frac{\partial \tilde{\eta}}{\partial \theta}=\frac{\theta}{c_v} \frac{\partial p}{\partial \theta}.$$

From (13.30) one also gets $\partial \tilde{\varepsilon}/\partial v=a/v^2$ for the van der Waals equation of state, so that in this case $c_v=c_v(\theta)$ by (H. 2), and

$$\tilde{\varepsilon}=-\frac{a}{v}+\int c_v(\theta) d\theta. \tag{13.33}$$

13.3 The Hugoniot Curve

The van der Waals equation of state possesses non-monotone isotherms for $\theta<\theta_c$, where θ_c is the critical temperature, and monotone decreasing isotherms for $\theta>\theta_c$ (see Fig. 13.2). The unstable region for this equation of state in the $v-\theta$ plane is

$$\Omega_u=\left\{(v, \theta) \in \Omega: \frac{\partial p}{\partial v}(v, \theta)>0\right\}. \tag{13.34}$$

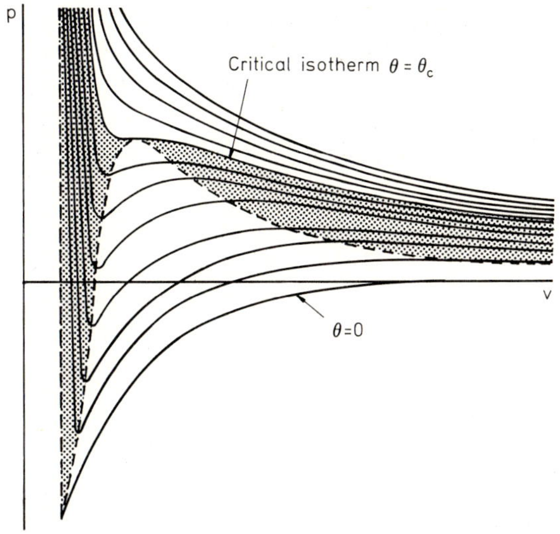

Fig. 13.2. Isotherms for a van der Waals fluid

Thus Ω_u is bounded by

$$\theta = \theta_u(v) \equiv \frac{2a}{R}\,\frac{(v-b)^2}{v^3}, \qquad b < v < \infty\,. \tag{13.35}$$

We have

$$\frac{d\theta_u(v)}{dv} = \frac{-2a(v-b)(v-3b)}{Rv^4}$$

and hence $\theta_u(v)$ has only one stationary point $v_c = 3b$ on (b, ∞). Clearly $\theta_u(v)$ takes on its maximum value $\theta_c = 8a/27Rb$, at $v = v_c$.

We now define

$$\begin{aligned} \Omega_s &= \{(v,\theta) \in \Omega\colon \theta_c < \theta < \infty\} \\ \Omega_l &= \{(v,\theta) \in \Omega\colon \theta_u(v) < \theta < \theta_c,\ b < v < v_c\} \\ \Omega_g &= \{(v,\theta) \in \Omega\colon \theta_u(v) < \theta < \theta_c,\ v_c < v < \infty\}\,. \end{aligned} \tag{13.36}$$

Here Ω_s is the super-critical vapor region, Ω_l is the liquid region, and Ω_g is the vapor (gas) region (see Fig. 13.3).

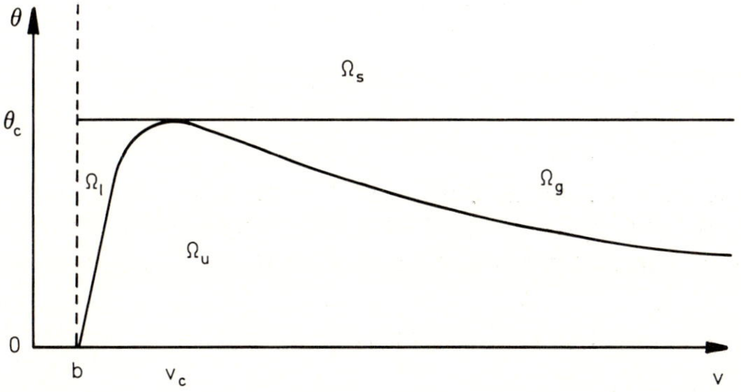

Fig. 13.3. Critical regions for a van der Waals fluid

We shall frequently refer to the fluid state in terms of v and p instead of v and θ. On such occasions it is useful to introduce the notation

$$\hat{\Omega}_\chi = \{(v, p(v,\theta)) : (v,\theta) \in \Omega_\chi\} \qquad \text{where } \chi \text{ is l, s, u, g, or empty}.$$

Note that the map $(v,\theta) \to (v, p(v,\theta))$ is a homeomorphism of Ω_χ onto $\hat{\Omega}_\chi$ since $\partial p/\partial\theta > 0$.

We now state and prove some useful lemmas concerning the Hugoniot curve.

Lemma 3.1. Let $(v_0, p_0) \in \hat{\Omega}$ and let I be any compact interval in (b, ∞) containing v_0. Then the Hugoniot curve generated by (v_0, p_0) approaches the isotherm passing through (v_0, p_0), uniformly on I, as $\inf\limits_{0<\theta<\infty} c_v(\theta)$ approaches infinity.

Proof. We have

$$H(v,p;v_0,p_0) = \hat{\varepsilon}(v,p) - \hat{\varepsilon}_0 + \tfrac{1}{2}(p+p_0)(v-v_0). \tag{13.37}$$

Hence

$$\frac{\partial H}{\partial p}(v,p;v_0,p_0) = \frac{\partial\hat{\varepsilon}}{\partial p} + \frac{1}{2}(v-v_0).$$

But

$$\frac{\partial\hat{\varepsilon}}{\partial p} = \frac{\partial\tilde{\varepsilon}}{\partial\theta}\bigg/\frac{\partial p}{\partial\theta} = \frac{c_v(v-b)}{R}$$

and so

$$\frac{\partial H}{\partial p} = \frac{c_v(v-b)}{R} + \frac{v-v_0}{2} \to \infty \quad \text{as} \quad \inf c_v(\theta) \to \infty. \tag{13.38}$$

Therefore we can solve $H(v,p;v_0,p_0) = 0$ for p as a function of v on I, say $p = h(v)$, if $\inf c_v(\theta)$ is sufficiently large. Assuming that this is so, we have

$$\left(\frac{\partial\hat{\varepsilon}}{\partial p} + \frac{1}{2}(v-v_0)\right)\frac{dh}{dv} + \frac{d\hat{\varepsilon}}{dv} + \frac{1}{2}(p+p_0) = 0.$$

Also

$$\frac{\partial\hat{\varepsilon}}{\partial v} = \frac{\partial\tilde{\varepsilon}}{\partial v} + \frac{\partial\tilde{\varepsilon}}{\partial\theta}\frac{\partial\theta}{\partial v}\bigg|_p = \frac{a}{v^2} - \frac{c_v(v-b)}{R}\frac{\partial p}{\partial v}.$$

Thus

$$\frac{dh}{dv} = \left(\frac{c_v(v-b)}{R}\frac{\partial p}{\partial v} - \frac{a}{v^2} - \frac{p+p_0}{2}\right)\bigg/\left(\frac{v-v_0}{2} + \frac{c_v(v-b)}{R}\right), \tag{13.39}$$

and hence $dh/dv \to \partial p/\partial v$ uniformly on I as $\inf c_v(\theta) \to \infty$. In turn

$$h(v) - h(v_0) \to \int_{v_0}^{v} \frac{\partial p}{\partial\theta}(z,\theta_0)\,dz = p(v,\theta_0) - p_0$$

uniformly on I as $\inf c_v(\theta) \to \infty$. Therefore $h(v) \to p(v,\theta_0)$ uniformly on I as $\inf c_v(\theta) \to \infty$, since $h(v_0) = p(v_0,\theta_0)$.

Let us define

$$\gamma = \frac{R + c_v}{c_v} \tag{13.40}$$

(note that γ is not the same as the ratio of the specific heats c_p/c_v, except in the case of a perfect gas).

Lemma 3.2. If c_v is constant, then $H(v, p; v_0, p_0) = 0$ can be solved algebraically for p as a function of v:

$$\begin{aligned} &p\left(\frac{\gamma+1}{2}\,v - \frac{\gamma-1}{2}\,v_0 - b\right) + \frac{a}{v^2}\,((2-\gamma)\,v - b) \\ &= p_0\left(\frac{\gamma+1}{2}\,v_0 - \frac{\gamma-1}{2}\,v - b\right) + \frac{a}{v_0^2}\,((2-\gamma)\,v_0 - b)\,. \end{aligned} \tag{13.41}$$

Equation (13.41) can be rewritten in the alternate form

$$(v - v_s)\left(p + \frac{\gamma-1}{\gamma+1}\,p_0\right) = \frac{2a}{\gamma+1}\,f(v)\,, \tag{13.42}$$

where

$$v_s = \frac{\gamma-1}{\gamma+1}\,v_0 + \frac{2}{\gamma+1}\,b \tag{13.43}$$

and

$$f(v) = \frac{2\gamma}{\gamma+1}\,\frac{p_0}{a}\,(v_0 - b) + (\gamma-2)\left(\frac{1}{v} - \frac{1}{v_0}\right) + b\left(\frac{1}{v^2} - \frac{1}{v_0^2}\right). \tag{13.44}$$

Proof. We have $p = \dfrac{R\theta}{v-b} - \dfrac{a}{v^2}$, and hence

$$\theta = \frac{(v-b)}{R}\left(p + \frac{a}{v^2}\right)$$

$$\tilde{\varepsilon} = c_v\,\theta - \frac{a}{v} = \frac{v-b}{\gamma-1}\left(p + \frac{a}{v^2}\right) - \frac{a}{v}\,.$$

But $H(v, p; v_0, p_0) = \tilde{\varepsilon} - \tilde{\varepsilon}_0 + \frac{1}{2}(p + p_0)(v - v_0)$, and the results now follow at once.

It is clear from (13.42) that $p = h(v)$ has exactly one singularity at $v = v_s \in (b, v_0)$. If $c_v \uparrow \infty$ then $\gamma \downarrow 1$ and $v_s \downarrow b$; if $c_v \downarrow 0$ then $\gamma \uparrow \infty$ and $v_s \uparrow v_0$.

Lemma 3.3. Suppose c_v is constant. Then

$$\lim_{v \to \infty} h(v) = -\frac{\gamma-1}{\gamma+1}\,p_0\,. \tag{13.45}$$

Proof. The result follows at once from formula (13.42).

In the next lemma we shall examine the behavior of $p = h(v)$ as v approaches v_s. To this end we define

$$\alpha = \frac{\gamma+1}{2\gamma} \cdot \frac{ab}{(y-1)} f(v_s) \tag{13.46}$$

so that

$$\alpha = b^2 p_0 + \frac{\gamma+1}{\gamma[(\gamma-1)y+2]^2} \left\{ (\gamma-2)(\gamma-1) + \frac{4(\gamma-1)}{y} + \frac{2}{y^2} \right\} a , \tag{13.47}$$

where $y = v_0/b$ (>1).

Lemma 3.4. Suppose c_v is constant and $(v_0, p_0) \in \Omega$. Then

$$\lim_{v \uparrow v_s} h(v) = \begin{cases} -\infty & \text{if} \quad \alpha > 0 \\ +\infty & \text{if} \quad \alpha < 0 , \end{cases} \tag{13.48}$$

$$\lim_{v \downarrow v_s} h(v) = \begin{cases} +\infty & \text{if} \quad \alpha > 0 \\ -\infty & \text{if} \quad \alpha < 0 , \end{cases} \tag{13.49}$$

$$\lim_{v \to v_s} h(v) = \frac{-2a}{\gamma+1} \left\{ \frac{\gamma-1}{2a} p_0 + \frac{(\gamma-2)}{v_s^2} + \frac{2b}{v_s^3} \right\} \quad \text{if} \quad \alpha = 0 . \tag{13.50}$$

Lemma 3.5. If c_v is constant, $p_0 \geqslant 0$ and $\gamma \geqslant 2$, then $\alpha > 0$ and

$$h(v) < -\frac{\gamma-1}{\gamma+1} p_0 \quad \text{for} \quad b < v < v_s .$$

Proof. We see from formula (13.47) that $\alpha > 0$ whenever $p_0 \geqslant 0$ and $\gamma \geqslant 2$. Thus (13.46) implies $f(v_s) > 0$ and from (13.44) we have

$$f'(v) = -\frac{\gamma-2}{v^2} - \frac{2b}{v^3} < 0 \tag{13.51}$$

since $\gamma \geqslant 2$. Hence $f(v) > f(v_s)$ for $b < v < v_s$. It now follows from (13.42) that

$$h(v) + \frac{\gamma-1}{\gamma+1} p_0 < 0 \quad \text{for} \quad b < v < v_s$$

and the lemma is proved.

Lemma 3.6. If the hypotheses of Lemma 3.5 are satisfied then the Rankine-Hugoniot jump conditions *cannot* be satisfied if $v_1 < v_s$.

Proof. In order for the Rankine-Hugoniot conditions to be satisfied, we must have

$$-m^2 = \frac{h(v_1) - p_0}{v_1 - v_0} ,$$

and so $h(v_1)-p_0>0$ since $v_1<v_s<v_0$ (by hpothesis and by (13.43)). But by Lemma 3.5 we have $h(v_1)-p_0 \leqslant -2\gamma p_0/(\gamma+1)<0$ and the conclusion follows.

13.4 Existence of Compressive Shock Layers

Theorem 4.1. Assume $(v_1,p_1),(v_0,p_0)\in\hat{\Omega}\backslash\text{Closure }(\hat{\Omega}_u)$, that $v_1<v_0$ and $H(v_1,p_1;v_0,p_0)=0$, and that

$$\frac{p_1-p_0}{v_1-v_0}=-m^2<0\,. \tag{13.52}$$

Furthermore, suppose that the chord connecting (v_1,p_1) to (v_0,p_0) lies above the graph of $H(v,p;v_0,p_0)=0$ on the interval $v_1<v<v_0$ and is not tangent to the graph at either end point. Assume finally that the straight line extension of this chord does not intersect the graph of $H(v,p;v_0,p_0)=0$ when $v>v_0$ and $p>p(v,0)$. Then there exists a unique compression shock layer connecting (v_0,p_0) to (v_1,p_1). Furthermore, $\theta_1>\theta_0$, $\tilde{\eta}_1>\tilde{\eta}_0$, the flow is supersonic at the state (v_0,p_0) and is subsonic at the state (v_1,p_1).

The proof of Theorem 4.1 will be carried out with the help of a series of lemmas.

Lemma 4.2. Each of the equations $M(v,\theta)=0$ and $L(v,\theta)=0$ uniquely defines θ as a function of v, say $\theta=\theta_M(v)$ and $\theta_L(v)$. Furthermore under the hypotheses of Theorem 4.1 the curve $L=0$ intersects the v-axis in exactly two points $v=b$ and $v=\bar{v}$, where $\bar{v}>v_0$. Moreover the curve $L=0$ lies above (below) the curve $M=0$ when $v_1<v<v_0$ $(v_0<v\leqslant\bar{v})$. That is

$$\begin{aligned}\theta_L(v)>\theta_M(v)\quad&\text{for}\quad v_1<v<v_0\\ \theta_L(v)<\theta_M(v)\quad&\text{for}\quad v_1<v<\bar{v}\,.\end{aligned} \tag{13.53}$$

Proof. We have

$$\frac{\partial M}{\partial\theta}=\frac{\partial\tilde{\varepsilon}}{\partial\theta}=c_v>0 \tag{13.54}$$

and

$$\frac{\partial L}{\partial\theta}=\frac{\partial p}{\partial\theta}>0 \tag{13.55}$$

so that both $M=0$ and $L=0$ can be solved for θ as a function of v. In particular by (13.22) and (H1)

$$\theta_L(v)=\frac{v-b}{R}\;\frac{a}{v^2}+p_0-m^2(v-v_0)\,. \tag{13.56}$$

It is clear from (13.56) that $\theta_L(b)=0$. In the $v-p$ plane the image of the line $L=0$ is the straight line $p_1-p_0=m^2(v_1-v_0)$. This line intersects the lower boundary of the region Ω, given by $p=p(v,0)$, $b<v<\infty$, only once, since $L=0$ has negative slope and $p=p(v,0)$ has positive slope (Fig. 13.4).

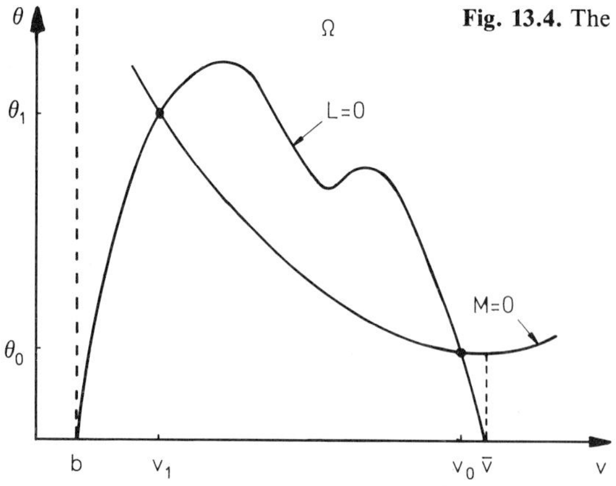

Fig. 13.4. The curves $L=0$ and $M=0$

Therefore in the $v-\theta$ plane the curve $L=0$ intersects the v-axis (i.e. $\theta=0$) only once in the interval (b, ∞) say at $v=\bar{v}$.

The curves $L=0$ and $M=0$ intersect in Ω if and only if the line $L=0$ and the curve $H=0$ intersect in Ω, as follows immediately from the relation

$$H=M+\tfrac{1}{2}(v-v_0)L\,. \tag{13.57}$$

Thus it is sufficient to prove that the curve $L=0$ lies above the curve $M=0$ for at least one value of v in the interval (v_1, v_0). We have by (13.38)

$$\left.\frac{\partial H}{\partial p}\right|_v=\frac{c_v(v-b)}{R}+\frac{(v-v_0)}{2}>0 \tag{13.58}$$

in some neighborhood of (v_0, p_0). Hence, by our assumption, in this neighborhood $H>0$ on $L=0$ if $v<v_0$. Thus by (13.57) in this neighborhood we also have $M>0$ on $L=0$ when $v<v_0$.

From Lemma 4.2 and equation (13.56) we see that the graph of $\theta_L(v)$ lies in Ω only when $v\in(b, \bar{v})$. In the next lemma we shall show that the graph of $\theta_M(v)$ lies in Ω when $b<v<\infty$.

Lemma 4.3. Under the hypotheses of Theorem 4.1 there is a constant $\bar{\theta}>0$ such that $\theta_M(v)\geqslant\bar{\theta}$ for all $v>b$, with equality holding only at $v=\bar{v}$.

Proof. The curve $M=0$ lies above the curve $L=0$ for $v_0<v\leqslant\bar{v}$ by Lemma 4.2, and the curve $L=0$ lies above the v-axis for $v_0<v<\bar{v}$. It is sufficient to show that the curve $\theta=\theta_M(v)$ takes on its minimum value in the interval (b, ∞) at $v=\bar{v}$. Now

$$\frac{\partial M}{\partial v}=\theta\frac{\partial p}{\partial \theta}-L \tag{13.59}$$

and by (13.54), since $c_v=c_v(\theta)$,

$$\frac{\partial^2 M}{\partial\theta\,\partial v}=0\,. \tag{13.60}$$

Therefore

$$\frac{\partial M}{\partial v}(v,\theta)=\frac{\partial M}{\partial v}(v,0)=-L(v,0)\,. \tag{13.61}$$

From (13.56) and Lemma 4.2 we see that $L(v,0)<0$ when $b<v<\bar{v}$ and $L(v,0)>0$ when $\bar{v}<v<\infty$. But

$$\frac{d\theta_M(v)}{dv}=-\frac{\partial M}{\partial v}(v,\theta)\Big/\frac{\partial M}{\partial \theta}(v,\theta)=\frac{L(v,0)}{c_v(\theta)} \tag{13.62}$$

and so $\theta_M(v)$ takes on it minimum at $v=\bar{v}$ and $\theta_M(v)\geqslant\theta_M(\bar{v})$ for all $v>b$, with equality holding only at $v=\bar{v}$.

According to the discussion in Sect. 13.2 a shock layer is a solution of (13.24) satisfying (13.25); the points $(v_0,0,\theta_0)$ and $(v_1,0,\theta_1)$ are of course critical points of (13.24). In the next lemma we examine the nature of these critical points.

Lemma 4.4. Under the hypotheses of Theorem 4.1 the critical point $(v_1,0,\theta_1)$ is a saddle point with a one-dimensional stable manifold and a two-dimensional unstable manifold. Furthermore the fluid velocity is subsonic at the back state (v_1,θ_1) and is supersonic at the front state (v_0,θ_0).

Proof. The acoustic speed c is given by

$$c^2=\left.\frac{\partial p}{\partial \varrho}\right|_\eta=-v^2\left.\frac{\partial p}{\partial v}\right|_\eta . \tag{13.63}$$

Thus

$$u^2-c^2=v^2\left(\left.\frac{\partial p}{\partial v}\right|_\eta+m^2\right) \tag{13.64}$$

since $u=mv$. We need to show that

$$\left.\frac{\partial p}{\partial v}\right|_\eta+m^2>0 \quad \text{at} \quad (v_0,\theta_0) \tag{13.65}$$

and

$$\left.\frac{\partial p}{\partial v}\right|_\eta+m^2<0 \quad \text{at} \quad (v_1,\theta_1)\,. \tag{13.66}$$

Now by hypothesis the chord connecting (v_1,θ_1) to (v_0,θ_0) has slope $-m^2$ and lies above the Hugoniot curve ($H=0$) in Ω. Thus the slope of the curve $H=0$ is greater than the slope of the chord at (v_0,θ_0). From (13.28) and the Gibbs relation (13.29) we now get

$$dH=\left(\theta\left.\frac{\partial\eta}{\partial p}\right|_v+\frac{1}{2}(v-v_0)\right)dp+\left(\theta\left.\frac{\partial\eta}{\partial v}\right|_p-\frac{1}{2}(p-p_0)\right)dv\,. \tag{13.67}$$

Thus

$$\left.\frac{dp}{dv}\right|_H=-\left.\frac{\partial\eta}{\partial v}\right|_p\Big/\left.\frac{\partial\eta}{\partial p}\right|_v=\left.\frac{\partial p}{\partial v}\right|_\eta \quad \text{at} \quad (v_0,p_0) \tag{13.68}$$

and so

$$\left.\frac{\partial p}{\partial v}\right|_{\eta} + m^2 > 0 \quad \text{at} \quad (v_0, p_0) .$$

This proves (13.65). We next show (13.66).

By Lemma 4.2 the curve $L = 0$ lies above the curve $M = 0$ for $v_1 < v < v_0$ and so

$$\frac{d\theta_L}{dv}(v_1) > \frac{d\theta_M}{dv}(v_1) . \tag{13.69}$$

The strictness of the inequality can be shown to follow from the non-tangency hypothesis and equation (13.55). Now

$$\frac{d\theta_L}{dv} = -\frac{\partial L}{\partial v} \bigg/ \frac{\partial L}{\partial \theta} \tag{13.70}$$

and

$$\frac{d\theta_M}{dv} = -\frac{\partial M}{\partial v} \bigg/ \frac{\partial M}{\partial \theta} ; \tag{13.71}$$

thus by (13.69)

$$D \equiv \frac{\partial L}{\partial v}\,\frac{\partial M}{\partial \theta} - \frac{\partial M}{\partial v}\,\frac{\partial L}{\partial \theta} < 0 \quad \text{at} \quad (v_1, \theta_1) \tag{13.72}$$

since $\partial L/\partial\theta$ and $\partial M/\partial\theta$ are positive. Substituting (13.54), (13.55), (13.59), and

$$\frac{\partial L}{\partial v} = \frac{\partial p}{\partial v} + m^2 \tag{13.73}$$

into (13.72) gives

$$D = c_v\left(\frac{\partial p}{\partial v} + m^2\right) - \left(\theta\frac{\partial p}{\partial \theta} - L\right)\frac{\partial p}{\partial \theta} \quad \text{at} \quad (v_1, \theta_1) . \tag{13.74}$$

But $L(v_1, \theta_1) = 0$ and (see (13.22))

$$\frac{\partial p}{\partial v} - \frac{\theta}{c_v}\left(\frac{\partial p}{\partial \theta}\right)^2 = \left.\frac{\partial p}{\partial v}\right|_{\eta} \tag{13.75}$$

so that

$$D = c_v\left(\left.\frac{\partial p}{\partial v}\right|_{\eta} + m^2\right) \quad \text{at } (v_1,\ \theta_1) . \tag{13.76}$$

Since $D < 0$ by (13.72), condition (13.66) now follows at once.

To show that $(v_1, 0, \theta_1)$ is a saddle point we linearize (13.24) about $(v_1, 0, \theta_1)$ to obtain

$$\begin{pmatrix} v' \\ w' \\ \theta' \end{pmatrix} \begin{matrix} = \\ = A \\ = \end{matrix} \begin{pmatrix} v - v_1 \\ w \\ \theta - \theta_1 \end{pmatrix} \tag{13.77}$$

where

$$A = \begin{bmatrix} 0 & 1 & 0 \\ -\dfrac{1}{\sigma}\dfrac{\partial L}{\partial v} & m\dfrac{\lambda+2\mu}{\sigma} & -\dfrac{1}{\sigma}\dfrac{\partial L}{\partial \theta} \\ \dfrac{m}{\varkappa}\dfrac{\partial M}{\partial v} & 0 & \dfrac{m}{\varkappa}\dfrac{\partial M}{\partial \theta} \end{bmatrix} \tag{13.78}$$

the entries being evaluated at (v_1, θ_1).

Now (at (v_1, θ_1))

$$\det A = \frac{m}{\varkappa\sigma}\left(\frac{\partial L}{\partial v}\frac{\partial M}{\partial \theta} - \frac{\partial L}{\partial \theta}\frac{\partial M}{\varkappa v}\right) = \frac{m}{\varkappa\sigma} D < 0 \tag{13.79}$$

by (13.72). Also by (13.54)

$$\operatorname{trace} A = m\left(\frac{\lambda+2\mu}{\sigma} + \frac{c_v}{\varkappa}\right) > 0\,. \tag{13.80}$$

Let ϱ_1, ϱ_2 and ϱ_3 be the eigenvalues of A. Suppose first that these are all real and that $\varrho_1 \leqslant \varrho_2 \leqslant \varrho_3$. Then $\varrho_1\varrho_2\varrho_3 = \det A < 0$ by (13.79) and so either $\varrho_1 < 0$, $\varrho_2 < 0$ and $\varrho_3 < 0$ or $\varrho_1 < 0$, $\varrho_2 > 0$ and $\varrho_3 > 0$. But $\varrho_1 + \varrho_2 + \varrho_3 = \operatorname{trace} A > 0$ by (13.80), and hence $\varrho_1 < 0$, $\varrho_2 > 0$ and $\varrho_3 > 0$. Next suppose that ϱ_1, ϱ_2 and ϱ_3 are not all real. Let $\varrho_2 = \alpha - i\beta$ and $\varrho_3 = \alpha + i\beta$, where α, β and ϱ_1 are all real. Then $(\alpha^2+\beta^2)\varrho_1 = \det A < 0$; hence $\varrho_1 < 0$ and $2\alpha + \varrho_1 = \operatorname{trace} A > 0$, so $\alpha > 0$. Therefore the critical point $(v_1, 0, \theta_1)$ is a saddle point with a one-dimensional stable manifold and a two-dimensional unstable manifold.

Define

$$\Phi(v, w, \theta) = m\left(\tilde{\eta} - \eta_1 + \frac{\sigma w^2}{2\theta} - \frac{M}{\theta}\right). \tag{13.81}$$

Lemma 4.5. Let $(v(x), w(x), \theta(x))$ be any solution of (13.24). Then

$$\frac{d}{dx}\Phi(v(x), w(x), \theta(x)) \geqslant 0\,. \tag{13.82}$$

That is, Φ is nondecreasing along the trajectories of (13.24).

Proof. We can use (13.13), (13.23), and (13.24c) to write (13.81) as

$$\Phi = m(\eta - \eta_1) - \frac{\varkappa\theta'}{\theta}\,. \tag{13.83}$$

Thus

$$\Phi' = m\eta' - \left(\frac{\varkappa\theta'}{\theta}\right)' \geqslant 0 \tag{13.84}$$

by the Clausius-Duhem inequality (see also (13.99)), completing the proof.

We have $\Phi(v_1, 0, \theta_1) = 0$ and, by (13.54) and (13.31),

$$\frac{\partial\Phi}{\partial\theta}(v, 0, \theta) = \frac{m}{\theta^2} M(v, \theta) \tag{13.85}$$

so that $\partial\Phi(v_1, 0, \theta)/\partial\theta > 0$ (<0) for $\theta > \theta_1$ $(\theta < \theta_1)$. Hence

$$\Phi(v_1,0,\theta)>0 \quad \text{for} \quad \theta \neq \theta_1 . \tag{13.86}$$

The function $\theta_L(v)$ has a maximum on the interval $(b,\bar{v})$, say $\bar{\bar{\theta}}$, while $\theta_L(b)=\theta_L(\bar{v})=0$ and $\theta_L(v)>0$ for $b<v<\bar{v}$ by Lemma 4.2. Now by (13.59) and (13.31)

$$\frac{\partial\Phi}{\partial v}(v,0,\theta)=\frac{m}{\theta}L(v,\theta) \tag{13.87}$$

and so

$$\frac{\partial\Phi}{\partial v}(v,0,\bar{\bar{\theta}})\geqslant 0 \tag{13.88}$$

since $L\geqslant 0$ for $\theta=\bar{\bar{\theta}}$. Thus

$$\Phi(v,0,\bar{\bar{\theta}})>0 \quad \text{for} \quad v\geqslant v_1 \tag{13.89}$$

since $\Phi(v_1,0,\bar{\bar{\theta}})>0$ by (13.86).

The curve $M=0$ intersects the line $\theta=\bar{\bar{\theta}}$ for some $\bar{\bar{v}}>\bar{v}$, as follows at once from the relation (see (13.62), (13.22) and (H1))

$$\frac{d\theta_M}{dv}(v)=\frac{1}{c_v}\left(m^2(v-v_0)-p_0-\frac{a}{v^2}\right). \tag{13.90}$$

From this fact and (13.85) and (13.89) it is clear that $\Phi(\bar{\bar{v}},0,\theta)>0$ for $0<\theta<\bar{\bar{\theta}}$.

Put

$$\bar{w}^2=\max_{\substack{v_1\leqslant v\leqslant\bar{\bar{v}}\\ \bar{\theta}\leqslant\theta\leqslant\bar{\bar{\theta}}}}\left\{-\frac{2\theta}{m\sigma}\Phi(v,0,\theta)+1\right\}.$$

We define a box B in phase space by

$$B=\{(v,w,\theta)\colon\ v_1<v<\bar{\bar{v}}_1,\ -\bar{w}<w<\bar{w},\ \tfrac{1}{2}\bar{\theta}<\theta<\bar{\bar{\theta}}\}.$$

Note that with the exception of the bottom of the box and the point (v_1,θ_1) we have $\Phi>0$ on ∂B.

Lemma 4.6. Under the hypotheses of Theorem 4.1 every trajectory which intersects the bottom of the box B leaves the box.

Proof. We have from (13.24)

$$\varkappa\theta'=m\left(-\frac{1}{2}\frac{\partial(\theta\sigma)}{\partial\theta}w^2+M(v,\theta)\right). \tag{13.91}$$

From Lemma 4.3 the curve $M=0$ does not drop below the line $\theta=\bar{\theta}$ in Ω. But $\partial M/\partial\theta=c_v>0$ and so

$$M(v,\theta)<0 \quad \text{for} \quad b<v<\infty \quad \text{and} \quad 0<\theta<\bar{\theta}. \tag{13.92}$$

Also

$$\frac{\partial}{\partial\theta}(\theta\sigma)=\sigma+\theta\frac{\partial\sigma}{\partial\theta}. \tag{13.93}$$

and by hypothesis $\sigma>0$ and

$$\lim_{\theta \to 0} \theta \frac{\partial \sigma}{\partial \theta}(v, \theta) = 0 \quad \text{for} \quad b < v < \infty .$$

Hence

$$\frac{\partial}{\partial \theta}(\theta \sigma) > 0 \quad \text{for} \quad b < v < \infty \tag{13.94}$$

and for θ sufficiently small and positive. We can choose $\bar{\theta}$ smaller if necessary so that (13.94) holds when $0 < \theta < \bar{\theta}/2$. Thus $\theta' < 0$ on the bottom of the box B and every trajectory which intersects the bottom of the box at some point leaves the box at that point.

Lemma 4.7. Under the hypotheses of Theorem 4.1 one of the (two) trajectories of the stable manifold of $(v_1, 0, \theta_1)$ enters the box B while the other never does. Thus there can be at most one trajectory of (13.24) connecting $(v_0, 0, \theta_0)$ to $(v_1, 0, \theta_1)$.

Proof. We need to show that the line tangent to the stable manifold is transverse to the plane $v = v_1$ in phase space. This line is parallel to the eigenvector associated with the *negative* eigenvalue ϱ_1 of (13.78). Let $(\xi_1, \xi_2, \xi_3)^{\mathrm{T}}$ be this eigenvector. Then we have

$$A \begin{pmatrix} \xi_1 \\ \xi_2 \\ \xi_3 \end{pmatrix} = \varrho_1 \begin{pmatrix} \xi_1 \\ \xi_2 \\ \xi_3 \end{pmatrix} . \tag{13.95}$$

We assert that $\xi_1 \neq 0$, which is the required transversality condition. Suppose for contradiction that $\xi_1 = 0$. Then from (13.95) and (13.78) we have

$$\begin{aligned} \xi_2 &= \varrho_1 \xi_1 = 0 \\ (\varkappa \varrho_1 - m c_v) \xi_3 &= m \frac{\partial M}{\partial v} \xi_1 = 0 . \end{aligned} \tag{13.96}$$

But $\varkappa \varrho_1 - m c_v < 0$. Hence $\xi_1 = \xi_2 = \xi_3 = 0$, which is impossible.

Now suppose one of the trajectories forming the stable manifold of $(v_1, 0, \theta_1)$ crosses the plane $v = v_1$ at some point. Then $\Phi > 0$ at this point since

$$\Phi(v, w, \theta) = \Phi(v, 0, \theta) + \frac{m \sigma w^2}{2\theta} , \tag{13.97}$$

and $\Phi(v_1, 0, \theta) > 0$ for $\theta \neq \theta_1$ by (13.86). By Lemma 4.5, Φ is nondecreasing along trajectories of (13.24); thus $\Phi \leqslant 0$ on the stable manifold and hence any trajectory forming the stable manifold cannot cross the plane $v = v_1$.

If we replace x by $-x$ in the system of ordinary differential equations (13.24) then the direction of each trajectory in (13.24) is reversed. We shall refer to (13.24) with x replaced by $-x$ as the *reversed system*. Note that Φ decreases along any trajectory of the reversed system and hence Φ is a Liapunov function in the sense of LaSalle for the reversed system.

Let us denote by T the trajectory of the reversed system that leaves $(v_1, 0, \theta_1)$ and enters B.

Lemma 4.8. Under the hypotheses of Theorem 4.1 the trajectory T is bounded.

Proof. We shall show that T is contained in B. For the reversed system, Φ is non-increasing along T and hence $\Phi \leqslant 0$ on T since $\Phi(v_1, 0, \theta_1) = 0$. But $\Phi > 0$ on ∂B with the exception of the point $(v_1, 0, \theta_1)$ and the bottom face of B. Thus T cannot cross ∂B except possibly at the bottom face. But by Lemma 4.6 and the fact that the system is reversed, every trajectory which intersects the bottom face enters B. Thus T cannot cross ∂B and so it is bounded.

To complete the proof of Theorem 4.1 we need to show that T enters $(v_0, 0, \theta_0)$, that is, the ω-limit set W for the reversed system consists of the single point $(v_0, 0, \theta_0)$. Since Φ is a Liapunov function in the sense of LaSalle for the reversed system and T is bounded, W is contained in the invariant subset of $S \equiv \{(v, w, \theta) \in \text{Closure}(B)\colon \Phi' = 0\}$, by LaSalle's invariance principle [13.12].

From (13.84) we have

$$\Phi' = m\eta' - \frac{(\varkappa\theta')'}{\theta} + \frac{\varkappa\theta'^2}{\theta^2} . \tag{13.98}$$

Substituting (13.24c) and (13.13) into (13.98) yields (after some calculation)

$$\Phi' = \frac{m^2}{\theta}(\lambda + 2\mu)w^2 + \frac{\varkappa\theta'^2}{\theta^2} . \tag{13.99}$$

Thus on S we have by (13.24c)

$$w = 0 , \qquad M(v, \theta) = 0 .$$

Comparing this with the basic equations (13.24), and particularly with (13.24b), it is clear that the points of S where $L \neq 0$ cannot be part of the invariant subset of S. Hence S consists at most of the two points $(v_1, 0, \theta_1)$ and $(v_0, 0, \theta_0)$ in Closure (B) where $w = L = M = 0$. But the first of these points cannot belong to W, since by (13.99) we have $\Phi' < 0$ somewhere on T so that $\Phi < 0$ on W. Hence W, which is of course non-empty since T never leaves B, must consist of the single point $(v_0, 0, 0_0)$. *Thus we have proved that the trajectory T connects $(v_1, 0, \theta_1)$ to $(v_0, 0, \theta_0)$, as required.*

It remains to show that $\eta_1 > \eta_0$ and $\theta_1 > \theta_0$. The first of these inequalities follows from (13.81) since $\Phi < 0$ on W, that is at $(v_0, 0, \theta_0)$. The second follows from (13.62) since $L(v, 0) < 0$ for $v_1 < v < v_0$.

Remark. Theorem 4.1 also remains valid when the Korteweg terms are suppressed in the constitutive formulae (13.9 – 14), that is, when $\sigma = 0$. The basic shock layer equations (13.29) then take the form

$$m(\lambda + 2\mu)v' = L(v, \theta)$$
$$(\varkappa/m)\theta' = M(v, \theta) ,$$

as in the classical paper of *Gilbarg* [13.13]. Proof of existence of shock layer connections in this case can be carried out either by using the method of Gilbarg or by noting that Φ continues to be a Liapunov function and then restricting the previous construction of the three dimensional box B to the two dimensional cross section in the (v, θ) plane.

Example 1. Suppose that c_v is a large constant, so that by Lemma 3.1 the Hugoniot curve generated by (v_0, p_0) is near the isotherm $p(v, \theta_0)$. Let us suppose that (v_0, p_0) is in the vapor region and (v_1, p_1) is in the liquid region. Furthermore, suppose that the straight line through (v_0, p_0) and (v_1, p_1) intersects the Hugoniot curve at only these two points (see Fig. 13.5).

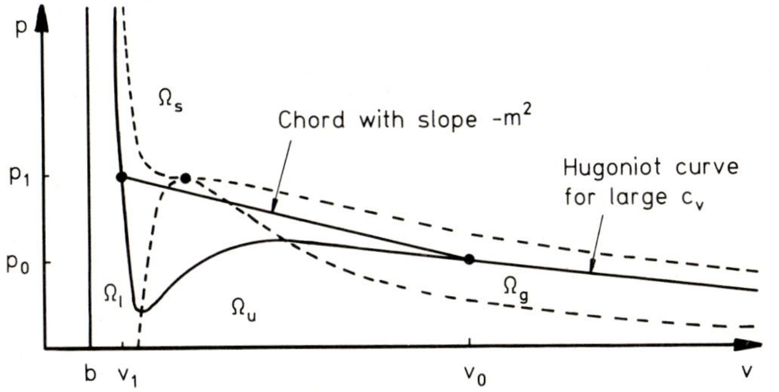

Fig. 13.5. The Hugoniot diagram for large specific heat

In this example the hypotheses of Theorem 4.1 are satisfied and hence there is a shock layer connecting (v_0, p_0) to (v_1, p_1). The shock layer converts vapor in the equilibrium state (v_0, p_0) into liquid in the equilibrium state (v_1, p_1).

The phenomenon of liquefaction shocks was first discovered in 1979 by *Thompson* and his co-workers [13.14], who produced such shocks experimentally as supersonic, strongly temperature dependent reflection waves in a specially designed shock tube apparatus. In view of our hypothesis for this example that c_v should be large it is interesting to note that the fluids studied in [13.14] all had specific heats with c_v/R at least 50 or greater. In this regard, both the experiments in [13.14] and the theory developed in this paper indicate that Landau & Lifshitz's remark, "It should be emphasized that condensation discontinuities are a distinct physical phenomenon, and do not result from the compression of gas in an ordinary shock wave; the latter effect cannot lead to condensation, since the increase of pressure in the shock wave has less effect on the degree of supersaturation than the increase of temperature..." (quoted from [13.14]), is incomplete and somewhat misleading.

Example 2. We keep the same set-up as in Example 1, but with (v_1, p_1) also in the vapor region and $v_1 < v_0$. Then there is a gas-gas compressive shock layer converting gas in the more rarefied equilibrium state (v_0, p_0) to gas in the hotter denser equilibrium state (v_1, p_1).

Example 3. Let us suppose that c_v is constant and $0 < c_v \leqslant R$, so that $\gamma \geqslant 2$. We choose (v_0, p_0) in either the vapor or super-critical region, with $v_0 \geqslant 7b$ and

$p_0 > 0$. Then $\alpha > 0$ by (13.47), and the only attainable states on the Hugoniot curve are to the right of

$$v_s = \frac{(\gamma - 1)\, v_0 + 2b}{\gamma + 1}$$

by Lemma 3.6. Since $v_0 \geqslant 7b$ and $\gamma \geqslant 2$ we have

$$v_s \geqslant \frac{(\gamma - 1) 7b + 2b}{\gamma + 1} = 7b - \frac{12b}{\gamma + 1} \geqslant 3b\,.$$

But the liquid region lies to the left of $v = 3b$, and hence no complete liquifaction shock is possible in this case.

References

13.1 D. J. Korteweg: Sur la forme que prennent les équations des mouvements des fluides si l'on tient compte des forces capillaires causées par des variations de densité. Arch. Néerl. Sci. Exactes Nat. II, **6**, 1–24 (1901)

13.2 C. A. Truesdell, W. Noll: *The Non-Linear Field Theories of Mechanics*, Handbuch der Physik, Vol. III/3, ed. by S. Flügge (Springer, Berlin Heidelberg New York 1965)

13.3 J. Serrin: Phase Transitions and Interfacial Layers for van der Waals Fluids. In *Recent Methods in Nonlinear Analysis and Applications*, ed. by A. Canforo, S. Rionero, C. Sbordone, C. Trombetti (Liguori Editore, Naples 1980)

13.4 M. Slemrod: Admissibility criteria for propagating phase boundaries in a van der Waals fluid. Arch. Rational Mech. Anal. **81**, 301–315 (1983)

13.5 R. Hagan, M. Slemrod: Viscosity-capillarity admissibility criteria with applications to shock and phase transitions. Arch. Rational Mech. Anal. **83**, 333–361 (1983)

13.6 M. Slemrod: Dynamic phase transitions in a van der Waals fluid. Univ. of Wisconsin, Mathematics Research Center, Technical Report #2298 (1981)

13.7 M. Gurtin: Thermodynamics and the possibility of spatial interaction in elastic materials. Arch. Rational Mech. Anal. **19**, 339–352 (1965)

13.8 J. E. Dunn, J. Serrin: On the thermodynamics of interstitial working. Arch. Rational Mech. Anal. **88**, 95–133 (1985)

13.9 E. C. Aifantis, J. Serrin: The mechanical theory of fluid interfaces and Maxwell's rule. J. Colloid Interface Sci. **96**, 517–529 (1983)

13.10 R. Hagan, J. Serrin: One dimensional shock layers in Korteweg fluids. In *Phase Transformations and Material Instabilities in Solids,* ed. by M. Gurtin (Academic Press, New York, London 1984)

13.11 R. L. Pego: Nonexistence of a shock layer in gas dynamics with a nonconvex equation of state. To appear, Arch Rational Mech. Anal. (1986)

13.12 J. K. Hale: *Ordinary Differential Equations*, 2nd ed. (Krieger, Huntington, New York 1980)

13.13 D. Gilbarg: The existence and limit behavior of the one-dimensional shock layer. Am. J. Math. **73**, 256–274 (1951)

13.14 G. Dettleff, P. A. Thompson, G. E. A. Meier, H.-D. Speckman: An experimental study of liquefaction shock waves. J. Fluid Mech. **95**, 279–304 (1979)

The **Institute for Mathematics and Its Applications** was established by a grant from the National Science Foundation to the University of Minnesota in 1982. The IMA seeks to encourage the development and study of fresh mathematical concepts and questions of concern to the other sciences by bringing together mathematicians and scientists from diverse fields in an atmosphere that will stimulate discussion and collaboration.

Hans Weinberger, Director
George R. Sell, Associate Director

Yearly Programs

1982–1983
Statistical and Continuum Approaches to Phase Transition

1983–1984
Mathematical Models for the Economis of Decentralized Resource Allocation

1984–1985
Continuum Physics and Partial Differential Equations

1985–1986
Stochastic Differential Equations and Their Applications

1986–1987
Scientific Computation

1987–1988
Applied Combinatorics

Springer-Verlag
Berlin Heidelberg
New York Tokyo

Springer